ROME BEFORE AVIGNON

ROBERT BRENTANO is Professor of History at the University of California at Berkeley. He is the author of *Two Churches: England and Italy in the Thirteenth Century* which was awarded the Hafkins Medal of the Medieval Academy of America and the John Gillmarry Shea Award of the Catholic Historical Association.

**BRENTANO, Robert. Rome before Avignon; a social history of thir-
teenth-century Rome. Basic Books, 1974. 340p il map 75-
75306. 15.00. SBN 465-07125-2**

Brentano (Berkeley) has again placed the historical profession in his
great debt; following upon his *Two churches* (1968) is *Rome before
Avignon,* perhaps his best study. In seven sections, Brentano treats
of the physical Rome, the ideas of Rome, the Roman polity, the popes,
natural and spiritual affiliations and bonds, and of wills and testaments
as social history. This study will not please those who believe that
history is a list of events in the order in which they happened, but it is a
rich, complex, evocative, sensitive, and humane product of a fine mind
at work. The plates are chosen to illustrate specific allusions made in
the text, the footnotes copious and so full as to make a formal bibliography unnecessary; spot-checking of the index proved it accurate. With
this brilliant social history, Brentano joins that small pantheon of superb scholars doing municipal history of the Middle Ages and Renaissance in Italy: such men as Frederic C. Lane, David Herlihy, Gene
Brucker, Werner Gundersheimer, William Bowsky. Highly and warmly
recommended for freshman to senior faculty member.

ROME
BEFORE
AVIGNON

A Social History of
Thirteenth-Century Rome

BY

ROBERT BRENTANO

Basic Books, Inc., Publishers
NEW YORK

© 1974 by Basic Books, Inc.
Library of Congress Catalog Card Number: 75-75306
SBN 465–07125–2
Manufactured in the United States of America
DESIGNED BY VINCENT TORRE
74 75 76 77 10 9 8 7 6 5 4 3 2 1

THIS BOOK WAS WRITTEN FOR

AND IS DEDICATED TO

MY SON

James August Burke Brentano

Contents

Contents

viii

Acknowledgments

BEFORE I thank some of the kind and scholarly people to whom I am indebted for their help to me in writing this book, I must explain to them a little, in order to disturb them less, what sort of book it has turned out to be. When years ago Bill Gum first suggested that I write a book on thirteenth-century Rome, we both thought that it would be quick, light, and relatively popular. I did not then intend, for example, to include any footnotes to work in languages other than English except for a few minimally necessary ones to things in Italian and Latin; and I intended to do very little research in unprinted sources specifically for this book. I found in writing that I had to and wanted to do a great deal more than I had originally intended. Increasingly the book came to be a scholar's ordinary academic book—but not completely. The result is clearly a hybrid, a chrysalid book and thus likely to repel equally "two classes of readers." I would rationalize to both sides that from beginning to end the book was written with the purpose of exposing what thirteenth-century Rome was like, of making the reader see and feel it. Sometimes I have moved fast over the surface and sometimes worried a lot with the sources, with how and how much we can know. This latter sort of knowledge seems to me a crucial part of seeing and feeling. I hope that to some readers the book will look like a three-dimensional picture with areas of diverse depth and intricacy; and I further hope that the footnotes will make the nature of each part of the book obvious.

By far my greatest debt in writing this book I owe to my wife Carroll Winslow Brentano. It was she who first really introduced me to Rome and induced me to live in it and study it. She still knows it much better and likes it more unreservedly and at the same time with more perception than I. Her unpublished work, "The Church of S. Maria della Consolazione in Rome," exposes as does little else the exact nature of a nexus of Roman social and artistic life in one period of Roman history. Next I should like to thank my friend Gene Brucker whose *Renaissance Florence* is what a book about a city should be; I wish that my book could more adequately repay his generous encouragement over very many years. I should like to thank Bill Gum not only for the idea of the book but also for not having lost faith in it over the years.

I owe a very specific debt of gratitude to Julian Gardner who has given me on extended loan one of his copies of "The Influence of Popes' and Cardinals' Patronage on the Introduction of the Gothic Style into Rome and the Surrounding Area, 1254–1305"; it is ironic, and Roman, that this, one of the very best books that has been written about thirteenth-century Rome, is formally an unpublished thesis. Certainly not all of Charles Till Davis's luminous work on Rome is unpublished, but his most recent work on Ptolemy of Lucca, Nicholas III, and Roman republicanism is. He has been kind enough to talk to me about it and to lend me work in progress. It is certainly work that will change and enhance our view of late thirteenth-century Roman history. Agostino Paravicini Bagliani's recent book, *Cardinali di curia e 'familiae' cardinalizie* has been a great help to me; I am sure that mine would have been a better book if his had been written earlier. One should await impatiently his forthcoming work on cardinals' wills in the late thirteenth century; he has been very kind in lending me material which he has gathered for it.

To Carlo Bertelli and Ilaria Toesca I am grateful for their generous gifts of many extremely helpful plates and offprints and for much more, and particularly in the last year for having shown me so much of what remains in Rome from the thirteenth century and is very beautiful. In this connection I should also like to thank Fra Corrado, O.S.B.C., of the abbey of Tre Fontane, who let me look and look again at the wonders of his abbey's cloistered frescos. I should like to thank Lucia Nonis Stefanello who explained to me Roman things with Venetian clarity. And for various ideas and explanations I should like to thank Anthony Luttrell, Peter Herde, Father Leonard Boyle, O.P., Richard Mather, Milton Lewine, Randolph Starn, and Gerard Caspary. I should like to thank Peter Partner for an early copy of his book *The Lands of St. Peter;* and I should very particularly like to thank my friend Miriam Brokaw whose extremely perceptive criticisms of the manuscript I was only partly able to utilize.

I am indebted to a number of keepers of archives and libraries. I should like to offer very special thanks to Virgilio Valeri who for years (over twenty that I knew) made the manuscript library at the Vatican a particularly pleasant place to work and who was especially helpful to me. But I am also much indebted to those who have helped me in the printed book library and the archives, and to those institutions themselves. Monsignor Emidio de Sanctis's kindness in admitting me to the

Acknowledgments

capitular archives at Rieti proved surprisingly helpful to my work on Rome; and I am also indebted to the efficiency and kindness of the keepers of the Archivio di stato in Rieti. The Archivio Capitolino at the Vallicelliana in which the Archivio Orsini is kept proved an unusually attractive place in which to work and its archivists very gracious. I have been shown much kindness by the librarians of the American Academy, the British School, the German Historical Institute, and (in the years in which it was open) the National Library at Rome. It may seem odd in connection with all this very sincere gratitude to thank, as one must, the Archivio di stato in Rome. But in fact in the overwhelming majority of instances the Archivio was able to find requested documents and to produce them within twenty-four hours. A number of its employees were exceedingly gracious; and its cataloging and indexing of collections of documents is in many ways distinctly superior to that even of the Vatican Library. Furthermore, it is now clear that its lassitudinous years are in the past.

I should like to thank Adrienne Morgan for having worked on the map of Rome and Giuseppe Zigrossi for having taken pictures of Matteo the Scribe. I should like to thank my editor Regina Schachter for her kindness and patience. And finally I want to thank all three of my children, Margaret and Robert as well as the dedicatee, for taking photographs of and drawing and generally appreciating Rome.

R. B.

Acknowledgments

xi

Illustrations

Illustrations

Illustrations

Introduction

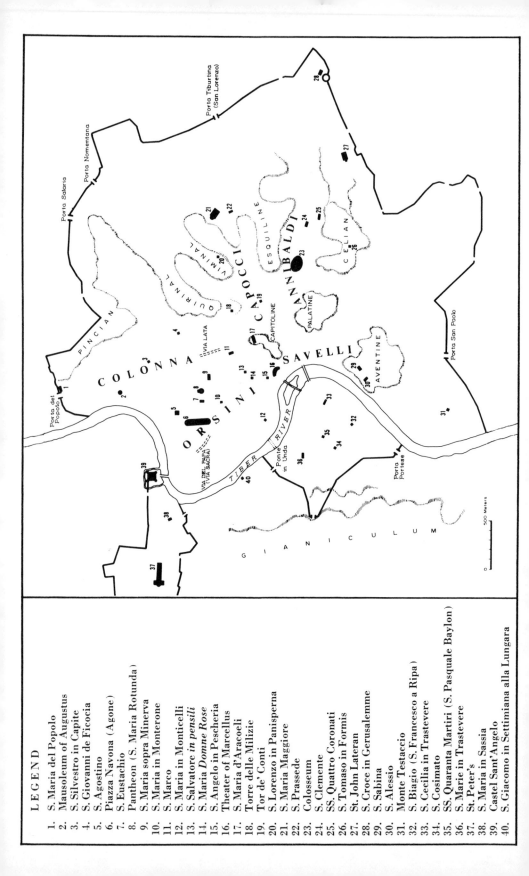

LEGEND

1. S. Maria del Popolo
2. Mausoleum of Augustus
3. S. Silvestro in Capite
4. S. Giovanni de Ficocia
5. S. Agostino
6. Piazza Navona (Agone)
7. S. Eustachio
8. Pantheon (S. Maria Rotunda)
9. S. Maria sopra Minerva
10. S. Maria in Monterone
11. S. Marco
12. S. Maria in Monticelli
13. S. Salvatore *in pensili*
14. S. Maria *Donne Rose*
15. S. Angelo in Pescheria
16. Theater of Marcellus
17. S. Maria d'Aracoeli
18. Torre delle Milizie
19. Tor de' Conti
20. S. Lorenzo in Panisperna
21. S. Maria Maggiore
22. S. Prassede
23. Colosseum
24. S. Clemente
25. SS. Quattro Coronati
26. S. Tomaso in Formis
27. St. John Lateran
28. S. Croce in Gerusalemme
29. Sabina
30. S. Alessio
31. Monte Testaccio
32. S. Biagio (S. Francesco a Ripa)
33. S. Cecilia in Trastevere
34. S. Cosimato
35. SS. Quaranta Martiri (S. Pasquale Baylon)
36. S. Marie in Trastevere
37. St. Peter's
38. S. Maria in Sassia
39. Castel Sant'Angelo
40. S. Giacomo in Settimiana alla Lungara

THE CITY OF ROME IS UNLIKE other cities in many ways. Its past is unique. It is "the broadest page of history," the "city of all time, and of all the world." For many centuries, including the century which this book observes, it was "The City" (*Urbs*) and needed no qualifying, identifying description. For some of those centuries, it, a city, ruled the Western world. Unlike other cities which might compete with it for ancient greatness, Rome's physical past has not been severed from it. It is in this not at all like Athens. Rome is not a museum of ruins set within a modern town. It is a crumbling mixture of all its pasts, jumbled together and still living, never dead but never freshly alive, all covered and covering in "casual sepulchre." The very thought of an opposite condition, of unsophisticated freshness, has been unthinkable of Rome at least since the time of Livy, and repulsive to it at least since the time of Tacitus.[1]

Perhaps one should say that there are two historical Romes: one, the plasticine city that classical scholars re-created, or created, particularly in the nineteenth century, a city of vast colonnades, perfect arches, perfect hexameters, stretching into that unbearable dullness that provoked orgy, murder, and satire; the other, the broken city, the foundation for new things to be broken in their turn—with the lovers in each house among these ruins dreaming in part the dreams of their predecessors, all hopelessly interwoven, like their houses, with their own.

Anyone particularly attracted by the Rome that has existed at least since about the year 410 is attracted by ruins. This ruined Rome, whatever else it is, is romantic Rome. Everything in it speaks of something else, of some broken echo—echoes of things which may have been dully, physically prosaic when they were first spoken, but that have been transformed by time. There is always (as Rome's most brilliant novelist, Nathaniel Hawthorne, said of Perugia) "the life of the flitting moment, existing in the antique shell." There is always the taste of which Webster was the master, the taste of the thing only per-

Introduction

fect, only mature, when it is crushed. Rome of the time between Augustine and Moravia is like the cassia. "Everywhere, some fragment of ruin suggesting the magnificence of a former epoch; everywhere, moreover, a Cross—and nastiness at the foot of it." [2]

The love which Rome inspires is not like that inspired by Florence or Perugia, Siena or Ravenna. It is not a gentle thing, Fra Angelico, holy picture, looking down from San Miniato, E. M. Forster, a perfect incapsulated past age. No one could say of Rome, as Margaret Fuller did of Florence, that it is like Boston. Rome, as it is ravaged, ravishes—the orange walls, the slackly ringing bells, the sinuous streets flowing like rivers over waves of buried shards and garbage, the succession of domes in which Giacomo della Porta and Carlo Maderno compete for some unimaginable prize, the endless debris of countless ancient columns. In their sleep, lovers of Rome can feel its contours with their fingers, imagine its whole shape, centering on its river and island and the old church surrounding its Ottonian well. Rome's great historian, Ferdinand Gregorovius, wrote, in 1856 when he was having strange dreams: "One night I found myself in the theatre; instead of actors the walls of Rome appeared on the stage, where they gave a magnificent dance. At the end Iphigenia appeared and addressed a speech to me, I being the only spectator in the building." (Gregorovius wrote this in April, when, as Romulus and Remus found, birds come to Rome.) Ernest Renan, who was staying at the Minerva, wrote in 1849, "This city is an enchantress; . . . I am no longer French; I am no longer the critic. . . ." [3] Lovers of Rome like to look down upon it, from its hills, from its city of roofs—the roofs which its inhabitants fight for and which gave pleasure to Filippo Neri and which bound to themselves Sixtus V, the good saint and bad pope from one of the city's great building centuries. Rome is not a sweet city. It is loose and florid. The bottoms of its trousers are rolled. It can frighten, as it did thirteenth-century English proctors and Gide's Amédée Fleurissoire and, perhaps, even Saint Bernard.

All the great historians of postclassical Rome have been in love with the city. And, with the possible exception of the greatest among them, Ranke, they have been in love with its decay, been in love with it because it was a great thing decaying. Decay is, in a way, a form of paradox. So Rome attracts men not only with its wanton beauty, but also with the almost absurd obviousness and variety of its paradox. Rome's history is the history of paradox; it localizes a universal attitude, a way of seeing things. In the thirteenth century "The City" was also a little town with little farms; the *caput mundi* was a ruin, the reliquary of the apostles a sink of corruption, the Vicar's house a counting-house. Rome's Christian history is a history of distance between things and ideas, between different kinds of, and opposing, ideas and incongruously juxtaposed things.

All this complexity and opposition is at the same time summed up and romanticized, softened and seen in a different way, through the concept of decay. In the twelfth and thirteenth centuries the concept coincided with a particularly pleasing intellectual cliché, the contempt for the world, the realization of the transitory quality of all earthly, physical things. This is clearly seen in the most intellectualized of Rome's twelfth- and thirteenth-century guidebooks. It is a work written about the middle of the two centuries by a Master Gregory, perhaps an Englishman, supposedly at the request of his associates, Master Martin, Dom Thomas, and others. Gregory begins his work by reflecting, with Hildebert of Lavardin and Lucan, upon the true meaning of what the city has to teach, as he recalls his first seeing it from afar, from the slopes of Monte Mario, as he looked down with wonder upon the place, with all its memories, with all its towers bristling like a field of grain. "Its ruins, I think, make it terribly clear that all temporal things soon come to ruin, since this head of all things temporal, Rome, slips daily languishing away." [4]

Most obvious, of Roman historians, in his taste for decay is Gibbon. Gibbon did not in the end really write the history of the city, but rather the history of its declining empire. But he had started to write of the city itself. His attachment was probably insufficiently physical— and he knew Rome very little (a few months, including eighteen weeks with a Scottish guide, he says)—or insufficiently physical in a specifically local way, to keep him within actual walls at an actual place. Gibbon's physicalness defines itself in his famous description of his finishing, far from Rome, the *Decline and Fall:* "The air was temperate, the sky was serene, the silver orb of the moon was reflected from the waters, and all nature was silent." The acacias, the garden, the summerhouse, and Gibbon himself are there in the passage of which this is the climax, and so is the moon, but the moon is a "silver orb," the moon of empire not of specific ruins—Gibbon was more Giannone than observation. Still, of a beginning even more famous, Gibbon had written what are probably the most familiar phrases ever written about the city of Rome. He had been caught in passing by the physical city, at least as it entwined itself about an idea. Even his slow senses were a little inflamed: "It was at Rome, on the 15th of October 1764, as I sat musing amidst the ruins of the Capitol, while barefooted friars were singing vespers in the temple of Jupiter [what Gibbon thought the temple of Jupiter], that the idea of writing the decline and fall of the city first started to my mind." [5]

Gibbon's words are the device of all historians of the city of Rome. Gregorovius came upon Rome a century after Gibbon: ". . . on October 2, 1852, at 4:30 in the afternoon, I entered Rome by the Porta del Popolo. . . ." [6] In the conclusion to his *History of Rome in the Middle Ages,* he wrote: "The idea of the work was suggested by the overwhelming spectacle of the monumental grandeur of the city, and perhaps by the presentiment which made itself obscurely felt at the time, that the history of the Roman Middle Ages would soon reach its perfect close in the downfall of papal dominion. . . ." But in a footnote three volumes earlier one finds Gregorovius being more exact: "A tower Anguillara still stands on the Lungaretta. Seen from the Ponte Cestio, Trastevere and the bank of the river form a curious picture. Here and there a grey baronial tower rises by the river among modern houses. It was while looking at Rome from this bridge that I conceived the thought of the present history." And to Gibbon's telling of his finishing his long, absorbing Roman work on June 27, 1787, Gregorovius echoes, "I began my task in 1855, and ended it in 1871."

Gregorovius's is still the major history of medieval Rome. No one who tells any part of that history can help admiring Gregorovius's book and being thankful for it, because Gregorovius worked so broadly and so deeply and put his great pattern together so well, sometimes so brilliantly. He has been wretchedly underrated. His excellences were precisely those that proved invisible to historians in the age of Tout and Petit-Dutaillis, and he was reticent about almost all the things that they considered important. But read on the death of Boniface VIII, on Cola, on the beauties of Santa Maria Maggiore, Gregorovius can now be seen in his full strength immediately. Unlike Gibbon he knew physical Rome well, and this knowledge gave a different substance to the decay for which he had a taste—took it farther from Tacitus. But decay was the theme of his elegy, as the yellow (as he thought it) Tiber flowed sadly among his ruins. [7]

Gregorovius expressed his love for Rome straightforwardly: "My work is merely an imperfect fragment, but such as it is, I gratefully and reverently lay it as a votive offering at the feet of Rome." Hawthorne was more perplexed. Did he hate Rome? Not really. Did he love it? Probably. As his Hilda says: "I sometimes fancy that Rome—mere Rome—will crowd everything else out of my heart"; and he himself adds: "It is very singular, the sad embrace with which Rome takes possession of the soul." But with hate or love, Hawthorne essentially saw Rome, in streets and buildings and gardens and surrounding hills, so that, as the structure of Augustus Hare's *Walks in Rome* indirectly but persuasively points out, once having seen Rome through Hawthorne's eyes it is impossible to see it again with one's own. In *The Marble Faun,* Hawthorne reproduced physical Rome for all the senses, its colors, its "evil smells," its temperatures, the feel and taste of its water, its sky—the masses of white cloud floating over the "ever-open eye" of the Pantheon, or as in the golden sunset "the bells of the churches suddenly ring out, as if it were a peal of triumph because Rome is still imperial." [8]

Gibbon sat on the Capitol, and Gregorovius stood on the Ponte Cestio. Ranke moved from side to side, from height to height, and caught the city in a connecting web. Ranke was the most sensual of Roman historians; like Hawthorne, he saw. In his *History of the Popes,* even the documents are "actually sensible." In his description of Paul v's building and his bringing water to the huge fountain of the Acqua Paola, which, from the Gianicolo, looks across all Rome to the surrounding hills, Ranke enclosed the city and inhabited it with sound:

Opposite to the fountain and the Moses of Sixtus v, but at a distance, the stream five times as powerful as the Aqua Felice, divides into four copious branches. From this far-famed hill, the scene of Porsenna's attack, now covered with vineyards and orchards and ruins, the traveller looks across the city and the country to the distant mountains, over which the evening hangs its many-coloured mist, like a transparent veil. The solitude is sublimely broken by the noise of the gushing abundance of water, the multitude of fountains, and this charm it chiefly owes to the Aqua Paolina.

Ranke's movement of confronting two opposing places or monuments with each other and thus capturing the distance between them could also happen within the smaller space of a great church. So in Santa Maria Maggiore, again opposing Paul v to Sixtus v, Ranke makes the Pauline Chapel look across to, and overawe, the Sistine Chapel. Most brilliantly of all, he ties the city and its civilizations together in his discussion of Sixtus v and the columns of Trajan and Marcus Aurelius (the Antonine Column), on the former of which Sixtus placed a statue of Saint Peter and on the latter of Saint Paul, "and from that time the statues of the two apostles have stood, confronting each other [although not, at least now, quite looking at each other] on that airy elevation, overlooking the dwellings of men." Ranke ties a Christian and modern ribbon to ancient and pagan bases and ties together the Piazza Colonna and the Forum of Trajan. Ranke's city has volume, and he looks at it, as every man should, from one hill to another, from airy elevations like the towers.[9]

In the continuous history of Rome, even of papal Rome, the thirteenth century, the period that stretches between the pontificates of Innocent III and Boniface VIII, is only a short fragment. It is, however, a significant fragment, that which precedes the medieval papacy's abduction, or self-abduction, to Avignon, that which crowns the high pretensions of the great medieval popes. Insofar as thirteenth-century Rome was ruled by its popes, it had an unusual set of rulers, extraordinarily well educated, articulate, and visible. The thirteenth century is the century of the old church's high experiment, of the efforts of men like Saint Francis and Innocent III to produce a new sort or new sorts of Christian total society. Their efforts left trails back and forth through the walled enclosure of the papal city. The thirteenth century is also the century when pope and Hohenstaufen emperor fought for the last time up and down the peninsula of Italy and across Rome, the century in which the popes' Angevin allies turned into their Angevin enemies as Angevins inherited, bloodily, Hohenstaufen interests. In the century's last years a concentration of patrons and artists produced in it one of the West's renaissances. Papal, curial, and family money, intended to glorify those who spent it, found a style in which a sort of naturalism and a breath of Gothic decoration enlivened a studied return to Rome's ancient past. A series of masters and their shops painted, carved, and constructed their frescos, mosaics, tombs, and ciboria. This work reached a final statement, before Rome went to Avignon, in Giotto's magnificent altarpiece which Cardinal Jacopo Stefaneschi gave to Saint Peter's and in the bronze Saint Peter of the Vatican which was cast by Arnolfo or some of his followers. Rome just before Giotto and Arnolfo, like the popes, went North was a, perhaps the, artistic capital of the Western world. The pageantry and propaganda of this painted, glittering Rome reached a final trumpeting crescendo in Boniface VIII's Jubilee of 1300; but from that jubilee's brilliance the Florentine historian Giovanni Villani, as he said, went home taught by falling Rome to write of rising Florence.

Thirteenth-century Rome may still be seen. To see it one must first take a painful "flight of imagination," although not one so terrible as that through which Freud leads his readers early in his *Civilization and Its Discontents,* when he forces them to imagine a totally reconstructed Rome "in which all the earlier phases of development continue to exist alongside the latest one" so that in "the place occupied by the Palazzo Caffarelli would once more stand—without the Palazzo having to be removed—the Temple of Jupiter Capitolinus; and this not only in its latest shape as the Romans of the Empire saw it, but also in its earliest one, when it still showed Etruscan forms. . . ." [10] To see thirteenth-century Rome it is necessary first to tear away that which is later. This means not just the sort of pleasurable tearing away that removes the hideous, dark, killing cap that a modern American-style hotel has placed on the Monte Mario (where that hill once made green the end of the Via Giulia) and that removes the street's poisonous traffic, but also tearing away the Via Giulia itself. Painful as it is even to think of Rome without Borromini, Bernini, or Michelangelo—limbs seem carved off—it is necessary. And it is necessary to rebury, or half bury and encumber, a great many relics of antique Rome, and to reestablish farms, gardens, and towers, and also a number of pieces of medieval Rome. This is a complicated process which requires study, planning, description. But something of thirteenth-century Rome as it remains now, its characteristic colors and shapes, can be identified immediately. Thirteenth-century Rome (extended through the classicizing end of the century to the death of Boniface VIII in 1303) is the Rome of the Aracoeli's façade, of the shapes and patterns and purposes of the prelates' tombs in San Lorenzo, Santa Prassede, the Aracoeli, Santa Balbina, the Lateran, and of the great Arnolfo ciboria in San Paolo and Santa Cecilia, of the Torriti and Cavallini mosaics, of the paintings at Tre Fontane, of Rome's two great towers, the Conti and the delle Milizie, of the (restored) fortified Cecilia Metella on the Via Appia. But the Roman thirteenth century shares its two most constantly apparent man-made physical characteristics with its predecessor century. The thirteenth century continued (although not exactly in the same style), the twelfth century's great series of brick campanili and cosmatesque floors (echoing echoes). And it went on living amid the birds and the lizards, the figs, the capers, and the acanthus, which always decorate Rome. Look at the campanile of Sant'Eustachio at ten o'clock on a March morning, or look across the back floor of the Aracoeli to the great west door when it is thrown open, so that you can see straight over Rome, across all the repeatedly rearranged porphyry and serpentine, the tufa blocks and reused ancient bricks, the pieces of column and frieze, to the campanile of Santa Maria in Trastevere and to the Gianicolo, and you sense something of the Rome which the Florentine painter Cimabue saw and went north to paint in Assisi, something of what the papal Rome of this book was, at its best, like.

Introduction

It will be obvious to the reader that this book means to suggest
the physical city of Rome and is in part addressed to the actual visitor
to the city. There are some books about which that visitor should know.
If he has the faintest taste for romance he should read and use Augustus J. C. Hare's *Walks in Rome*. There are also two helpful, interesting, and different modern guidebooks. Anyone who can read Italian,
even slightly, should use *Roma e dintorni,* in the series "Guida
d'Italia" of the Touring Club Italiano; anyone who cannot will find
Georgina Masson's *The Companion Guide to Rome* attractive and fascinating. Finally, the visitor should get himself a copy of Father Leonard Boyle's *A Short Guide to St. Clement's Rome* (Rome, 1968); he
should take it with him to, or get it at, Saint Clement's, and there, in
the church, use it to work out the succession of medieval Roman centuries.

CHAPTER

I

The Physical City

HE TOTAL POPULATION OF THE
city of Rome in the early thirteenth century has been thought to have
been about thirty-five thousand. If this estimate is even close to correct,
all the Romans of Rome could have sat down in the Colosseum, if it
too had not crumbled and gradually fallen, if it had not turned itself
into an exotic wildflower garden (still perhaps nurturing seeds from
the paws and fur and fodder of ancient beasts) and into a center of
dreams. The dreams—or nightmares—about the Colosseum were
old by the thirteenth century, and their direction fixed. As long as the
Colosseum should stand, Rome would stand; when it should finally
disappear, so would Rome, and with Rome the world. So when for the
year 1231 the Cassino notary and historian Riccardo of San Germano
recorded a great and terrible earthquake which lasted for over a month,
he told of its horrors, of the earth quivering around San Germano, of
the clear water of fountains turning color, of the campanili falling from
churches, of towers and houses, and the whole thing stretching itself
out over the land from Capua to Rome, and terrifying people into pen-
itence and flight, and the abbot of Cassino into ordering barefoot processions over all his lands. And Riccardo told very particularly of the
Colosseum in Rome: *Et tunc de Coliseo concussus lapis ingens eversus
est* ("And then, shaken from the Colosseum, a huge stone was dis-
gorged"). The Colosseum and its stones were a center of men's attentions. Their fall was particularly watched.[1]

But at the same time, this romantic and symbolic object was very
casually used. In it, prominent Roman families fortified themselves as
if it were a caved cliff. And in 1263 Fra Giordano, a Cistercian monk
from Fossanova (whose brother, according to papal sources, was ille-
gally holding papal lands) was living in a house at the Colosseum with
his own grandmother.[2] In the middle of the thirteenth century the old
Colosseum-controlling family of Frangipane (too loyal, it has been
thought rather oddly, to the emperor Frederick II) was replaced, but
not too quickly and decisively, by the newer family of Annibaldi.

The Physical City

13

There in the Romans' Colosseum the Annibaldi might pretend their descent from great Hannibal. From there they could guard the roads to their country estates, and launch their sons on profitable careers at the curia, and immure their daughters in neighboring convents. The Colosseum, and its wildflowers, its dreams, its Annibaldi, and its renegade Cistercian with his grandmother, is a fair symbol of thirteenth-century Rome.

But although it is a fair symbol, the Colosseum is not a unique one. On the Celian Hill, in the ruins of part of the Acqua Claudia, one of the aqueduct courses that had brought fresh water from the distant countryside to the heart of the classical city, there lived, in the last four years of his life, until December 1213, a curious saint named John of Matha (or Giovanni or Jean of Mathe). John was one of the founders of the order of Trinitarians, the friars who wear a blue and red cross on the front of their white habits, not the most famous but the first papally approved of the great new orders of the very end of the twelfth and the first half of the thirteenth centuries.

John's biography has been romantically, and perhaps interestedly, elaborated in modern times, and much of his actual life is unsure. John is thought to have been born in Provence, to have studied in Paris, and, having grown increasingly pious, to have become a priest. While celebrating his first Mass, he saw a vision in which the Lord God held by His two hands two slaves, one black and the other white. The message that John saw into this vision, that he should do something to help free Christian slaves in Moslem captivity, molded his mind and formed his purpose. But at the same time John felt increasingly attracted to a life of eremitic withdrawal, and he withdrew to a country hermitage. Some time later John founded his order with the approval of Innocent III and began the serious work of redeeming slaves and rehabilitating them in hospitals. Although the initial center of enthusiasm for the new order seems to have been Marseilles, a point of debarkation from Africa, John himself eventually came again to Rome. There a house, convent and hospital, was founded around a piece of the Acqua Claudia, the Arch of Dolabella (built in the consulate of Cornelius Dolabella and Junius Silanus), and the old connected monastery of San Tommaso in Formis (Saint Thomas in the Watercourse). There John lived, a saint in an aqueduct, a hermit in a city, a recluse of the old French sort next to his new order's very modern, active work of Christ, saving the bodies as well as the souls of the destitute. John was a proto-Francis; and if the Claudia did not work as it once had, it gave Rome, and the world, a new sort of life-giving water.[3]

The active hermit in the aqueduct (who is there remembered in thirteenth-century mosaic) is a second symbol of thirteenth-century Rome. The market of Sant'Angelo in Pescheria is a third. There, near the vast ruins of the Theater of Marcellus, ruins that formed the base of a family fortress for the Pierleoni and then for the Savelli, and actually within the ruins of the portico which Augustus had refurbished in honor of his sister Octavia and that had held a temple of Jupiter and Juno, the collegiate church of the Holy Angel offered for hire to the sellers of fish the antique marble stones in its "temple" portico. Sant'Angelo was the hub of Rome's great fish market, and there the busy sort of petty mercantile activity that gave Rome much of its surface vivacity was strikingly apparent. At Sant'Angelo, as at the Colosseum and San Tommaso in Formis, the life of thirteenth-century Rome was sewn into the ruins of classical Rome; the life of the present used the artifacts of the past to its own purposes. Rome was a place in which flourished, or at least existed, places with names like the Church of the Holy Savior of the Arch of Trajan, where branches of families identified themselves with the names of ancient monuments, by or in which they lived, and where a man could be unremarkably named Gregory son of Caesar.[4]

Surrounding all the separate, smaller, used ruins of Rome were the great ancient walls, the walls built by the emperor Aurelian in the third century, with the Vatican additions erected during the ninth century by Pope Leo IV. Medieval Rome used the walls of ancient Rome (made Christian with crosses), but it used them in its own way, careless of their ancient purposes. Although thirteenth-century Rome burst through its gates and streamed, in suburban farms, fortifications, and churches, out along the roads leaving Rome, it did not fill at all densely the whole area, well over three thousand acres, enclosed by its walls. The greater part of this enclosed area was lightly inhabited, divided into small farms and even papal and monastic vineyards, decorated with churches and basilicas, punctuated by the towers and fortresses of powerful Roman families. The heavily populated part of Rome lay on the east side of the river between the bridge to Sant'Angelo and the bridge to the island. It touched parts of Trastevere and the Vatican, and to the east the Trivio and the Suburra; but for the most part it lay in the horn between the southwestern side of the Corso (then the Via Lata) and the river.[5]

Any observer must notice repeatedly the contrast within the walls of medieval Rome between this crammed urban space and the other, vineyard city. The sense of the tightly filled part of the city is caught in

The Physical City

a chronicler's account of Boniface VIII's death: "The awareness of Boniface's death was brought to the people suddenly by rumor, quick in its enjoyment of bad news, rushing through the alleys and down the blind streets and making the whole shocked city reverberate with the sound of wailing." [6] This old cramped city is not completely lost from modern Rome. Although much of medieval Rome has, even within the last century, been destroyed, there is in many parts of continuously occupied Rome a sort of crusty hive of medieval and early Renaissance (and sometimes ancient) undervaulting, the whitewashed workshops and storerooms of artisans, upon which Rome has built and rebuilt itself, allowing an occasional old wall or tower to remain, protruding through the modern building. Some streets and alleys, cut through by more modern regularizations, maintain in part their old, independent waywardness as do the Arco degli Acetari and the Arco di Santa Margherita off the Via del Pellegrino, and the Vicolo Savelli, the Vicolo del Babuccio, and the Vicolo della Cuccagna. One also feels the old city buried under the hillocks and humps of streets like the Via di Santa Maria in Monticelli and the Via di San Paolo alla Regola (although there was still in 1245 at least some little arable land in the Arenula near San Paolo), and like the streets that are now the Torre Argentina and the Funari, with their cobblestones undulating over centuries of life and garbage. One feels it too in the visible distance between Rome's present surface and its old surfaces, as at the Pantheon, in those Roman pits and pocks that cut down through the centuries. But it is also true that a better sense of the town life of medieval Lazio and perhaps even of Rome (if a provincial town can ape a great city) is found at Anagni where the middle ages have been less tampered with and the gray houses still lie one upon another.[7]

Of the other Roman city, the vineyard one, less is left. One has some sense of it, particularly in autumn, on the Gianicolo, some around the Savelli fortress on the Aventine, or in the Forum. But rustic Rome is now too antique and romantic to recall very sharply the old practical farms.

The two different cities within the walls, different, but, as should become apparent, intricately connected, were of course not isolated from the great world outside. Beyond their connection with their immediate suburbs and short of their connection with all the provinces of the great church whose center they formed, they, as the composite city of Rome, had a very special, although certainly not always amicable, connection with a fairly large group of provincial towns. The papal curia itself was only intermittently at Rome. Perugia, Orvieto, Rieti,

Viterbo, Tivoli, and Anagni, as well as more distant places, were also papal capitals. Denizens of these towns, as well as Romans, became particularly involved with the papal court and, as scribes or butchers or spicers, moved about with the curia. In the other direction, Roman nobles who were at home in Rome in the Campo dei Fiori or the Mausoleum of Augustus or the Theater of Marcellus had major holdings and palazzi in smaller towns, and they had great country estates and fiefs throughout central Italy in places like Palombara, Palestrina, Terracina, and Ninfa. Roman monasteries and religious institutions controlled large portions of smaller towns like Sutri and Campagnano. So, in a way, a realistic map of Rome would not be bounded by the walls, but rather it would stretch out into the surrounding provinces in a series of superimposed networks following personal connection and the interests of real property.[8]

Within the city there were physical divisions other than urban and rustic. The Tiber divided the city, but at the same time it, and its bridges, joined the city's parts, its three or four, not just two, parts, because the Vatican, with its Leonine walls, and Trastevere proper were not joined within the city walls. The river also gave the city at its center an island with a tradition of healing and a healing well. On the island and along the river's sides were landings, mills, and crowded and jealously guarded fisheries. The city had its seven hills. They were not always identified as the same seven hills, but the important number was seven. Numerology was more important than geography. Some of the hills, particularly the Capitoline and less so the Aventine, were important as hills in the thirteenth century.

Above topography there were man-made spaces and volumes, relatively inflexible and angular, ancient and new, which gave the city its internal physical structure, shape, and texture, gave it areas of physical resistance and extension—and sometimes paved them with mosaic. Beside the small squares and rectangles of towers, the cloisters of religious houses placed their great, measured squares and rectangles among the unruly streets and the irregular fields. The effect of these ordered spaces in the country is still visible and easily imaginable. Their effect in a crowded, disordered medieval city (even one like Rome, with at least small piazzas) must have been much more impressive; it must have been like the effect that mosques still have in the disorder of the medina at a place like Rabat.

Thirteenth-century Rome came to be divided into thirteen districts, or *rioni*. To the extent that these *rioni* defined voting places (and that extent is very unclear), their boundaries must have had some pre-

The Physical City

cision. But, in general, people talking of their *rioni* in contemporary documents seem only to be trying to tell the neighborhoods in which they live. There seems to be no consistency about when a man felt he should identify himself by his *rione;* sometimes he certainly used it as casually as he would a street or a hill, like Antonio of the Via Lata or Pietro of the Pincio. Further, most late thirteenth- and fourteenth-century *rioni* had double names (recalling their double pasts); people sometimes spoke of the *rione* Campo Marzio and sometimes of Santa Maria in Aquiro. There does not seem always to have been a clear distinction in contemporary minds between *rioni* and *contrade,* their smaller divisions, which may merely mean that *rione* (*regio*) was not a completely fixed formal term. So in the first half of the fourteenth century it is possible to describe a man as being of the *rione* Ripa or Sant'Angelo and another man as being of the *rione* or *contrada* Campo Marzio.

It is possible that if better records of thirteenth-century Roman games and contests in the Piazza Navona (then the Agone) or on the Testaccio survived, one could, with some effort, imagine a colorful picture of mutually exclusive and patriotic *rioni* battling for some sort of intraurban prize or glory. But the surviving evidence makes the *rioni* seem in themselves neither important nor colorful. They do sometimes more or less coincide with neighborhoods which do have some significance as centers for certain sorts of artisans or merchants. One can say of thirteenth-century Rome, for example, that the *rione* Sant'Angelo was the *rione* of fish sellers, that the *rione* San Marco (or Pigna) was particularly the *rione* of merchants and their *botteghe,* or shops. But the neighborhood of the Lombards, on the other hand, would seem to have been around the Pantheon (Santa Maria Rotunda), itself the intersection of three *rioni:* Pigna, Colonna, and Sant'Eustachio.

The evidence does not really suggest, either, that in this part of the middle ages the *rione* Trastevere was considered to be another city, as it may have been at other times, as in the mid-twelfth century, and as it has been thought to be by some historians. In the later thirteenth and early fourteenth centuries, Trastevere does not seem to have been separated from the rest of the city in any peculiar or artificial way— except in the case of the double senators of 1306. So, for instance, in 1290 Angelo Malaspina is described in an ordinary way as a doctor of laws and a citizen of Rome of the *rione* Trastevere. What does seem important about *rioni* is that there were, or came to be, thirteen of them, a number which, with its multiples, played an important role in the organization of corporations and of the commune during parts of the later middle ages. Thirteen *rioni* could mean all of Rome, as four corners has sometimes meant all of the world.[9]

If the sense one has of Rome divided into *rioni* is palely abstract, this is not at all the sense one need have of a Rome divided into parts and looked at in other sorts of pieces. One of the most exciting religious corporations and religious ventures of thirteenth-century Rome was the combination of the old church of Santa Maria in Sassia and the new hospital of Santo Spirito in Sassia. Sassia was the old Saxon or English *borgo,* one of that series of *borghi,* or quarters, near the Vatican which maintained the names of the foreigners who had once clustered there. Santa Maria still retains, in paint, romantic memories of Anglo-Saxon kings; and in the thirteenth century it still had distinct English memories and at least slight English connections—King John was a donor to it, and in 1290 a Thomas Anglicus was a member of the community.[10]

The hospital was organized and supported by Innocent III ("Holy Scripture teaching him"), who connected it with the new order founded by Guy of Montpellier for maintaining hospitals. Innocent arranged a yearly liturgical station at Santo Spirito, to which the Veronica, the great relic of the towel on which Christ's suffering face (according to the then increasingly accepted interpretation) had painted itself as He went to His crucifixion, should be carried by the canons of Saint Peter's from its home in the basilica. Innocent provided that on the annual occasion as many as one thousand paupers from outside the hospital and three hundred from inside it should each be given three *denari,* one for bread, one for wine, and one for meat. Innocent ordered that the station should occur on the first Sunday after the octave of the Epiphany when the gospel, until recently, was read: "Jesus went with his disciples into Cana in Galilee." The Veronica was to be carried in a reliquary of gold and silver and precious jewels and then shown to the people. The papal gift to the poor was meant to encourage others to give. And because man does not live by bread alone, but by the word which comes from the mouth of God, the pope with his cardinals was to come to the station and celebrate a solemn Mass and preach a sermon. Innocent's audience was to be fed in three kinds —carnally with food, sacramentally with the Eucharist, and spiritually with the food of instruction. Indulgences were to be granted to those who were present.

Innocent found the day's gospel text appropriate, as it in fact seems, but particularly because in the hospital of Santo Spirito the six corporal works of mercy echoed the six vats of water turned into wine,

The Physical City

vices turned into virtues, evil into good. Just as, moreover, Jesus came with his disciples to the marriage at Cana, so, preached Innocent, the effigy of Jesus Christ carried by the ministers of the church (the canons of Saint Peter's) came to Santa Maria and Santo Spirito in Sassia; just as the Mother of Christ was invited to the wedding, so here is the church dedicated to her. The establishment of the station involved the whole pious paraphernalia of thirteenth-century modernism—the face of Christ, His remembered life, the sacrament in the Mass, the corporal work, the active pope—in this new hospital set near the river, a little to the south, between the Vatican and the river.[11]

Its location near the river is important at least to Santo Spirito's legendary history. By the end of the middle ages it was said that Innocent had been driven to the foundation of the hospital by a terrible dream about the Tiber, that fishermen who went to the river brought back nets full not of fish (or of men, exactly), but of the little bodies of dead babies unwanted by their sinful mothers. Innocent, as followers of Francis will recall, was a pope who attracted dream legends, but there are parts of the hospital's earliest rule which lend some substance to the story. Ejected orphan infants were to be nursed and cared for by the house, and poor pregnant women were to be freely received there and ministered to with charity. Moreover, sinful women who promised to remain chaste were to be given a haven, if they wished it, at Santo Spirito, through Holy Week and the octave of Easter. There were also to be little cribs in the hospital, so that the infants born of pregnant pilgrims there could lie commodiously and alone. But the hospital was clearly not designed solely for infants. One of the rule's most impressive chapters demands that once a week infirm paupers should be sought out through the streets or alleys and piazzas of the city (*per vicos . . . et per plateas,* phrases constantly used about the city), and brought back to Santo Spirito and diligently nursed and cared for. And in another place the reception of the sick is described. They were to be confessed, given Communion, and then taken to bed, and fed and cared for each day in charity. When the bell rang each day for dinner, each unoccupied brother was to rush to serve the poor, and each Sunday the priests and clerks, brothers and sisters of the hospital were to go through the hospital taking the Mass and the gospel and epistle to the sick. In this hospital dedicated to the Holy Spirit, where the *Veni Creator Spiritus* was sung, the illumination of the Spirit was meant to encourage a various physical charity.[12]

Innocent gave his physical piety a physical setting. There is much more that is physical, particularly in instruments of gift and rent and sale tied to specific places, which remains from Santo Spirito. Some of it makes strange contrast with Innocent's preached homily and regulating letter. In 1306, ninety-eight years after Innocent's act, the Masters of the Buildings of the City were forced to take action in response to complaints about conditions in the neighborhood of Santo Spirito. On the street to Saint Peter's, much frequented both by Romans and by foreigners, across from the gate of Santa Maria and Santo Spirito and opposite some of their houses, were some vacant lots or gardens and a charcoal pit. Into these vacant places so many people had thrown so much garbage and filth that the smell was horrible, revolting to the passersby on their way to and from Saint Peter's, and dangerous, it was thought, as a source of infection to the inhabitants of the hospital. (The chapter of Saint Peter's had repeated difficulties with the cluttered approaches to their basilica. In 1233 and 1279 they brought cases before the Masters of the Buildings against the denizens of the Leonine city in an effort to keep the streets, including the Ruga Francigene—"French Street"—and the piazzas leading to the basilica clear of projections added to buildings. At one point they were permitted to project into the piazza no farther than the portico of one of the houses of Santo Spirito and Santa Maria.) In the 1306 garbage case the Masters investigated, collected evidence, sent their submasters, one a mason, and a notary to the spot, and finally themselves went to find out who was responsible for the vacant land and who should close it and keep it clean. The answer was clear. Marking the boundaries of the vacant land were three pieces of marble, of which one was noted to be antique and one a little column that was marked with that cross which was the sign of the hospital of Santo Spirito. It identified the hospital as the responsible owner. This sign was also on the hospital's wax seal, and, as a medieval archival mark, it still identifies some of the hospital's fourteenth-century documents. It was the mark, "the banner of the cross" (*vexillum crucis*), which the rule, without describing it, ordered brothers of the hospital to wear on their capes above their chests and on their cloaks on the left side. On the hospital, the cross with two transverse bars, the *crux gemina*, still marks Santo Spirito's part of Rome.

The business documents, the instruments dealing with real property, preserved by the hospital of Santo Spirito construct a detailed, but incomplete, map of the hospital's local neighborhood: who lived next to whom, in the *borgo* of the Frisians, on the corner of the streets leading to San Michele and to San Lorenzo "de Piscibus" (an intersection which was in a place just off the eastern end of Bernini's south colon-

The Physical City

nade); how the lot which was being transferred was held and how much it was worth; where its gardens were; and that, in this case, there was an attached piazza in front of it and that it itself was an inhabited house with a hut without a roof—or, in the fourteenth century, that a man from Gubbio rented a house from the hospital for seven florins a year, that the house was placed in the hospital's portico, that it had a sign of four faces on it. But the point about the real neighborhood of a place like Santo Spirito is that, although it had a local node, it was not in a physical sense restrictedly local. It is true that even outside their closest home (in the Arenula, outside the Porta San Pancrazio, or deep in the country) the brothers of the hospital found neighbors in the family Papareschi and the churches of Santa Rufina and Santa Maria in Trastevere, local neighbors in the sense that they were from Santo Spirito's side of the river. Still, Santo Spirito, like other corporations and families in the city, built up subsidiary neighborhoods far from the central sign of the *crux gemina,* as far away, in Santo Spirito's case, as Gallese and Nepi. The brothers, if we can trust their names, themselves came from as far away as Trani and Offida and even England. The center of their complex of neighborhoods of property was the chamber of the hospital in which the rents were received.[13]

Around this center, if the fears predicted in the rule for the house and order were realized, the action was sometimes disturbingly picturesque—by it the neighborhood was enlivened. The house and order had of course to face the possibility of trouble, to be prepared to be charitable but firm, forgiving but cautious, to be careful in the weighing of evidence. They had to be prepared to treat properly the arrival of heresy, or leprosy, or magnates in a house—all to be treated appropriately, the first with vigor, the second with practical charity, the third with deferential propriety. The house had to beware of becoming a post exchange or a bank; the order had to be concerned with the problems of elections and buildings and burial and of brothers' wandering through the streets alone (for which they could be punished through their food and where they ate it). The order had to avoid for itself the sort of sexual problems that might have arisen had married couples been admitted together, had the old and young been placed too closely together in dormitories, had men and women been allowed to serve each other too intimately (to wash, for example, one another's heads). Fornication committed by brothers and sisters of the house was a special problem, and for it, when it was a matter of public knowledge, there was a special remedy: public, very public, penance to be performed in the same town or vill where the fornication had been performed, on Sunday after Mass, as the people were coming out of church. Rules were established for dealing with brothers who beat servants to the point of drawing blood—and the difference of blood from the nose and other blood was recognized. Further rules were established for dealing with any servant who argued with a brother and called him *latronem* or *fornicatorem* or *malefactorem*—he was to be beaten through palazzo and cloister and ejected from the house; for dealing with any servant who had stolen (and been proved certainly to have stolen) goods of the house, and thus of the paupers—with the stolen goods tied around his neck, he was to be beaten through the palazzo to the *portone,* and then with bread in his hands he was to be thrown from the house; for dealing with servants who fought with each other clamorously to the drawing of blood and the use of weapons (and about whom this had been established by witnesses)—they were to be dragged through the hospital to the *portone* nude, beaten very hard with straps, and thrown from the house. These punishments of servants, more startling than those prescribed for the brothers of the house, but in a way comparable to those prescribed for the monks of English Eynsham convicted of its bootless crimes (sodomy, robbery, heresy), who were shot forth from its gate called the Wiket, must, if in fact instances of them occurred, have proved as arresting to pilgrims passing on the way to Saint Peter's as the smell of garbage.[14]

Santo Spirito, in spite of its retained crosses and the popping servants of its rule, may seem an insufficiently evocative example of a local Roman neighborhood, lying too inconspicuous and undefined, too built-over, in the flat land between the Vatican hill and the river. The most conspicuous remaining medieval neighborhood in Rome is the greater Aventine, with its clear remnants of the domination of a single thirteenth-century family, the Savelli, with Honorius IV Savelli's remaining curtain wall, with Savelli tombs within the Aventine cluster of medieval churches of the newer orders, Premonstratensian, Dominican, with Saint Dominic's orange tree. The Aventine is distinct.

On the Aventine stood the monastery of Sant'Alessio, or more properly and formally Santi Bonifazio e Alessio. It still stands. From some angles, from the river, from across the city on the roofs of the Piazza Mattei, it looks much as it must have in the thirteenth century. In certain lights and in certain seasons, on April mornings as the birds fly about the trees that now stretch down from its heights to the river, Sant'Alessio is oddly, very oddly, reminiscent of Durham hanging above the Wear. Durham proctors in the later thirteenth century on their way to a Savelli pope on the Savelli hill, encrusted with the

The Physical City

Rocca Savelli and dominated by the churches that the Savelli patronized, must have felt a momentary illusion of enclosed, homely holiness before the strange monastery.

Whatever the thirteenth-century or the modern impression of the Aventine and its houses, they were not really self-contained or restrictedly local. Nor was Sant'Alessio, for example, without a very heavy burden of guarded real property. Sant'Alessio held the gardens around the monastery. It also held gardens, lots, and crypts under the monastery stretching down to the Tiber at that part of the Tiber's bank called, from its ancient ruins, Marmorata, with five fishponds or fisheries there, and the Palazzo Eufemia, and the lands around it down to the road that led to the Porta San Paolo and to the Testaccio and the lands around on the left hand back to the church of San Nicola de Aqua Salvia and to a place before Santa Prisca and around to the road that led back to the monastery, between it and Santa Sabina—a huge local swathe still suggested by roads in maps as late as Falda's of 1676. And because Euphemiam was Saint Alessio's father, in the Palazzo Eufemia and Falda the fourth and the seventeenth centuries seem joined. But Sant'Alessio holdings swept beyond this central conglomeration out the city gates—Porta Appia, Porta San Paolo, Porta Maggiore, between the Tuscolana Vecchia and the Tuscolana Nuova—and into the country and the distance, to mills, vineyards, towers, fisheries, saltbeds in Ostia, and shipwrecks far away. Properties brought goods and money back to the monastery. Three of the named fisheries on the Tiber ("Palatellum strictum," "Columpnellam," "Farum") were in 1278 rented for five years to two men from the *rione* Sant'Angelo and the *rione* Ripa for forty *soldi* a year and fourteen "lovely, fat shad with their roe" each April. And in 1271 one-half *pezza* of vineyard outside the Porta Appia was rented to a widow, the guardian of four children, for three and one-half *soldi* and one-half *rubbio* (perhaps four bushels) of onions to be paid each feast of the Assumption, and for one-quarter of the new wine at the vintage and one-half canister of grapes, and for food and drink for the monastery's representatives when they came to collect the dues. There were to be Tiber fish and Appian onions for the canons of the Aventine. The interests of Sant'Alessio were hardly more bound by their church and cloister than were the actions of Honorius IV by his Aventine wall. These Roman neighborhoods were not, for institutions, neat little mosaic tiles with sharp edges which could be fitted together to make a whole pattern.[15]

Within the walls, across the neighborhoods, who lived in Rome, and how? It is quite conceivable that this might be the sort of question which could not be answered at all—or which could only be answered in terms of the great itinerant curialists or the more dramatic members of the greatest Roman families who spread themselves across the pages of chronicles and political history. Fortunately this is not the case for thirteenth-century Rome. One can, in a way, look inside their houses and see the Risabellas and the Clarastellas, the Giovanni di Pietros and the Pietro di Giovannis, James Shortarm called Four Feet and Francesco di Orlando, the man in the *rione* Colonna in 1269 with his century's name.[16] One cannot see them in the intimate acts of their daily lives—pouring water from their pottery pitchers, eating their chickpeas, and perhaps their green sauce, their fried flowers, their *torte di Re Manfredo* of fresh *fave,* and drinking all those gallons of suburban wine—unless they appear as actors in the cast of a saint's life, a chronicle, in the testimony of a trial, or as models for a painting or mosaic.

What one can see of ordinary Romans, repeatedly and in detail, is any act or condition (and there are very many of them) connected with one characteristic Roman and human activity, the transfer of property. It sounds like a limiting phrase, and it is. But at best, one can be told even startling things in property transfers: that as a woman watched the Elevation in a Roman church (the Augustinian Hermits' San Trifone) she was so moved by religion that she vowed to give a great fortune to God and his paupers (but to God at Santo Spirito—where there were paupers—not at San Trifone), a fortune so great that the four thousand florins reserved from it for the nuns of San Lorenzo in Panisperna seemed a minor reservation. This moment of revelation is from what is essentially a charter—although it must be admitted that it is late, from the 1350s, and not the act of an ordinary woman, but rather of Margherita Colonna, the daughter of the Magnifico Stefano Colonna and the widow of the Magnifico Giovanni Conti.[17]

The spiritual elation of a Roman widow at the Elevation of the Mass is not of course the sort of information that charter evidence gives in abundance about Roman society and topography. The abundant evidence makes clear that Romans lived with their mothers-in-law, that is, that extended families of various generations lived together in one house, or half-house, although they might in fact hold various other properties as a group in other parts of Rome or the suburbs. It makes clear that they paid their rents to their landlords, or received rents as landlords, in both money and produce, and particularly in wine. It makes clear that tenements could be sublet, with the permission of the landlord and his retention of his right to ground rent, but that in the actual area of Rome there was seldom a ladder of more than two proprietors on a piece of land, and that often there was only one.

The simplicity of Roman vertical tenure makes contrast with its horizontal complexity. Reversions and usufructs for life were common. So in 1220 a woman who held a close with trees near San Biagio (San Francesco a Ripa) in Trastevere next to her own uncle's property granted its use for her son to an appointed administrator, reserving it for her own life and for her mother's, but also granting that should her descendants fail the property should go to the monastery of San Cosimato in Trastevere. In spite of the fact that, in some ways at least, the mixed Roman-Germanic law of medieval Rome was less favorable to women than that of classical Rome and that rural feudal tenements might be withheld from women—*quia femine in feuda venire non possunt* (in 1207)—the positions of women and minors were important and protected both in theory and practice in Rome; and this in itself encouraged horizontal tenurial complexity.

Inheritances, moreover, were frequently not divided. Extended families frequently held properties jointly so that they were tied together by their possessions as well as by their conventional living conditions, and sale or gift sometimes replaced part of the family and injected a stranger, or strange institution, into the joint holding. Individual men and women in thirteenth-century Rome were at least as much held together in groups (most normally in family groups) as they were divided by their real property. And much of a Roman's future life and future tenure was decided while he, or she, was a minor and thus a particularly restricted member of the family group. The actual pieces of land (often in half-forgotten, named places—"the Plain of the Palms," "the Three Columns") were of irregular size and shape, and of curiously various extensions above and below the ground (up into lofts and down into crypts and metal deposits), but they were regular enough so that the absence of a fourth side in the case of a triangular piece of land, for instance, was sometimes worth noting, so that it would not seem that one boundary had been left out of a description.[18]

Roman instruments of transfer also tell a lot about who lived next to whom and what their occupations were—that rolling abundance of shoemakers and notaries. The impression one derives from the instruments is overwhelmingly of a society not at all without class, but in which the classes, rather loosely defined, were, geographically, all closely intertwined. The effect, in palazzi and in neighborhoods, is rather like that of those Roman witness lists which put noble and butcher, scribe and smith, next to each other, just as they do the man from Palombara and the man from France.[19]

The thirteenth-century Roman lived in a house with a garden beside and behind it, next to, or behind, the old wall of a monastery, San Silvestro in Capite, for example. He held his land with his family and paid a ground rent of two *denari* a year to the monastery. One of his neighbors was a notary or scribe, and the other was a set of four heirs. On the fourth side of his property passed the public road. Or he was a shoemaker living with his wife across from the portico of the Pantheon, with, for neighbors, a notary, a butcher, and a tower of the Sant'Eustachio family. If he had a last name, an agnomen turning surname, it began (surprisingly frequently) with Bucca or Bocca; every sort of mouth from beautiful to dark identified Romans, but whether they were called by the orifices of their persons or the doors of their houses (whether they were out of Lenore Kendall or Alfred Tennyson) is difficult to tell.

To get a real sense of what these documents of land transfer are like, and what sort of ordinary information they give, the sort of information from which these conventional thirteenth-century Romans can be built, one must look at the detail of a few ordinary documents. But it is of course true that these documents are less interesting and make less sense when they are seen in isolation than when they are seen in the context of a related known family or institution like the Orsini or San Silvestro in Capite. In 1245 a group of owners sold one *pezza* of arable land of five *rubbi* (one should think of a plot of no more than a few acres at most) in the Bravi outside Porta San Pancrazio at the Fontana Vulinna. The owners sold the land for thirty-seven and one-half *lire*. The neighboring property was held on two sides by the hospital of Santo Spirito, on the third side by the church of San Biagio (whose priest Gualterio was present for some part of the negotiations), and on the fourth by a man named Guido. The land was sold to a man named Accurimbono and his heirs.

The Physical City

The interesting part of this transaction is its sellers. They held the land jointly in five shares. One share was held by the church of Santa Maria in Trastevere (and nine clerks of the church, including three priests, consented to the action). The other four shares belonged to a collection of descendants of a man named Locrerengo, by then, 1245, dead. The document's description of the relationships of the sellers and their consenting relatives unveils a difficult genealogy of over thirty people in four generations and in the four collateral branches descended from Locrerengo's sons (that is, from the sons of Locrerengo the elder, for, although none of his sons was called Locrerengo, two of his grandsons were—a not uncommon Roman phenomenon). Locrerengo the elder had a son Giacomo, then dead, whose widow's name was Biancofiore. Giacomo and Biancofiore had had three daughters: Purpurea, Giacomina, and Buona, who herself had a child named Giacomo. Purpurea was also dead, and Biancofiore, the grandmother, was guardian of Enrico and Buona, two of Purpurea's children who were still minors. (Purpurea's other son, Bartolomeo, was not, for purposes of consenting to this transaction, considered a minor. Full majority arrived at twenty-five; but there were grades between the complete inability of the Roman infant and the freedom of the full adult.) Locrerengo's son Riccio, who had married a woman named Teodora, was also dead. One of Riccio's three sons was named Giovanni. Giovanni himself was dead in 1245, but his wife Giacomina survived, and she was the guardian of her three minor sons, Matteo, Petruccio, and Giovanni, who held and sold one of the five shares of the land in the Bravi. Giovanni's two brothers together held two shares. One share was held by Buonfiglio, a son of the older Locrerengo and husband of Oddolina. Locrerengo's other recorded son, Rufino, who had been married to Stefania, was also dead. From this marriage there appear to have come four children: Pietro, married to Adelascia; Locrerengo, married to Stefania; Rufino, married to another Stefania, which marriage had produced a daughter, Viola; and Giovanni.

From the actions of 1245 it is impossible to tell how this tangle of family had molded its interests in the Bravi property into their 1245 shape. One must suspect that this *pezza* of land outside the Porta San Pancrazio represented a fraction of a larger inheritance which was divided equally among a group of heirs and that if all the shares of the total inheritance were reconstituted this distribution would not look so strange. Nevertheless, the sense of intensely populated conglomerate ownership would survive. The repeated acts of consent that medieval Roman law advised, and also the joint holding, broadcast the propertied reality of the family group. The additional presence of the nine clerics of Santa Maria in Trastevere, representing their share, makes clear that this sort of tenure did not forbid the entrance into this joint-stock landholding of nonrelated men or institutions.[20]

Across Rome in the *rione* Trivio, or Trevi, almost twenty-five years later, in 1269, a matron named Angela was selling her tenure of a house with a garden to Rodulfucia, widow of Rainaldo Schibane, for eight *lire* less five *soldi*. The house is said to have been "by the church of San Giovanni de Ficocia," that is, just off what is now the Via della Panetteria where it is joined by the Via Maroniti, between the Trevi and the Tritone, in the Campo dei Arcioni. Angela held the property from the monastery of San Silvestro in Capite, which received two *soldi* for its consent to the transfer. The property was bounded on its four sides by the tenement of Angelo de Vento, who held from San Gio-vanni, by that of Lorenzo di Giacomo di Bartolomeo Caczocotti, who held from the monastery of San Silvestro, by the heirs of Tommaso di Giovanni Iaconi, who held the property behind, and, in front, by the public street. The witnesses included a Gottifredo de Monticello. The consenting prior of San Silvestro was Giovanni de Monticello, probably a man from the *rione* Arenula, but possibly from the prominent noble family that gave the church the antipope Victor IV. Of the other four witnesses, two were scribes or notaries (besides, of course, the redacting notary), and one was a shoemaker named Pietro di Benedetto. Angela herself was identified as the wife of Giovanni Macrapellis and the daughter and heiress of the dead Matteo da Sermoneta and his widow Sapia. Sapia, who was by this time remarried to Angelo di Stefano di Tebaldo, consented to the sale. It seems a small woman's world by the Ficocia.[21]

This area near the Trevi was one of the neighborhoods in which San Silvestro in Capite had a number of holdings, and neighbors and principals reappear in various instruments. Just a week before Angela's sale, on January 8 rather than January 15—a pair of Mondays— her neighbor Lorenzo di Giacomo di Bartolomeo Caczocotti had bought a house with a garden beside and behind it from Rainaldo di Egidio for seven *lire*, five *soldi*, with the promise of a yearly ground rent of four *denari* to the monastery of San Silvestro, to be paid on the feast of Saint John in the summertime. (The *capite* of San Silvestro is generally thought to be Saint John's head, which is San Silvestro's greatest relic.) San Silvestro was also paid a five *soldi* fee for its con-sent. The boundaries of this tenement were the tenements of Sapia (held of San Silvestro), Angela, Blasione, and the street. Rainaldo's

The Physical City

family appears—a wife Maria, a son Giovanni, and a daughter-in-law Rosa. Gottifredo was again a witness, as were Angelo de Vento (Angela's neighbor), the priest Donato of the church of San Salvatore in Onda (by the city side of the Ponte "in Unda," or "Ianiculensis," or "Agrippae"—now the Ponte Sisto), a marblecutter (*marmorario*) named Nicola di Lorenzo Romanelli, and a shoemaker (this time a *sutore* not a *calzolario* as before) named Bonagura.[22]

Lorenzo's document should be put next to the immediately preceding one (among those preserved by San Silvestro in Capite) in which a widow Romana, whose husband had been a shoemaker (or descended from a shoemaker or named Calzolario), is shown selling three *pezze* of vineyard and one-half of a wine-making vat to a man named Romano for fifty *lire*. The vineyard is described as being outside the Porta Nomentana or outside the Porta Salaria, at the watercourse called the Forma di San Silvestro, next to the *forma,* and bounded otherwise by a *fontana,* with the public road at the vineyard's head and the river at its feet. From this vineyard, instead of money San Silvestro, not to be robbed of its rights over the land, was to receive one-fourth of each year's wine and each year three canisters of grapes.[23]

In 1200 the abbey of San Silvestro had rented for twenty-nine years for a lump sum and an annual money rent a house with a garden behind it in the *rione* Colonna and the *contrada* Vigna to a mother and daughter, Altemilia and Hostisana. Their neighbors again form an interesting group: on one side, the heirs of a man named Gregorio di Lorenzo, who held their tenement from San Silvestro; on the second side, Oddone the Miller and the heirs of Benedetto Romano (or di Romano), who held from the monastery of Santa Maria in Campo Marzio; on the third side, Nicola and Bobolo, the bastard sons of Maccafora, who also held from San Silvestro; on the fourth side, a lane that ran to the street. The lane ran in and out of a little neighborhood of congested monastic tenements, with mother and daughter, heirs, bastards holding together.[24]

These four San Silvestro documents suggest that tenants paid money rents within the city and rents in kind outside the city walls. As a general pattern, this would seem to make sense, but it was certainly not always the case. Sant'Alessio's shad and roe were from within the city, so were some rents in wine. In 1232 the monks of San Cosimato in Trastevere were attempting to establish a vineyard on a vacant piece of land in front of their monastery and within the walls of Trastevere. For the first four years of the tenure, which was meant to be perpetual, the new tenant, Sassone Ferrario (or Sassone "the smith"), was to hold freely, but from the fifth year he and his heirs and assigns were to return each year to the convent one-fourth of the new wine and a measured canister (two *palmi* across and one *semisso* deep) of grapes. The vineyard-reestablishment, wine, and canister agreement ought not to be thought of as peculiarly monastic. In the 1270s Compagio di Giovanni Lucidi planned a similar rent on land which he himself had recently acquired outside the Porta Aurea in the Vatican near Santa Anna and which (probably) he was about to lose to the papacy. After four years of tenancy, the rent was to consist of one-fourth of the wine and two and one-half canisters from two *pezze*. In the agreement between San Cosimato and Sassone there is a clause which is usual in these Roman agreements; provision is made that if Sassone should find gold, silver, iron, lead, or any metal or stone worth more than twelve *denari* he should return half of it to the convent and keep half of it for himself.[25] This common clause is really very prominent in Roman contracts; the sense of the rich ground full of useful and romantic treasure pervades the business of transfer and lease.

An arrangement somewhat similar to the one between San Cosimato and Sassone was made with Radulfo Carbonis in 1248 for land in the neighborhood of San Giovanni in Lombrica; the monastery demanded one-fourth of the oil and fruits and eventually one-fourth of the wine. (Like the Sassone agreement, the Carbonis arrangement was in purpose and style of the general *emphyteusis* type.) But one cannot generalize to the point of saying that in Trastevere payments were in kind. In a transfer of a house with a garden beside and behind it in the *contrada* Santi Quaranta (Quaranta Martiri or San Pasquale Baylon, at the corner of what are now the Via di San Francesco a Ripa and the Via delle Fratte di Trastevere), Barone di Paolo di Viola and his wife Regimina di Biagio di Romano Mellini, acting with the consent of her sister Caracosa and her sister's husband Pietro di Gregorio, received four and one-half *lire* from the buyer, Prassete di Giovanni di Pietro, saving always the rights of the charter of renewal (presumably after nineteen or twenty-nine years, or three generations, the term is not stated here) to the monastery. The instrument of renewal may have of course insured a yearly rent in kind, but there the abbot and monastery receive money—six *denari*—for their consent to the alienation of this house "of their right" in a neighborhood in which their rights were thick. In fact, neighbors on two sides of the Quaranta Martiri house held their property from the monastery—and on its fourth side the house was bordered by the street.[26]

Wine and grapes were not the only rents in kind. In 1191 San

The Physical City

Silvestro in Capite had rented country property for money and an annual rent of one-half flagon of oil, the flagon to be three *palmi* in circumference, the *palmo* to be measured by the *palmo* marked in the stone at the door of the monastery. In 1198 the convent of San Cyriaco in Via Lata contracted beyond a money payment for a full and pleasant combination of annual rents in kind: twenty *rubbi* of good grain at harvest, one *rubbio* divided between beans and chickpeas, six *salme* of good wine, three *manci* of bundles of wood, all to be brought to, and inside, its cloister near the Via Lata (the Corso). In 1249 San Silvestro bargained to receive two one-pound *saculas* of wax or their money equivalent, in addition to its money rent for property outside the Porta Tiburtina.[27]

Nor were all Roman holdings composed of real property in the narrow sense. As elsewhere, property could be in rights. So in 1218 two inhabitants of Rome bought two days (with their nights) in a river mill at Gallese, for six *lire* of Lucca. Three years later, other Romans bought the other five days of the mill (with equipment—a property of mixed nature) for fifteen *lire* of Lucca. (The price for a day had not changed.) River mills were an obvious and important part of the economy both in Rome and in the surrounding countryside, and the clauses with which mill tenants protected themselves against the devastations of floods and armies indicate the conventional instability, natural and political, of the countryside.[28]

If one reads enough of these medieval contracts, the city and suburbs are repopulated with petty farmers in their gardens. The garden of the house with half a well by the church of San Nicola "de Melinis" is green again (where now there is only the broad, ugly grayness of the Arenula) in an instrument of the notary Nicola Mellinus, as three witnesses from the Giudea (or Giudia) and Cola Fusco watch a man from the *rione* Campo Marzio acquire property. And around San Nicola a little neighborhood of destroyed churches is returned (one might not guess in looking that there had once been so many more churches): San Valentino de Piscina, Santa Maria in Giulia (with its convent), San Salvatore in Giulia, San Nicola di Calcarario—and the Calcarario itself, the neighborhood where, in the middle ages, marble antiquities were, it is believed, remodeled or destroyed for modern uses, for lime, where now are cats, an underpass, where perhaps Caesar died, the enclosed, restored ruins of the republican temples, and the queerly rebuilt medieval tower of the Boccamazzi.[29] Besides all that, in these documents, all over Rome one hears again the coughing and scraping feet of hundreds of financially interested in-laws.

The recovered greenness and in-laws cluster in one particular set of documents connected with a major, and famous, recovery of property in the 1270s. A rather beautiful inscription preserved in the Palazzo dei Conservatori in the Capitol (where it was brought in 1727) says that in the first year of his reign Pope Nicholas III (who was consecrated on December 26, 1277) had enlarged buildings in the Vatican and in his second year built the walls of "this *pomerium*"—this open space—the *viridarium novum;* and one knows from a chronicle that he also made a fountain. Gregorovius wrote of this, rather excessively: "The feeling for nature thus again awoke, and for the first time for centuries, Rome saw a park laid out." Nicholas III's work at the Vatican required the acquisition of a number of small farms, like the vineyard of two *pezze* "next to the old walls of the Leonine city behind the sacred palace and the houses of the lord pope on which were built the latrines of the palaces." What the actual attitudes of sellers and buyers to one another was, it is impossible to say, but their negotiations caused two parts of Rome to meet face to face: on the one side were the clerks of the papal camera, professional accountants, and other hangers-on of the papal curia, like the papal scribe who was the rector of Sant'Egidio outside the gate (a repeatedly scribal church, which Boniface VIII would give to the basilica of Saint Peters); on the other side were the Romans who held the neighboring vineyards, and their families.[30]

The actual sales and consents to sale, which stretched over some months, were regularly in two parts. The principal seller was brought before the acting chamber clerk at the Vatican, and there he sold his vineyard, often before curial witnesses, who obviously were present in their professional capacities, and also before other sellers. Later, a chamber clerk with witnesses and notary would present himself at the house in which the seller lived in Rome (because the sellers did not live on these pieces of vineyard) to gather the consent of the seller's wife and other interested members of his family. So, in April 1278 Nicola di Giovanni di Angelo de Amatiscis of the *rione* Parione appeared with his nephew Federico in the camera, and on the same day Nicola's wife Angela gave her consent at home. On May 1 Giuliano di Lorenzo di Pietro di Lorenzo de Cerinis and his brother Pietro of the *rione* Ponte (across the river from Castel Sant'Angelo) appeared in the camera, and then consent was given by Giuliano's wife in front of the house where they lived (the brothers' father was by this time dead, but theirs was at least a ménage of three). On May 5 Pietro di Ugolino de Speculo, also of the *rione* Ponte, for himself and in the name of his

The Physical City

brother Ugolino sold, and Oddolina, Ugolino's mother, and Giacoba, his wife, consented—the wife through her father Don Giacomo Rosso—before the brothers' house. The next day the brother Ugolino appeared for himself at the balcony or windows of the chamberlain's palazzo in the Vatican.

So, one after another, they appeared, people from the Piazza of the Botteghe ("shops") of Castel Sant'Angelo, from the Arenula and Campo Marzio and San Lorenzo in Damaso, the important Compagio di Giovanni Lucidi from Sant'Eustachio, actually from right by the Pantheon. And the clerk and the notary went out around Rome to the sellers' composite homes, to Bartolomea, Compagio's wife, and Aldreda, his son Giovanni's wife, at Compagio's house, to Donna Comitissa and Donna Teodora, the wives of the brothers Pietro and Paolo. They went to the house of Giacomo Magalocti, who had sold in his own name and that of his three brothers, Lorenzo Magalocti called Locti, Ponzello, and Tommaso called Sucio; and the three brothers and Teodora, Giacomo's wife, consented at home. An uncle-guardian, Andrea Barbarubea (or Barbarossa), who was related probably to another seller, Pietro Bursa de Barbarubeis of San Lorenzo in Damaso, sold for his nephew; and two nieces and their mother, Andrea's sister-in-law, consented at the ward's house. A sale by Matteo di Bartolomeo Bavosi de Malioczis was consented to at home by Constanza, his mother, and Froga, his wife. Besides the sense of life in the city, and besides the bringing together of Vatican and Pantheon, which these documents provide, they make clear certain points: Roman extended families lived together even when the family held more than one property; they lived in the city and held farms in the suburbs; the people who held these little farms (although not always of little value, as Compagio's sixty- and ninety-*lire* vineyards show) were not themselves, in the normal sense of the word, farmers.

One can in some part recover, and see, the actual physical houses in which these people lived, which they owned or leased. (Observable leased houses were generally held on long-term leases like a rush(?) house in the *rione* Colonna leased in 1234 for three generations.) The descriptions of houses between the Colonna and the Porta del Popolo (or Porta Flaminia or San Valentino) belonging to San Silvestro in Capite imply a rustic simplicity that is only somewhat surprising. Earth, rubble, and rush, and sometimes tile, would seem to have been the materials for building. The actual structures were sometimes simple sheds and lean-tos (although these may not actually have been much lived in), sometimes built over or adjoining dugout cellars.

Sometimes there were solared lofts, occasionally a tower and a cellar. On the other hand, in heavily populated Rome there is constant talk of gables, towers, and reused classical marble. Everywhere people used convenient classical remnants.

A rich sense of the texture of the second Rome, jumbled, reused, heavily urban, and, at the same time, ruined, is provided by thirteenth-century documents preserved in the archives of the family Orsini (the most spectacular and complicatedly visible branch of the Boveschi family). The late thirteenth- and early fourteenth-century Orsini were a prolific family with many branches of their own, and it was necessary for those branches repeatedly to divide and reassemble great segments of the entire family's holdings. In the late thirteenth century, moreover, parts of the family were involved in the aggressive collection of power and property, in their own aggrandizement. To them, with money available partly from office, including the highest office in Christendom, the urban center of the old city offered one worthy and natural arena for the accumulation of holdings. With the background of their Boveschi inheritance and linked Boveschi holdings, they could and did collect and consolidate a group of neighborhoods stretching in an intermittent chain from the *rione* Sant'Angelo in Pescheria (near, for example, what is now the Piazza Mattei and Sant'Ambrogio, by the de Galganis holdings and the palazzo "of the chancellor," and near the Piazza Giudea), through the Piazza dei Satiri and the Campo dei Fiori, out into the *rione* Ponte by the bridge to Castel Sant'Angelo and the Vatican.

The mélange of property types which they actually collected and consolidated must have suited their purposes perfectly, because the small neighborhoods which formed their great swathe were thick with stones and ruins and towers, fortifiable and fortified, and often high, ideal for the pursuit of "family wars." At the same time and in the same places were scattered shops, granaries, lots, habitable houses, and tenements which produced rent and other income. So, in a nicely balanced pattern, the Orsini could simultaneously support their wars and protect their incomes (although it should not be assumed that different Orsini were always on the same side in a war). The Orsini, insofar as they were urban, and urban Rome fit together so neatly that it becomes an intriguing but difficult puzzle to try to decide which more forcefully shaped the other.

The most strikingly evocative of the Orsini neighborhoods, in the documents and on the site itself, is probably that around the Piazza dei Satiri, or Zatri as it was called in the thirteenth century, a piazza which

The Physical City

still exists just off the curve of the remains of the Theater of Pompey, behind the church of Santa Barbara, near the Campo dei Fiori. There one can see the Orsini with their neighbors and predecessors and landlords, the de Tartaris, the de Stincis, and the church of Sant'Angelo in Pescheria (a landlord from a neighboring *rione*). In the neighboring church of San Lorenzo in Damaso, in 1296, the noble Fortebraccio (di Giacomo di Napoleone) Orsini and four of his nephews acting also for two others (Orso, Leone, and Giovanni di Francesco; and Giacomo, Nicolò Comitis, and Fortebraccio di Napoleone) made proctors for themselves to lease or re-lease, at thirty-nine years, property from the church of Sant'Angelo in Pescheria. In this transaction, the church was represented by one of its canons, Malabranca de Galganis, and the vicar of its cardinal deacon. They intended to lease the *trullum,* or *trullo,* called the "trullum of Donna Maralda" (which Maralda or Marala, it has been suggested, was a wife of an Orsini of an earlier generation), and the place where the *trullum* was, and the *arpacasella* and the place where it was, together with the new tower "now" built there, and the palazzo next to it, and all the crypts and lots and properties of the other adjoining houses and piazzas and their appurtenances in the arc of possessions of the church. (The *arpacasella* has been connected with the *arpacasa,* the tower, according to Saba Malaspina, which Fortebraccio's "Ghibelline" father had among his ruins when he was supporting Arrigo of Castille and Corradino and opposing Charles of Anjou.) The old ruins of the arc of the theater had been joined perhaps by the new ruins of destroyed and confiscated "Ghibelline" fortifications, honeycombed with caves and fortified with a new tower. It is not really clear what in the description is tower and what ruined theater, what destroyed and what standing, but the texture and part of the shape of the used ruin are apparent. This complex of properties, caves, and heights was said to be in the *rione* San Lorenzo in Damaso, and to be bounded by a public road, by the road that went to Santa Barbara, by the Piazza Zatri, and on the fourth side by the garden of Francesco Tartari (or de Tartaris) and the heirs of Leonardo de Stincis. Two days later, the act of leasing took place at San Lorenzo in the presence of five members of the household of Francesco, the Orsini cardinal deacon of Santa Lucia in Selcis; the property's description was repeated, with the careful stipulation that the church of Sant'Angelo was not to lose whatever rights it had in the churches of Santa Barbara and San Martino (de Panarella).

Already in 1242 and 1263 Orsini had been engaged in collecting property around the *arpacasella,* acquiring from a Cenci one of the five parts of the whole *trullum,* called the *"trullum* of Gregorio de Trullo," from the ground to its peak, with its ascents and descents, and the whole shop next to the greater ancient gate to the property, and half a house and half a garden behind the *trullum,* and a whole shop under the *fondaco,* known as the *fondaco* of Leone de Trullo, and property above and below on the side of the church of San Martino, and the vacant area next to the public road. In 1290 some of the de Stincis heirs, holding as a *consortium,* sold their share of the *trullum* "which was once Donna Merala's," that is, one-third of two-fifths, to Don Francesco di Napoleone Orsini (described at this time as a notary of the apostolic see), for seventy *lire* of the senate. The property was described as being in the *contrada* (*in contrata*) San Lorenzo as it was again (and also "in the parish of Santa Barbara") when Francesco Orsini, the notary, bought another one-fifth of the *trullum* ("once of Donna Marala") and one-fifth of the *arpacasella,* for one hundred gold florins. In this last case the neighbors are described as the sons and heirs of Giacomo and Matteo Orsini, the Tartari, the de Stincis heirs, and the public road. Angela, the wife of the seller, who was a Roman citizen named Angelo di Giacomo Rubei Catellini, consented to the sale, in the public street before Angelo's house (although the property had been in the Catellini family); and Francesco Orsini's proctor, the priest Egidio, rector of Santa Barbara, went to the *trullum* and *arpacasella* with Angelo and was physically invested with this Catellini fifth.

Although the descriptions of the *trullum* of Donna Maralda-*arpacasella*-Piazza dei Satiri properties are particularly evocative, they are not really unlike others in the Orsini documents. There are, for instance, the descriptions of the Mannetti properties in the *rione* Cacabario (in a neighborhood with a garden and a house belonging to the little church of Santa Maria in Publicolis) some of which came into Orsini hands in 1294 and was divided into thirds for Fortebraccio and his two sets of nephews. These properties were a collection of tower, oven, garden, and house. One house had a shoemaker in it and one a smith. A related Mannetti document of 1270 talks of the "Palatium Merulatum" with a kitchen and a stable below it, and also a house underneath it where a smith lived, and a scribe's house, and a house "de capite crucis" where a spicer lived, and a barbican, and holes or loops for shooting arrows through, and the arch of a tower, and a columned palazzo, and *podia* and benches, and stones and walks around the place. The document also contains regulations to be followed in dividing the property that concern the height of fortifications, walls, and barriers, and the disposition of liquid waste.

The Physical City

There are also the ruins of towers and houses called the "Baroncina" in the *rione* Sant'Angelo in Pescheria, with a granary and a cloister at the foot of the tower, and a walled lot, near San Salvatore de Baroncinis (on the Piazza Giudea on the side toward the Piazza Mattei) which Orsini (Giacomo di Napoleone and Matteo Orso) bought from Giovanni Cenci in 1271. At the same time the Orsini also got the ruins of the houses called "Quinque Palaria," houses with Jewish neighbors, and a house called "Hosterium," where Matrona Hebrea, Matrona the Jewess, lived next to a spicer and property of the church of San Salvatore de Baroncinis. There are, too, the de Galganis properties around the "Palatium Cancellarii" (of the Malabranca?) in the *rione* Sant'Angelo in Pescheria, near the present Piazza Mattei; their neighbors included the monastery of Santa Maria Massima, Pietro Conti, and Manuel Beniamin de Vitali, the Jew. With part of this complex, Donna Constanza de Galganis was invested through and by means of the gate of the steps of the palace and the door of the enclosure called "de Galganis" in August 1275. The list could go on, as *consortium* after *consortium* (family group after family group), and men and women, buy and sell these mixed pieces of urban Rome, the surface of which was covered with many things including very variously tenanted houses.[31]

Some relatively simple late medieval houses survive in Rome, some not too fussily restored. There is a familiar house across from the church of Santa Cecilia in Trastevere, and another one fairly close to it on the corner of the Vicolo della Luce and the Lungaretta. These houses are very noticeably made of brick and other classical remnants. That this sort of house seems to have followed an ancient Roman vernacular style is particularly clear in the case of houses with open-fronted shops beneath them. Their materials are very like those of contemporary churches like the Aracoeli. They repeatedly use a constantly visible and characteristic arch, like those now to be seen embedded in walls on the Vicolo della Luce or holding overstructures and covering passages as in the Vicolo Sinibaldi or the Arco dei Cenci. If one maintains a certain amount of reserve and preserves oneself from too facile an antiquarian enthusiasm, one can look with profit at the restorations of tall medieval houses at San Paolo alla Regola. A characteristic complex of tower, antiquity, and confused building remains at the Torre Margana near the Capitol, and close by there is a clearly observable specimen of reused material on the Via della Tribuna di Tor de' Specchi (no. 3A), a building which has been tentatively identified as a Boveschi tower.[32]

A nice description of a fairly important house-complex survives from the early fourteenth century, from 1331, only a few years after the popes had gone to Avignon.[33] The describing documents, themselves preserved in the capitular archives at Rieti, are the result of a disputed inheritance (that of Giacomo di Don Giacomo) within the rich and important family of Ponte, a family in fact established around the city side of the Ponte San Pietro (now Ponte Sant'Angelo), the bridge which led from the city to Castel Sant'Angelo and Saint Peter's, and from which the family undoubtedly took its name (although, perhaps indirectly, through the *rione*). The Ponte family was of the very second importance (like, perhaps, the Sant'Eustachio, but without their known age) and was connected with both the Orsini and the Savelli. The 1331 description is included in a statement that invested with property Donna Andrea, the widow of Giacomo and the daughter of Giordano di Andrea Ponte. (She was doubly a Ponte woman.) The decision was based upon gifts *inter vivos,* dower agreements, and palatine and senatorial decisions stretching back more than thirty years.

The description of settlement required a description of Ponte urban properties. The properties described include a warehouse, an artisan's shop or storeplace in the fish market near the Capitol (and Sant'Angelo in Pescheria), a vineyard outside the Porta Castelli in the Prati near Castel Sant'Angelo, and a miscellany of places near the Ponte San Pietro: walled gardens, land with a house held by the church of Santi Celso e Giuliano, the holding of a tenant named Pietro de Palma, or la Palma (as in the local name of San Silvestro de Palma, the neighboring church), a tenement bordering on possessions of the Basilica of Saint Peter's, a piazza with a pergola, the river, the holding of Napoleone Orsini, the street to the Palma, and the water gate. One of these properties was a tower, the Torre de Rainone, with a garden next to it, and the gate between it and the adjoining palazzo.

But it is the adjoining palazzo that catches one's eye. It is described as a grand palazzo and house, colonnaded, with a garden and close, with solars and chambers above it, and a marble stone next to it, and with all its walls, at the foot of the bridge, and on the *via sacra* which led to the bridge.[34] It sounds a *signorile* city scene. But insofar as it is *signorile* the scene is incomplete. The description goes on to the palazzo's shops, that of the scribe (or the son of a scribe) and the widow of the barber, the portico or shop of a man named Giovanni di Jonta (a *sutore*), the house of a saddler and of a pruner, and the shop of a goldsmith, "which houses and shops and palazzo are all joined together in the *rione* Ponte San Pietro between the street of the bridge,

The Physical City

the street that goes to the Posterula (river-gate) Raynone, and the river Tiber." It is a beehive by the river, with the Ponte women, the barber's widow, the garden and goldsmith, all housed together by the tower, the ancient monument, and the "yellow" Tiber. It speaks, as do earlier Orsini documents, the physical involvement of various classes in medieval Rome. It tells clearly of the long heritage of those palazzi of modern Rome, the Costaguti, the Mattei, the Borghese, which, at least now that they have attained some age, shelter within themselves a similar variety. The Palazzo Costaguti, physically late Renaissance and after, is, as a way of life, very medieval Roman.

The picture of the Ponte palazzo comes from the beginning of the 1330s. There is not, I think, a similar picture of a relatively simple and domestic Roman neighborhood before the 1360's.[35] One can know a lot about the thirteenth-century neighborhoods of the Orsini and the tenants of San Silvestro in Capite and San Cosimato. But the way in which Sant'Angelo in Pescheria comes alive in 1363 makes earlier things seem black and white and two-dimensional. Sant'Angelo's brilliance is due to the accidental survival from that and following years of the cartularies of the notary Antonio di Lorenzo Stefanelli de Scambiis, who worked, for the most part, around the area of the fish market and its church.

There are obvious dangers in using a fourteenth-century document for explaining the life of thirteenth-century Rome. By 1363 the popes had been in Avignon for fifty years. By then Sant'Angelo had known intimately the commune of Cola di Rienzo. It had not been ten years since Cola had been killed on the Capitol and his dead body hanged near San Marcello. It had not been twenty years since the Cola allegory crying "Angel, angel" from the fire had appeared on the very wall of Sant'Angelo. Cola had been born nearby in the Arenula, by the river, within sound of the mills, near the synagogue and San Tommaso of the Cenci (in an Orsini as well as a Cenci and Jewish neighborhood), fifty years before the composition of the notarial cartulary of 1363; Cola was himself a notary, and an observer of ruins.[36] The scene at Sant'Angelo in 1363 is obviously the wake of Cola, not the world of Innocent III or Boniface VIII.

Still, the assembly, the total action at Sant'Angelo, which bursts from the notarial book, pulls together, as nothing earlier can, half-seen and half-understood earlier fragments. There is also every reason to believe that much was unchanged. The removal of the popes had of course made Rome poorer; and one must assume that Rome fifteen years after the Plague was a great deal emptier than it had been in the year of the Jubilee. There is, in fact, real evidence that Rome declined after the departure of the popes in the early fourteenth century, and the belief is not solely dependent upon common sense, often a surprisingly poor historical tool. There survives, for example, in the Vatican, a book of the properties and houses of the priorate of the city of the Order of Saint John of Jerusalem, from the 1330s.[37] The report starts with the property of San Basilio in the city itself, and with two shops in the *contrada* Tor dei Conti, in front of the palazzi of Don Stefano Conti; it continues through thirty-one other houses and shops in the area of the Torre (which of course still stands) and against the hill that was then, and is now, called Magnanapoli (although not all of the houses' and shops' locations are specifically stated). Of the thirty-three houses and shops, eleven were vacant at the time of the account, and seven were paying rents considerably reduced from their accustomed ones. Of an expected San Basilio urban rental income of something over 118 or 120 florins, over 35 florins were not at the time of the report being returned. It makes a dismal sounding list: "Item, the house where Paulucio the *ortolano* ("gardener," "truck farmer") lives returns one and one-half florins; it was accustomed to return four and one-half florins" —item after item.[38] It is a list that (although there were plenty of deserted vineyards in the thirteenth century) argues strongly against anything like a bursting local economic revival in early fourteenth-century Rome, one that would have made thirteenth-century Rome seem by comparison a sleepy little town; and this is well before the Black Death.

There is good reason to believe, from this sort of evidence, that Sant'Angelo in 1363 was not part of an unrecognizably more vibrant town than it would have been in 1300—and "vibrant" is the adjective that describes the distinguishing characteristic of life seen in Sant'Angelo's early notarial cartularies; in them Sant'Angelo sparkles and flashes like the fish it sold. The first cartulary exposes one of the characteristic qualities of the medieval city: its brisk, petty mercantile activity among the ruins. It makes clear that medieval Rome was not somnolent, not sleeping in its ruins; its decay was a decay of limited action and movement. This sort of action, it is true, may have been more apparent in 1363, more noticeable to an actual visitor, than it would have been in 1300, just because less else was going on. Similarly the jumble of classes may have been more relaxed in the Palazzo Ponte in 1331 than it would have been in 1300 because of the absence of curial

The Physical City

tenants. The removal of the curia had surely brought other people and types of people forward and made them more noticeable; again, it was Cola's time.

There was an important man, a *pezzo grosso,* in Sant'Angelo in 1363. He was Matteo de Baccariis, "noble" doctor of both laws, rich, omnipresent, with a complex of relatives present in and renting a complex of houses—like Mascio de Baccariis's house in the Ruga Judeorum ("the street—or passage—of the Jews") with Luca de Baccariis's house at its side and behind it the Porta de Baccariis. He was also a man who by the time of his death had collected a library rich in expensive books, particularly of both laws, the most expensive of which was a *Speculum iudiciale* (by that Guillelmus Durandus who lies in Santa Maria sopra Minerva) worth forty florins; he had, too, a valuable Bartolo on paper worth twenty florins. Matteo, although learned, was a man with a fat cigar—and he married well twice. Matteo brought pasturage in the country, arranged for his daughter's inheritance, acted as an advisor and an arbiter, as in a case between a Trasteverino and a Trastevere priest. As a canon of Sant'Angelo and a local innkeeper watched, Matteo's servant Francesco bought wine from a neighbor, wine stored in the neighbor's house in two vats, wine to be resold in the Piazza of the Lateran (out across the relatively empty spaces of the *rioni* Ripa and Monti).[39]

Cast rather in the shade by Matteo and his family, but present in the cartulary, is the great local noble family, the Savelli, under and against whose great house-fortress, the Theater of Marcellus, the church and neighborhood of Sant'Angelo stood (and stand). Savelli involvement in the life of the city (which did not exclude involvement in the city's usury) becomes apparent, for example, through a family sale that was officially notarized near the church of Santa Maria in Campitelli late in the year. The principal, Magnifico Nicola di Magnifico Bucio Savelli, acting also for his brother Guglielmo, sold part of a cluster of properties in the *rioni* Ripa, Sant'Angelo, and Campitelli for two hundred florins to the heir of a notarial family (and cousin of a canon of Santa Pudenziana). The sons of Bucio Savelli had held their property jointly with the Magnifici Antonio and Luca Savelli, with Luca holding one-half, the sons of Bucio, and Antonio each one-fourth. The sale is interesting partly because the property included one-fourth part of the place in the Portico di Ripa where vegetables were sold and the place where the money changers stayed.

This sort of Savelli involvement is seen again in a transaction of a member of the family Grassi (a family connected with the tower which still stands next to Sant'Angelo). This transaction, which took place at the monastery of Santa Maria de Massima (now Sant'Ambrogio), involved property in the Ruga Judeorum and also that "butcher shop or house with a place for keeping the beasts (*remettitorio*) in the *rione* Ripa under the houses of the Savelli." (One need not see the Savelli leading the beasts to slaughter; but one must see clearly that the beasts by whom the great families, Savelli as well as Conti, were surrounded were not just the beasts of chivalry.) Although Savelli had long replaced the Pierleoni as the dominant local nobles by 1363, Pierleoni were still around, among them Lello di Donna Laura Pierleone, who had a tenement by the cloister of the palazzo of the cardinal of Sant'Angelo, and Don Pietro di Domenico Pierleone, who was a canon of Santa Maria in Porticu.[40]

Families great and small move around Sant'Angelo, but the people who move most are the fishmongers and their families. A number of these fishmongers can be watched leasing the stones for them to sell fish on. The ancient marble stones in front of the church of Sant'Angelo belonged by right to the church and its canons, and the canons rented them on long leases (although in the thirteenth century the approval of the cardinal deacon of Sant'Angelo, or his vicar, might also be sought for other long leases). So on July 4, for a term beginning in the following year, the canons received eleven florins for the lease of half of one of the church's stones, with a promise on the tenant's part of an annual rent of two florins a year on the feast of Saint John, and on the canons' part of re-renting on the same terms to the fishmonger's male heirs. These stones, before or under the *templum* of the church, surrounded by other similarly leased stones, sometimes held by two mongers jointly, were serious and valuable properties and were discussed in much the same terms as were shops and even arable fields and vineyards. Privately held stones near the church's collection formed part of local dowers, and they were important parts of sales and property transfers, as in that from a Grassi mother-in-law to a Grassi daughter-in-law involving the properties around the Torre Soricara next to the church—in an area that can be mapped out from the cartulary and looked at in its inhabited ruins today.

The piquancy of the fishmonger-stone documents is increased by the stones' mutilated survival in the Portico di Ottavia, and their unmutilated and visible survival in familiar prints—the road, the Porta of the Ponziani (a family involved in fishmongering), the Ponziani holdings, the Grassi tower, the *templum* of the church, the Grassi stones.[41]

The Physical City

Fishmongers are constantly present, and in considerable number, witnessing all sorts of acts in and near the *templum*. A fishmonger returned the money that a Grassi woman had borrowed from a Trastevere lender. (They, reasonably, connect the two sides of their river.) In the church of Sant'Angelo two fishermen of Terracina accepted seven florins on deposit from a fishmonger named Pietro Corre, and they promised to send their fish to him in the accustomed way. Three other fishmongers, one the son of a scribe-notary and one from the adjacent *rione* Campitelli, witnessed the agreement. A Sant'Angelo man leased from Benedetto, the rector of the church of San Lorenzo in Piscinula, across the river in Trastevere, a fishery, or fish pond, in the Tiber, which belonged to the church, which was called the pond "in pede pontis Polçelle" of Trastevere, and which was next to another San Lorenzo pond. For ten years San Lorenzo was to receive on each Easter an annual rent of three florins and two shad with roe (*laccias oviatas*) and two tender male shad (*laccias lactinatas*).

The fishmongers' women appear, too; it is neither a cartulary nor a society that excludes women—a wife and mother is named future guardian of her children should her husband predecease her; if she should predecease him, his father is to be guardian, or if he has died, his wife, the children's grandmother. Women's (and not just fishmongers' women's) possessions appear—the ring of Paola, the fishmonger's widow, the chest of Sofia, the miller's wife. And three Renaissance-sounding nuns, Plasira, Laura, and Euphemia, appear in the medieval-sounding convent of Santa Heufemia.[42]

One of the marble stones for selling fish under the *templum*, to which the canons of Sant'Angelo held their "ancient" rights and which they let in their accustomed way, a stone next to that held by the heirs of Giacobello di Paolo Grassi, and before which was the street, was described as being near the wall of Sant'Angelo where the image of Saint Christopher was painted. (Rome was a painted city.) Sant'Angelo of the fishmongers was not only a fish market; it continued to be and to seem to be a church, and some of the business of the prior and three (or more) canons was ecclesiastical. (Like the canons of similar local churches, the canons themselves seem generally to have been local or relatively local men provided by Urban v or his predecessors.) They could, as they did in May, deal with marble (and mosaic in private chapels) over men as well as under fish—although, even in sepulture, talk of a neighboring counter and shop, the church's shop, could creep in. A fishmonger and a Ponziano left in his will, along with money for repairs at the Lateran, money for oil for the lamp of the image of the "glorious Virgin Mary" at the church of Sant'Angelo, as a neighbor named Ceccha left a striped tablecloth, or altar cloth, to the image of the glorious Virgin at Santa Maria de Massima, half a block away.[43]

Actually ecclesiastical bequests from the *rione* Sant'Angelo, with a center in the close neighborhood and a heavy extension into churches connected with the neighborhood, spread out over much of the city (from Sant'Angelo, the Campitelli with its chapel dedicated to Saint Nicholas, and San Salvatore de Pedeponte, to the Lateran, Santa Maria Nova, the Aracoeli, Santa Maria sopra Minerva, San Callisto in Trastevere, and San Pietro in Montorio). These bequests, like the great preponderance of evidence from both the thirteenth and the fourteenth centuries, argue against too rigid and narrow an interpretation of enclosing Roman neighborhoods.

Their argument against enclosure is reinforced by the actors in the scenes in the Sant'Angelo notarial book, and in the scenes themselves. They are not frozen in the *templum* of Sant'Angelo. They move to the courts on the Capitol and to the Campitelli with flashbacks and asides to places like the garden of Santa Maria domne Rose, where still or again there is an isolated stretch of green in the center of Rome between Santa Caterina and the Polacchi, to the fishery in the Tiber called "Locapraiello," to the Lateran. Although the fishmongers of Sant'Angelo dominate the scene, with a chorus of canons (and also an occasional solo by the chaplain) in the background, and the ceremonial progress of the indirectly vegetable-selling Savelli, there are plenty of strays: Sabba's daughter, the nun of Sant'Eufemia; Lello Buccabella, merchant of the Campitelli, a goldsmith from the same *rione,* and the merchant Gregorio di Enrico of Florence; the mobile population—two men from Marseilles now of Naples, Englishmen from the Biberatica, Constanza, a woman moved in from Ostia; the heirs of John de Montenoris (identified by his part of Rome) from the *rione* Sant'Eustachio; Cola the butcher of the Ripa, the butchers of Campitelli, and the Lady Symonetta, wife of Cola Conti butcher of Campitelli (and Cola seems to be a butcher's name as Nucio does a fishmonger's name); Antonio di Paolo de Corso from the *rione* Ripa; and millers, at least sometimes, from the Arenula, for the Tiber was a milling river; and clergy from Trastevere and the city side—canons of Santa Cecilia, of San Clemente, of Santa Maria in Porticu, from that order of canons which abounded in Rome. But at the center of these people, in

The Physical City

Sant'Angelo, were always the fishmongers, dominating the scene, like that member of the prominent family from beside the church, Paolucio di Lorenzo Ponziano called Capograsso, "the Fat-head." [44]

The question of how restricted neighborhoods were, or could be, in the thirteenth and fourteenth centuries can be asked most sharply and in a way answered in terms of the Jews and the ghetto. The main point is perfectly clear; thirteenth- and fourteenth-century Roman Jews generally lived in a specific neighborhood or neighborhoods, but not in a ghetto. The ghetto was created in the sixteenth century by Paul IV (and opened again in the nineteenth century by Pius IX, both deceptively purposed popes). Paul's ghetto enclosed a neighborhood that had already been partly Jewish for well over three centuries.

Lack of enclosure should not imply that the Jew's life in medieval Rome was as pleasant as the Christian's; and it would seem that, in spite of the protective edicts of popes like Innocent IV, Gregory X, and Nicholas IV, most of the nonsense about papal protection of the Jews should be forgotten. It is very unlikely that the papacy could have protected the Jews from a really hostile society if it had wanted to. Besides, the popes' constant and sometimes hysterical emphasis upon a pure Christianity (which often enough meant Christianity completely untainted by current heresy or by support of a current antipapal figure like Frederick II, rather than Christianity in the exact pattern of Christ) must not have had a very beneficial effect upon the position of the Jews. The medieval Roman Jew would seem to have suffered a slight, prejudiced, informal inequality before the law when he was involved in processes not subject to his own law. The thirteenth century itself was not, to put it mildly, a century of unmixed benefit for the Jews of Western Europe, as is shown by Innocent III's great council at the Lateran with its insistence on identifying pieces of costume, by Louis IX's attitudes, by the expulsion from England, and by Elijah de Pomis's death in Rome. That is to say, it is absurd to pretend that the position of the Jew in the thirteenth century was an ideal one, one free from persecution, and that bad things only came with the Renaissance.

This said, it must be made equally clear that Jews did not live as outlaws or outcasts in the thirteenth- and fourteenth-century Rome. Jews could be papal physicians, as Isaac ben Mordecai was to Nicholas IV. In the early fourteenth century Jews were rewarded with Roman citizenship. In August 1280 the Jewish mystic Abraham ben Samuel Abu'lafia, originally from Saragossa in Spain, went to the pope's palace to convert Nicholas III, but Nicholas had died a few days too early. Christians who wanted to be protected from what they chose to consider to be Jewish incursions upon their rights took their cases to court; Christians and Jews might succeed each other to the tenure of the same piece of property. Jews could make wills.[45]

The point is made quickly enough. In the thirteenth-century Orsini documents, Jews and Christians in the *rione* Sant'Angelo in Pescheria hold adjoining tenements casually and without comment. It is said of a heretic, admittedly, that his property was next to "the Jews," but this is only meant to describe his property's boundaries. In 1363 Luca de Baccariis, of the prominent Sant'Angelo family, rented a house of his in the Ruga Judeorum for a year to the Jew Sagaczolo di Bonaventura, also of the *rione* Sant'Angelo, for a rather high-sounding rent, perhaps, but not particularly high for this neighborhood. Luca himself held the property on two sides of the house; on the third side was the street, and on the fourth the Porta de Baccariis. The act of renting was performed before the house of the lawyer Matteo, the most prominent member of Luca's family. One of the witnesses was the priest Benedetto, rector of the church of San Lorenzo in Piscinula.

One hundred and twenty-five years earlier, in 1238, the Masters of the Buildings of the City made decision in a dispute between the church of Santa Maria "domne Berte," represented by the priest Gualterio, and four Jews, Acosiliolo, Nasaçolo his brother, Monayçolo, and Moscettulo. The priest had complained that the Jews were throwing dyes and dyed water out into the street in front of their house and that these then ran down to the church. The Masters condemned this practice. They ordered that if in the future the Jews should throw dyes and water dirty from dyes out in front of their house they should build an underground covered conduit that would not obstruct the street and that would carry the dye and water to the sewer (*clavicam*). The Jewish dyers lost their case (near a neighborhood that is still Jewish and still involved in the cloth and clothing trades). A Jewish renter a few blocks away paid a rather high rent, and in this case it was only rent, for his house. But none of them was excluded from the community and its normal traffic, presided over by the Masters, witnessed by a priest. In 1363, without any noticeable sort of dramatic explosion, a simple act of property transfer could bring together a resident of the Piazza

The Physical City

Giudea and property on Monte Mario, at the head of the Prato San Pietro next to the road to the Ponte Milvio.[46]

The Jewish neighborhood had once been in Trastevere. By 1309 the church of San Biagio de Cacabariis (near the Argentina) was in the middle of the Jewish quarter. A new synagogue had been built on the city side of the river after the ancient one was destroyed by fire in 1268—its anniversary became an annual fast day. The new neighborhood was established; the migration had been marked by the Ponte Giudeo ("Bridge of the Jews"). (The name had come to be applied to the Ponte Fabricio, the Ponte Quattro Capi, which still survives, crossing the left branch of the Tiber, connecting the *rioni* Sant'Angelo and Ripa with the medieval flank of San Bartolomeo on the island. The bridge still bears Fabricius's name, still carries its four heads, and still leads to, or from, the Jewish quarter.) The crucial movement to the new neighborhood could have happened between Benjamin of Tudela's visit to Trastevere in 1165 and the rebuilding of 1268, but it was probably earlier. The old neighborhood did not disappear completely. In 1300, the will of Tommaso da Ocre, cardinal priest of Santa Cecilia in Trastevere, left ninety-seven *lire* to the heirs of Salamone Tadei da Roma, who lived near the palazzo of Santa Cecilia. The money was intended to repay what Salamone had spent to buy a house next to the cardinal's kitchens for the church. As late as 1404 the record of a transaction which took place in the *curia judeorum,* a court or piazza near Santa Cecilia, talked of the neighboring house of Coymello the Jew. There is certainly nothing to suggest that the new neighborhood had any of the romantic horror it came to have for Hawthorne, his "confusion of black and hideous houses, piled masses out of the ruins of former ages." In fact, Hawthorne's former palaces, which "possessed still a squalid kind of grandeur," had not yet, with the most inconspicuous exceptions, been built.[47]

Another principle of forming neighborhoods and examples of neighborhoods so formed demand at least brief separate consideration. These are the neighborhoods of specific sorts of artisans. Of these one of the most noticeable on a map of Rome is a neighborhood which gave its name to a *rione,* a half-name to a double *rione,* the Ponte-Scortecchiaria (or *scorteclari, scorticlaro, scorteclariorum*). The Scortecchiaria was the neighborhood of men, or people, who worked with, and dealt in, leather and hides. Almost as apparent a neighborhood is the Calcarario, the area of the workers who turned ancient marble into modern chalk, whose powder seems sometimes still to dominate the Largo Argentina, where provincial antiquarians and conservationists like Henry of Winchester must have shopped in the twelfth and thirteenth centuries, and where for many centuries, but no longer except for Santa Lucia de Calcarario, a cluster of churches remembered the marble worker or chalkmaker in their names. The streets of this general part of Rome and the churches proclaim the neighborhoods of artisans, of candlemakers and keymakers and ropemakers and chainmakers. The beautiful twelfth-century campanile of San Salvatore alle Coppelle still identifies the place of the barrelmakers.

These names stretch back into an often specifically unidentifiable antiquity. The streets with which the names are connected have sometimes changed. This change is a general Roman phenomenon most famous in the case of the change of the Via Lata. In the middle ages the Via Lata was the name of the great Corso; it is now the name of the small street which goes along the side of the church of Santa Maria in Via Lata, itself named from the old name of the Corso where its façade is. But the convention of giving things names like "of the ropemakers" (*funari*), even if those names have sometimes changed their locations (as the Funari has), makes clear the fact that Romans were accustomed to thinking in terms of clusters of specific sorts of artisans. Quite surely in thirteenth-century Rome there was some concentration of artisans and dealers, but equally surely there was, as there is in modern Rome, a scattering of every sort of trade in every part of Rome.

The Ripa di Roma, the shipping *contrada* on the lower right bank of the Tiber, the neighborhood where Saint Francis is said to have stayed, was unsurprisingly the home of sailors. These men are revealed in the late thirteenth-century senatorial investigations of the lingering results of earlier Genoese piracy, abduction, and confiscation, which had adversely affected the lives and goods of the sailors and their neighbors. These Ripa sailors appear before senatorial notaries and talk of the common maritime knowledge, gossip, and memory of their Trastevere *contrada*. Their's seems to have been a unique neighborhood.[48]

Meat, like ships and fish, is a separate and special case. There was a great butchering center by the Torre de'Conti, in what had been the Forum of Nerva. By the thirteenth century the place was called the Ark of Noah, one hopes with deliberate macellary reference from a Bassano sort of vision of the ark and not just from a confusion of names (or standing water, as Adinolfi thought). The size of the Ark's marketing operations is suggested by a single transaction in May 1295, when the curia was in Rome.[49] This single sale brought to the Ark 124 sheep and 13 goats. The price that the seller, Pietro Cimino of

The Physical City

49

Rieti and his associates, charged the buyer, a man named Nicolucio di Leonardo of Perugia, who held properties in Rome, Rieti, and Perugia, for all this meat or potential meat was 116 *lire*. The transaction was enacted in Rome in *macello Arce Noe* in front of two witnesses from Orvieto and one from Viterbo (papal subcapitals which furnished men to the curia). One's romantic view of the Torre, the bastion of the Conti (the family of Innocent III, Gregory IX), a center of savage internecine war and human slaughter, must be modified to include its being surrounded by prosaic animal slaughterhouses functioning in order to feed people their lunches and dinners. Gregorovius's families who "only now and then burst forth with the wild din of arms" must have startled the sheep.[50]

Through thirteenth-century Roman documents there is a rush of artisans, who are joined by laborers (in the sense that gardeners and farmers are laborers) and masters of various skills. Among these masters, medical doctors, notaries, and judges hold a special place. They formed a literate and peculiarly skilled sort of middle class, a special ballast in the rather unstable Roman community—men like, in 1263, Matteo the notary, son of the deceased Benedetto, notary and medical doctor of the Calcarario of Rome. Scribes and notaries like Matteo, who are present for the writing of every recorded transaction (preserving and identifying themselves with their memorable notarial signs), and often, too, serving as witnesses, mingle in every neighborhood with very various sorts of neighbors. Although they and their professional colleagues gave Rome a needed class as well as needed services, they could be the descendants of butchers and fishmongers, as a judge could be the descendant of a blacksmith. These notaries were often men of property with vineyards in the suburbs. They left considerable legacies—in 1246 the three daughters of Rainerio, who had been a notary in Trastevere, sold four *pezze* of vineyard outside the Porta Portese, in the "Tertio," to the monastery of San Cosimato for fifty *lire*.[51]

That medical doctors should have been prominent inhabitants of thirteenth-century Rome is not surprising. Italy was one of the homes of medicine. Salerno was not too far from Rome. And in the thirteenth century malarial fevers were pressing upon the city (with its polluted rivers and shallow wells), threatening the swampy southern suburbs, darkening the hot summers as the Dog Star's madness gleamed from Orion's foot, and frightening neurotic and hypochondriacal curial clerks (although one sensible north German, at least, advised that mid-August was the best time for coming to Rome). The Tiber, particularly, life-giving, central, bringer of fish and trade, could be a killer river, as well as one that brought Rome's enemies, like Pisan ships in the twelfth century or the Provençal fleet under Charles of Anjou. Great floods brought death and filth to Rome, as in 1230, battering bridges, or in 1276, with a high-water mark that is still recorded in the Arco dei Banchi.

A town whose greatest denizen could make, or be believed to have made, a man a cardinal because he was a doctor good with fevers, would seem to provide a good market for physicians; Italian communes generally fought for doctors and offered them special privileges. When, in Rome, Nicola "One-Hundred Lire" (*Centum libre*), Octabiano or Romano Zampo, called "lo Regio," the widow Buonasera, Gregory Tasca and Lancelot his brother, or the Lady Africa living in the *contrada* Campo degli Arcioni in *rione* Trevi, or the mother and son, Rosa and Angelo, living in half their house with the garden behind it because they had sold the other half, when any or all of them felt the heavy misery of the hot sickness coming upon them, they could know at least that they lived in a town full (if not full enough) of doctors. There was even a tower with a doctor's name, the tower of Enrico Medico by Saint Peter's. Two of Rome's doctors who were brothers both had sons who became famous Talmudic scholars. Besides Martin IV's cardinal doctor in the 1280s, an Englishman named Hugh of Evesham ("the Phoenix") with offices at San Lorenzo in Lucina, lesser doctors were involved in papal administration. In the 1240s, for example, the medical doctor Bartolomeo was chaplain to the papal vicar. Doctors, like notaries, were involved in the total community. They could hold vineyards and be the neighbors of shoemakers.[52]

In the thirteenth-century lists of witnesses and neighbors in which these professional men join the ever present canons, priests, and other clerics, there is a great crush of identified workers—shoemakers and innkeepers and cooks; smiths and drawers of water; a combmaker; spicers in the Ponte; Francesco the tailor of the Via di Papa; Taliaferrus Ferrarius, whose trade seems clear: "Smith the smith"; cloth bleachers from Trastevere; abbatial knights and warehouse keepers; the painter Huguccio at San Silvestro in 1251; the miller in the cloister of Saint John Lateran; Pistoian merchant bankers at their shop; the son (Leopardo the clerk) and servant (Giovanni of Pisa) of the famous Pisan painter Master Giunta at San Clemente in May 1239; Cosmato, the (cosmatesque) marble worker, at Saint Peter's in the *sala* of the chamberlain in May 1279. They meet oddly—the shoemaker and the Capocci, the tailor and the Caffarelli. And they, again, like the notaries

The Physical City

and the doctors, do things and own things unconnected with their crafts or their neighborhoods: so the *muratore* ("builder") Matteo Vecclasolo buys a widow's vineyard outside the Porta del Popolo; an *ortolano* ("gardener") from the *rione* Ripa holds property outside the Porta at the Lateran. Crafts seem to appear out of the very ground, as when a widow and guardian of her son, on the one side, and the monastery of San Cosimato, on the other, fight over bake-ovens on Trastevere property—or as in a quite casual reference to the common charcoal pit.[53]

But in spite of all this activity and all these presences crushing together to present themselves, there is little firm record of the real existence or activity of specific craft gilds and organizations in thirteenth-century Rome. In the century of the rising gild, Rome is queerly quiet. The two great old trade organizations were those of the cloth merchants (*mercatores pannorum*) and the agricultural gild (*bobacterii*, more literally cowboys or perhaps ploughboys)—organizations which represented the purveyors of the first great commercial commodity, beyond bread and wine, and the major occupation, in a way, of a still very noticeably agricultural town. They were the crucial trades for people whose primary needs were still quite obviously eating and dressing. In the fourteenth century, by 1317, there were thirteen major gilds dominated by the two great old gilds. The gilds and their heads are said to have been important to the revolutionary "democracy" of Brancaleone d'Andalò, who came to power in 1252, and the heads of gilds were consulted in 1267. Representatives of corporations of merchants and shipowners had been particularly active in dealing with the Genoese in the twelfth century. Some historians have traced specific gilds and corporations to the thirteenth century—innkeepers, shoemakers, bakers, medical doctors, haberdashers, builders. It is perhaps easy to exaggerate the frailty of thirteenth-century Roman gilds, but only the corporations connected with the selling of cloth have any very persuasive, visible thirteenth-century reality.[54]

The cloth merchants had an early local presence in their neighborhood of the *mercato* underneath the Capitol by the church of San Marco, in the area now dominated by the Victor Emmanuel monument. When they can be seen meeting, they meet in the church of San Salvatore in Pensili, where now is the Polish church of San Stanislao on the Via delle Botteghe Oscure ("the Street of Dark Shops"). By the mid-thirteenth century the merchants seem to have controlled five lesser gilds (the *lanaroli*, the *bammacarii*, the *mercerii*, the *accimatores*, and the *cannapacioroli*), and to have controlled themselves by fairly elaborate constitutions or statutes.[55]

Statutes of this sort are almost always difficult sources for the actual behavior of the communities they purport to regulate. They are often derivative; they are often unrealistic. The statutes of the merchants of Rome are also, at least in their details, extremely hard to date. They come from a 1317 recension of a 1296 confirmation of mid-thirteenth-century enactments. Still, one can at least say that the gild of the merchants was in the thirteenth century building up an elaborate system of consuls and courts, regulations and almsgiving, a meeting place and a pattern of behavior. The statutes are most vivid in their regulations of how customers should be taken through the shops and how and where they should be shown cloth. If one merchant takes his cloth out to show it better to a customer, no other merchant is to try to show the customer his cloth until the first merchant's cloth has been fully looked at, rerolled, bound, and carried away. Only one merchant at a time was to move the customer from the dark shops to the light; cloths were not to be compared, and customers were not to be stolen. A greedy, competitive bazaar lives in the statutes, controlled by a community of merchants, so that the members of the community might profit more through regulated cooperation—but it cannot be clear exactly how this bazaar fits the bazaar of Rome.[56]

The cloth merchants aside, the actual existence of individual gilds is less secure than that of the organizations of servants who served pope and curia. The glaring oddity about the gilds of this important city full of artisans, bankers local and foreign, people and trade, is their unimportance. Like the governmental organs of the city, they seem to have been very poorly and primitively formed.

The reason for this oddity lies in the nature of the city's importance and the connected source of much of its income. Rome was the most important, and the name, capital of an important, but still itinerant, court, the center, but not the fixed center, of a great, worldwide government. It was also the cult center, and so the pilgrim and tourist center, of the Western world. Much of Rome's population was transient, and the transient population was the source of a great deal of Roman income. The chronicler who said that he saw, in the Jubilee year of 1300, two clerks at Saint Paul's outside the Walls, standing day and night with great rakes, raking in the coins left by pilgrims, wrote a bright metaphor of Rome's source of wealth.[57] In this sense every Roman was Saint Peter's man, and Saint Paul's, and a chorus of lesser saints'. Saint Peter made him rich and made his wine sell at a better price. Rome was very far from being a stagnant little place like sixteenth-century Lincoln, preserving its elaborate gild structure within its walls, and it was also very far from contemporary Florence, with its burgeoning commercial growth, channeled, it would seem naturally, into the form of gilds. Rome was already, or was about to be, a city in which salesmen at booths around the Vatican—booths on the stairs, in the portico, and beneath the Navicella—sold painted replicas of the Veronica, as well as figs and tooth extractions. Also at the Vatican, within the boothed periphery, pilgrims could procure lead or tin badges with the images of Saints Peter and Paul on them (like one found recently in Dublin); the pilgrims got these badges, which would show that they had been to the doorstep of the apostles, from the canons of Saint Peter's whose monopoly in their manufacture had been confirmed by Innocent III. Commercial activity at the tombs of Saints Peter and Paul and around the Lateran court would seem, again naturally, to have found patterns more fluid than those of Florence or, later, Lincoln.

The press of Roman pilgrims has left almost palpable memories. One can almost feel (perhaps particularly in the Sistine Chapel in the summer) the terrible day at Saint Peter's when an English Benedictine monk was crushed and fatally injured there. William of Derby, the monk, a man of importance but not of spotless reputation, had, in the great year, 1300, come to Rome with three other monks, two of them priors, as he had been, and the third a monk from his own house, of Saint Mary's, York. William of Derby went to see the great relic of the Veronica. As everyone pressed forward to be near the relic, William was caught in the crowd. He was injured. His leg was crushed. He never recovered. He did not live to go home.[58]

One can still feel the greed of the Romans as they watched these people falling into their nets. In April 1235 Angelo Malabranca, a Roman senator who had brought the commune back to peace with the pope, took pilgrims back under the direct protection of the senate and freed them from the trammels of the Roman secular courts. (This Angelo's son or grandson, Egidio, was by the late thirteenth century a San

Silvestro in Capite tenant of gardens in the *rione* Colonna—thirteenth-century Rome is an intricate tangle.) On September 15, 1235, Angelo acted to protect the peace and quiet of the city and all coming to it (his job, he said) and to save pilgrims from the violent avarice of innkeepers. The senator had heard that many inhabitants of the area around Saint Peter's had violently forced pilgrims and "Rome-seekers" to lodge in their houses and, even worse, had gone into other people's hospices and found pilgrims already quietly settled down there and then forced them to come out and lodge in their own inns. Innkeepers were forbidden this behavior by Angelo's edict. Pilgrims were to be allowed to stay where they wanted to, and to buy the things they needed where they wanted to buy them. Violators and molesters were to pay a fine, half of it to the canons of Saint Peter's for their daily commons, half for the repair of the city walls.[59]

Contemporaries were overwhelmed by the number of pilgrim tourists. In 1300 the Florentine Giovanni Villani thought that a great part of the Christians then living, men and women, came and went; he further thought that there were 200,000 pilgrims within the city at one time. And Guglielmo Ventura, who was there and stayed for fifteen crowded days, said he saw a crowd so big that no one could possibly count it, but that the Roman rumor was that there were "twenty hundred thousand" men and women. Guglielmo says that he often saw both men and women trampled under other people's feet and that he himself was often in danger of being trampled. Guglielmo found the price of bread, wine, meat, and fish quite reasonable, but the price of lodgings and fodder for his horse very expensive. A contemporary Roman, Jacopo Stefaneschi, explained the abundance of food elegantly; he remembered the miraculous, the five loaves and two fishes, and compared the natural: the Lord will give his goodness, the land will give its fruit. But no one had any doubt about the fact that Rome in 1300 was very crowded; and no one has ever doubted that Romans profited from the crowd. Villani said that the Romans got rich from selling their goods. Gregorovius wrote, after describing the Jubilee year, "Immense profits . . . accrued to the Romans, who have always lived solely on the money of foreigners." [60]

Less transient transients than the pilgrims were the members and followers of the Roman curia. Their presence or absence in a papal town was important enough; in Viterbo, papal and curial presence doubled rents.[61] Great as households of functionaries, cardinals, and, particularly, the pope were, with their swarm of scribes, advocates, and accountants, it was not just they who swelled the population of Rome

The Physical City

when the curia was present. There were also those business pilgrims, who in various pursuits followed the curia—proctors and representatives of churches as diverse as Canterbury and Bari and of laymen, bringing the business of their principals before the papal court.

There were great prelates and their households. The newly elected archbishop of Canterbury, Robert Winchelsey, for example, arrived in Rome in May 1293. Nicholas IV had died in April 1292, and the cardinals had not yet chosen his successor. In spite of the fact that eight cardinals had already fled the supposedly dangerous Roman summer heat and gone to salubrious Rieti, Winchelsey found three cardinals in Rome. He dealt with the cardinals, but he found to his horror that his letters of credit from merchants in London were not accepted by curial bankers. He was forced to reestablish his credit. He left Rome quickly to seek other cardinals. Early in 1294 there was still no pope; at Eastertime Winchelsey and his group were back in Rome. In July 1294 the cardinals finally chose a pope, who was to be Celestine v. Celestine was consecrated near L'Aquila in late August. Winchelsey's election was confirmed, and he was consecrated near L'Aquila in early September. He then rushed home to Canterbury. But for two summers, with the intervening winter, Winchelsey and his itinerant household had helped to swell the group of curial followers, crowding, when in Rome, the crowded parts of Rome, drinking from the local vineyards. There were many households of Winchelsey's sort lodged about the curial towns. They complemented the more permanent foreign households, like that of Hugh of Evesham (the English physician-cardinal from 1281 to 1287) whose full, ambassadorial household was established near San Lorenzo in Lucina (or the household of Ottobuno Fieschi, of which fifty-three members can be named and identified).[62]

But even the realization of the presence of these foreign households at the curia does not give an adequate notion of its drawing power. There were also all those tradesmen who followed it and its followers, people like Vanni di Nicola di Bruno of Viterbo, a butcher "who followed the curia," and Fico of Perugia, a poulterer "who followed the curia," whom we see, each in his Roman house, as Robert of Selsey, a Canterbury proctor, pays them their bills on December 23, 1266, pays them for chickens and capons, game and meat.[63] Butchers and poulterers, like clerks, joined the curia in papal towns like Perugia, Rieti, Orvieto, and Viterbo and moved with it. Witness lists, like the one from the large sale of Rieti sheep at the Ark of Noah, are a litany of the towns which the curia visited. These local foreigners moved in and out of Rome, making its actual population rise and fall, bringing to the city queer tempestuous periods of crowded, heavy trade and then of relatively slack emptiness.

The vineyard-owning residents of Rome and the shoemakers who profited from the presence of the transients seem pale beside them. Denizens, together with the less important transients, seen against the papal court itself, fade even more. They become a chorus without faces. And a chorus was probably what both pope and emperor wanted the populace to be as long as it sang approvingly. For the pope, particularly, the whole city probably often seemed no more than a setting for triumphal processions—and that could still reasonably be argued to be the city of Rome's greatest importance in the thirteenth century, at least if one accepts something of thirteenth-century values. Rome was the place which surrounded the pope's procession from the Vatican to the Lateran on the day of his coronation. These processions, of which the coronation procession is the greatest, in fact, like ribbons, tie together the neighborhoods and neighbors of the city of Rome. It is as if Ranke had invented them to do in physical fact what his own ribbons, tying together the two apostles on their columns across the city, the two fountains facing each other, did in fancy. They tie together the bundle with its disparate contents; and they emphasize the fact that all these people, whether local or transient, are economically, if not spiritually, Saint Peter's men—men of "lu baron san Piero."

One of the most brightly described of thirteenth-century papal processions occurred on November 15, 1215, while Innocent III's great Fourth Lateran Council was in session. The Western world, at least through proxy, was then present at the Lateran, listening to the establishment or clarification of the rules for its own behavior. The procession took the papal court across the crowded city and the river to the great old church of Santa Maria in Trastevere—great particularly in legend, for there, it was believed, a sacred spring of oil had flowed to announce the birth of Christ. Innocent III's procession went to Santa Maria to consecrate the church. Santa Maria had been rebuilt in the years after 1139 by the Trastevere pope Innocent II (a Papareschi), but it had not been very surely reconsecrated, and Innocent III took upon himself, at the height of his career, the duty of reconsecrating "Our Lady of the Flowing Oil." The procession moved to the blare of music, to a sound that "trumpeted" like an elephant. It was led by brilliantly dressed nobles. They were followed by a multitude of clerks and laymen. The chronicling contemporary who wrote of the procession moves in and out of and around Biblical phrases; he has Roman children join the procession, as he follows the familiar antiphon for Palm Sunday,

The Physical City

and he has them bear branches of olives as they shout their praises to the Lord unceasingly, saying, "as is the custom," *"Kyrie eleison, Christe eleison."* As the procession crossed the Tiber it found a Trastevere festa in the piazzas and the little streets, with lanterns, innumerable lanterns, hanging on cords, in their brightness competing with the light of day. Banners flew from towers and houses.

Guessing part of the itinerary, one can rewalk the procession any Saint Martin's week and look at the November morning haze romantically swathing the island and incompletely hiding the Aventine, covering the river, making pale the campanile of Santa Maria in Cosmedin. Some of the old towers remain on the island and in Trastevere—the Anguillara, Santa Rufina, part of the Mattei. The Gianicolo, Trastevere's enclosing hill, with its leaves changing and falling and heightening the color of the brick, bursts into the ends of Trastevere streets and over the corners of piazzas. At the end, still with the Gianicolo hanging above, is Innocent II's great church, with its façade half in light, its later campanile shining. Inside one finds more than Innocent III could find. In the apse Innocent would have seen only the twelfth-century mosaics above the level of Christ's lambs (and only the lambs, the city, and the feet of the saints on his far right as he approached could have sparkled in the morning light); Innocent could not have seen the lower mosaic panels, those by Pietro Cavallini, from the later thirteenth century.

In looking at the two sets of mosaics—the ones Innocent saw and the ones his century produced—one sees in style and line much of the whole nature of thirteenth-century history, the case within which the thirteenth century moved and what it did with its memories, particularly with its Byzantine memories, specifically, here, in its capital city. In the more restrained of the majestic, imposing twelfth-century mosaics (that is, not the mobile and fantastic Evangelists' symbols and prophets), under a patterned welkin, Christ and his mother, formally enthroned, are flanked by solemn, static saints. They fill, in a line, the cup of the apse. Beneath them the thirteenth century is busy with one of its sorts of interpretation of nature and life. In a series of frames, the human story of the life of Mary is told, but in a style in which the twelfth century, here, would not deign to, or could not afford to, talk, catching at the echoed life and artifacts of the observer, calling him with graceful angels' wings.

But the naturalism of the style, the seeming reflection of life lived, is in a way misleading. Almost every image in Cavallini's panels comes from some "established iconography." Even in the panel of the Nativity, which seems particularly freshly drawn from life, with its Joseph, its animals, its trees, its charming little dog sitting, every piece except the *Taberna Meritoria* itself (Santa Maria's miraculous ancestor, applicable really only here), seems to be in fact a rather standard piece of iconography. Much of it—the pipe, the tree by the stream—is Byzantine, and much liturgical. If the dog, who is at least relatively rare, could bark, he would bark to you in Greek, and probably something from the Orthodox Christmas rite. But this too is in a way misleading. The assemblage, the arrangement, the style are Cavallini's or of his time, or they are in part. He, like the Rome in which he lived (and in this he is its emblem), used the remaining, visible past to its own purposes.[64]

There were lots of Roman processions, short ones and long ones, annual ones and occasional ones: the Candelora from Santa Maria Maggiore to San Martino ai Monti, only a few blocks wending through the Capocci towers and the buildings of Santa Prassede; the king of the Romans' pretentious descent from Monte Mario on his way to be made emperor at Saint Peter's, with ceremony, sword, stirrups, and games, crosses and thuribles, and the singing of "Behold I have sent my angel."

The historian Saba Malaspina, to whose showy, patterned style the description of procession was, he must have felt, suited, wrote grandly twice of high royal reception-processions in the 1260s, one for Charles of Anjou, one for his opponent Corradino. Charles's fleet entered the Tiber, and then, "as is the custom," it was pulled to Rome. At Saint Paul's, Charles disembarked, and there he was met by a huge crowd of the Roman people (*plebs*) of every sort, men and women, old and young, lay and clergy, and of various religious orders, carrying palms in procession and shouting "Osanna." The dancing of feet, singing of songs, and playing of instruments circled through the piazzas and streets of the city (and the old crowded city is recalled). And the nobles on horse played at their festive games.

A few years after Charles's triumphal entry, however, Corradino's entrance from the other side of the city evoked from Saba, to his Guelf confusion, an even more glorious description, rich in gems and drapery, of Corradino's reception by the fickle Roman people: *Ecce venalis gens Urbis, plebs Gebellenia,* the people of a city which had too frequently violated the honor of its ancient liberty with its harlot's games. As Corradino approached from the Prati of Saint Peter's, under Monte Mario, a wonderful sound of women playing on a multitude of musical instruments came from within the city, of women playing on cymbals and

The Physical City

drums and trumpets and viols. The city itself, along the path that Corradino took toward the Capitol, was decked almost unbelievably. Ropes and cords were thrown back and forth from opposing houses (and again one senses the crowded city—although the houses may well have had courts inside and gardens behind); and from these were hung not the customary laurels and branches but the richest of cloths of every description—in a society whose rich vocabulary of textiles and draperies reflects its major national commercial activity and, perhaps, concern. To the goldens and purples, the silks, the rich mantles and coverlets, towels and curtains, of work foreign and domestic, were added furs and jewelry with sparkling stones, all draping and swathing the path which Corradino would follow.[65]

Rome was a city of processions. And its greatest repeated processions were papal. One can see some of them recorded still—by an eleventh-century painter at San Clemente, by a thirteenth-century painter at Quattro Coronati. Of these papal processions the greatest, again, was that of the pope's coronation. That procession tied together the two sides of the city, the Vatican's northwest and the Lateran's southeast. It went from wall to wall. It also tied together the two greatest churches in the West, the two basilicas that each claimed for itself a towering preeminence. The Vatican held in its crypt the precious relics of the greatest of apostles; it held the hand which had held the keys from which all Rome's modern power was seen to stem. It was Saint Peter's church. The Lateran was the world's cathedral. It was the most ordinary home of the pope. It was the reliquary of great, fantastic, and mysterious relics, from the ark of the covenant and from the life of Christ, Aaron's rod, the relic of Christ's circumcision, strange objects in thirteenth-century Rome, perhaps less compelling than they had once been, but compelling enough still to add to the holiness of the Holy of Holies with its portrait of Christ and with Pilate's stairs. Late in the thirteenth century, under the Roman builder pope, Nicholas III (Orsini), the Lateran's Holy of Holies (*Sancta Sanctorum*) was rebuilt, and its holiness was reshaped and given the vitality of a current idiom. Both basilicas, both the Vatican and the Lateran, Saint Peter's and Saint John's, had essays written in their praise, each by one of its own canons, in the twelfth century. In them, great names and great memories (Constantine, the apostles) swirl around the *"omnium ecclesiarum caput"* at the Vatican and the *"Caput . . . mundi"* at the Lateran.

Silver and ceremony, gold and porphyry, liturgy and incense decorate the coronation procession from its beginning to its end—and grotesqueries. It is not simply pretty. As the description in the "Deeds of Innocent III" says, the whole city was crowned; there were incense, palms, flowers, songs. The "Life of Gregory X" talks of Gregory's going to the Lateran through a city that was like a bride, adorned with golden necklaces and shining in silken clothes.[66]

The new pope moved across the city after his consecration by three cardinal bishops at the Vatican and his public wearing of both episcopal miter and royal crown. The pontificals of the thirteenth century describe the procession as being in seventeen parts, or layers. The first layer was that of the papal horse caparisoned, the second, a subdeacon carrying a cross, the third, twelve flagbearers with red banners and two with cherubim on lances. There followed groups of officials, scribes, advocates, judges, singers, and the deacon and subdeacon who had read the epistle and gospel in Greek, abbots from outside the city, bishops, archbishops, and abbots from the city, patriarchs and cardinal bishops, cardinal priests, cardinal deacons. Finally in the seventeenth group, the small, final, elevated one, came the pope himself, a subdeacon carrying a ceremonial cloth, and a servant carrying an umbrella —and also, in late century at least, the prefect of the city carrying more apparatus.

The procession moved past the great fortress of Sant'Angelo across the river through the *rione* Ponte and into the *rione* Parione. There, near the Orsini fortress in the Campo dei Fiori, the pope was met by the leaders of the Jews of the city who offered him their laws for his approval. The procession passed on through the towered, campaniled city to San Marco by the Mercato (the market still recalled by San Biagio al Mercato, peering out in restored ruin from beneath the Victor Emmanuel Monument) and on to Sant'Adriano (now again the, restored, curia of the ancient senate). Thus the procession that moved along the Via di Papa had passed under the Capitol from thickly inhabited commercial Rome to the "archaeological zone," to the ruined forum with its half inhabited but still potently symbolic arches (like the Arch of Septimius Severus with its tower) and its churches, by the Torre de' Conti, within view of the Palatine (a ruin of monastic farms and of the sinister pile of the Septizonio), around the Colosseum, near the Torre degli Annibaldi, and by San Clemente. At five stations the pope's chamberlains scattered coins to the crowds. Durandus says too that from Saint Peter's to the Lateran there were triumphal arches for the pope and that clergy stood censing with censers of incense and that

The Physical City

for this the Romans were paid thirty-five *lire* for the arches and the clergy thirteen and one-half *lire* for the censing—as they say in Rome, *chi paga passa.*

At the Lateran the new pope proceeded through the elaborate rites of taking over that part of his domain and its symbols to the final royal banquet where he sat alone. They were, some of them, curious ceremonies. The pope was seated upon an antique marble privy seat and then lifted up from it "like the simple out of the dust and the poor from the mire." He took coins from the lap of his chamberlain and threw them to the crowd as he uttered words that must occasionally have stunned, if not choked, a speaker or an auditor, "Silver and gold are not for me"—but it is the Roman, perhaps the Italian, genius (perhaps particularly in quotation) to avoid at the same time subtlety and candor.[67]

The processions which crossed back and forth over the face of thirteenth-century Rome caught it only tentatively within a specific century. Their movements were filled with ancient gesture. They crossed a surface littered with curiously half-absorbed pieces of the past. This temporal complexity, this tentative grasp of the momentary present, this lack of chronological definition in the processions is like that exposed in the relatively modern language of some thirteenth-century contracts, as when in 1238 some baths (*termas*) on the Palatine in the *rione* Biberatica were sold for seventy *lire,* baths from whose peaks, the document of sale specifies, war and peace could be made, against and with the right men (new wars and new alliances amid the ruins of ancient conveniences).[68] Thirteenth-century Romans moved to their own purposes across a very used city. The antiquity of their ruins was embellished by the repetitive timelessness of nature. In the summers of every century, presumably, the birds skim the river by the island. In every spring daisies turn the grass around Sant'Angelo white; a rose blooms; the Gianicolo yellows with wildflowers. Year after year the capers shape their leaves, and blossom on the rocks. They must have blossomed as Innocent III and Gregory X listened to ancient, but not unchanging, liturgies.

There is a strange timelessness, too, about Roman games and

Roman songs. The most familiar Roman games took place on the Testaccio (the hill near the Porta San Paolo built of the shards of ancient amphorae) and in the Piazza dell'Agone (the Piazza Navona). Gregory X wrote of the Testaccio, and young Conradin's representative in the city saw its games. By the early fourteenth century Rome had turned into a place of elaborate courtly ritual; in one of its tourneys a Roman courtier courting a Lady Lavinia could appear carrying the device *Io sono Enea per Lavinia* ("I am Aeneas for Lavinia") and a Pietro Capocci, *Io di Lucrezia Romano sono lo schiavo*. Gregory X found the ribaldry and roughness of Carnival troublesome. It was a custom for roughly celebrating Romans to get alms in the papal palace in the evening on the last day of Carnival; each got a half piece of meat, a piece of bread, and a cup of wine. These ribalds behaved so coarsely that they kept genuine paupers from receiving alms and in fact forced them out of the palace. In order to subdue all this perverse merrymaking, Gregory tried to put more order into the almsgiving, to insure that one hundred platters of bread, one beef, and fifty casks of wine be properly distributed.

It was during Carnival that the Testaccio games occurred. They involved the pope and the prefect, drinking and demonology, men on horse and men on foot. On the day of the games, horse and foot rose, and after *pranzo* they drank among themselves. Then the foot went to the Testaccio, and the prefect took the horse to the Lateran. The pope, if resident, came from the palace and joined the horsemen. Together they rode to the Testaccio, and there they watched the games. Signs were made of the coming abstinence of Lent: "They kill a bear; the devil is killed, the tempter of our flesh. Bullocks are killed; the pride of our pleasure is killed. A cock is killed; the lust of our loins is killed —so that we may live chaste and sober in suffering of spirit, that we may be worthy to receive the body of Christ on Easter day." The formula describes a rite of animal sacrifice—and before the sacrifice the beasts were elaborately carted, with red cloth, through the city; and the Romans and their lusts were, in symbol, parted.[69]

It is hard to tell what parts of the festivals of students, recorded from Rome's past, survived into the thirteenth century, what, for example, of the New Year's festival. On New Year's Eve, boys went out in the evening wearing masks and carrying shields. They drummed and whistled and danced about, went to houses, sought gifts, and ate. They ate beans that day (as still, in parts of the American South, one eats black-eyed peas on New Year's Day). The next day, on January 1, some of them went from house to house carrying olive branches and

The Physical City

salt, throwing handsful into the hearth fire, wishing joy and happiness, and with the constant implication of desired fertility, saying, "Tot filii, tot porcelli, tot agni"—for children, piglets, and lambs. And they were advised before the sun rose to have eaten honeyed beans or something sweet and all the year they'd find it sweet—without contention, case in court, or much hard work.[70]

The boys' songs may be too early for the thirteenth century, the wiles of the country too late. It is difficult to know how early the surviving country saws and charms started, to know when men around Rome first believed that they should look for the weather to the peak of Monte Cavo or to the lightning over Maccarese:

> Quando lampa a Maccarese
> pija la zappa e torna al paese
> [When there's lightning over at Maccarese,
> take your hoe and go home to your *paese*.]

It is equally difficult to decide when men from the Roman countryside first fought witches in the dark by making a sign of the cross in the ashes or fought the evil eye (*malocchio*) by putting three hairs and seven pairs of pips of grain in a half-glass of water, while saying something like the *cantilena*:

> Sette fratelli carnali in vigna stava
> tre zappava, tre vangava, e l'invidia schiattava
> si prega la vergine Maria
> la mandi via

[So that of the seven sets of grain brothers in the vineyard of hair, three are hoed, three dug, one, the invidious, pair cracked, and it, pray the Virgin, carries the *malocchio* away.]

Some of these verses are of course late; presumably the pea blossom, the *fior di pisello,* that tells of the beloved prisoner in Sant'Angelo is of the period of Scarpia rather than of the Orsini. But many of these songs and poems that once filled the Roman air are like the brick wall along the Via in Selcis or the ground-level rubble at the head of the Vicolo della Spada d'Orlando ("the alley of Roland's sword"), or near the top of the Via di Santo Stefano del Cacco, like old Roman houses and, particularly, like the farmhouses from which the lightning over Maccarese was observed. They are old; they are composite. It would take a lifetime of specialized learning or a laboratory of chemicals to date any specimen exactly, and that would not date the type.[71]

Popular religious drama, related both to the liturgy of the processions and to the charms, is more fixed in time. Dialect plays, in Romanesco, about the Baptist, Saint Christopher, and the end of the world survive from the fourteenth century. By the end of the fifteenth century the Corporation of the Gonfalonieri was holding its sacred plays in the Colosseum. The sort of enthusiasm that both the plays and the Gonfalonieri activities represent is thought to have grown particularly from the religious excitement which reached a sort of crest in central Italy in the demonstration of 1260. It is a sort of enthusiasm intimately connected with heresy. Heresy was not absent from thirteenth-century Rome; it provoked a sharp reaction from Gregory IX which led to a disgusting scene of condemnation conducted by Gregory's senator, Annibaldo degli Annibaldi, in front of Santa Maria Maggiore in February 1231. And in 1266 heretics from the *rione* Sant'Angelo in Pescheria, a father and his son, Don Riccardo de Blancis and Pietro, were ordered to wear yellow crosses, two palms long and four digits wide, on their chests and shoulders, and to have their house torn down, a house that had "the Jews" (the Piazza Giudea?) on one side and the street to Sant'Angelo in Pescheria on the other. In a way, the sacred plays and the observable Inquisition which Franciscan friars acted out in the Aracoeli at the very end of the thirteenth century are twins, and twins partly in their lack of spiritual elaboration. The plays' actions seem very simple, and the great figures whom the Inquisition attacked were Colonna, whose heresy had to do with their reservations about Boniface VIII.[72]

All the wisdom (and superstition) that floated through the Roman air was not folk wisdom. Thirteenth-century Rome gives every evidence of having been, in the simple modern sense, a very literate city. A man called a "master of grammar" can turn up quite casually as a witness to a document, as does, for example, the *maestro* Giacomo, *doctor grammatice,* in 1283.[73] Rome was not a learned city in the sense that Bologna, Paris, and Oxford were learned; but much of the world's learning came to, and was employed at, its court. A *studium* was organized at the curia itself, and in 1303 Boniface VIII attempted to organize a university for the city. Roman cardinals sometimes had great and expensive libraries (one worth 1,787 florins). They also gathered around themselves households that contained not only men of diverse national and regional origin but also men with various sorts of skills and learning. Cardinals' colleges of chaplains were, insofar as they were institutions, learned institutions; and in this the cardinals emulated their papal masters. Late-thirteenth-century Rome was a suit-

The Physical City

able setting for a great scholar like the English Franciscan John Pecham. It was a city, an Italian city as well as a cosmopolitan one, where a great many people wrote, and not all of them were notaries or chancery scribes.[74]

Although much medieval Roman building cannot be dated and, because of that, seems timeless, some major Roman monuments speak quite firmly in describing not just what the thirteenth century saw but what it built (and what it borrowed from the past as well as from Venice and the East). They also, in spite of the destruction of almost all of the Vatican and Lateran, give some real sense of the scale and quality and, in its way, classical dignity of particularly late thirteenth-century Roman civilization. Its greatness is clear at Santa Maria Maggiore. There, before its apse mosaic, even Gregorovius, who had a great many at least superficially unflattering things to say about the visual quality of papal Rome, was overwhelmed. He wrote: "The mosaic fills the building with a solemn golden splendour that is more than earthly. When illumined by the sunlight falling through the purple curtains, it reminds us of that glowing heaven, bathed in whose glories Dante saw SS. Bernard, Francis, Dominic, and Bonaventura. Then the spell of the work seizes us with its radiance like the music of some majestic anthem." [75] If one sees the apse reflecting the glow of a great ceremony as on the annual August festa when a snow of white petals falls in the basilica, it is hard to react less ecstatically than Gregorovius did— although perhaps with a little less of the color and idiom of "O love, they die, in yon rich sky." The mosaic is a major work on a great scale. It is majestic and, as its sister in the Lateran must once have, it handles with majestic grace different pasts and different sorts of imagery (part copied or revisualized from the old fifth-century mosaic, part new)— the classical frolics in the rivers at its base, the central, then very modern, coronation of the Virgin, the old saints, and the new friar saints, the vines and birds, and the great winged triumph of supporting angels. The mosaic has portraits of its two patrons, Pope Nicholas IV and Cardinal Giacomo Colonna, and is signed *Jacobus Torriti Pictor* (Jacopo Torriti). It was presumably begun before Nicholas IV's death in 1292 and was essentially finished before Boniface VIII's war with the Colonna, probably before 1296. It is not the only great mosaic remaining from its period in Rome: on the façade of Santa Maria Maggiore itself are the "Rusuti mosaics"; and there are the Cavallini mosaics in Santa Maria in Trastevere.

Cavallini worked in fresco as well as in mosaic at a time when Roman fresco reached a stunning zenith at, for example, Cistercian Tre

Fontane. At its most frivolously, but not unpleasingly, progressive, Roman painting incorporated Northern Gothic decoration as it did, for example, in canopies over saints at Cavallini's Santa Cecilia in Trastevere. More impressively it recalled, as did contemporary mosaic, old Roman style and matter. In brilliant and sophisticatedly employed hues it presented (and presents) to its viewer the natural limbs of men and animals moving in a visually understood nature. To the monks working in the fields of Tre Fontane, their painter, at the end of the century, presented his didactic series of upper storey frescos, with the ages of man, the senses, and biblical and moral exempla, in forms like those of a real, although bright, eagle teaching a real eaglet to fly and a real fisherman fishing for real fish.

Cavallini's own fresco of the Last Judgment (again with a flourish of angels' wings) is still in large part visible in the inner façade of Santa Cecilia in Trastevere. At Santa Cecilia, as at San Paolo before it, the pictorial creativity of Cavallini was joined (in 1293) by the sculptural and architectural creativity of Arnolfo di Cambio. At Santa Cecilia the two were brought together by the taste, imagination, and extravagance of a magnanimous patron, almost surely the French cardinal priest, Jean Cholet. At San Paolo in 1285 Arnolfo had built his first Roman ciborium, the piece of architecture with sculpture which forms a canopy over the high altar and thus over the celebration of the Mass and the Eucharist. The ciborium at San Paolo is, in the words of Julian Gardner, "the first comprehended reflection of French Gothic architecture in the Roman region." The succeeding ciborium at Santa Cecilia, as Gardner has explained, shows both a deeper comprehension of Gothic and a more serious Roman classicism, a combination which is itself one of the distinguishing characteristics of Roman art in the last two decades of the thirteenth century. Emblematic of this combination is the mounted saint (perhaps Tiburzio) who rides from the southwest corner of the Santa Cecilia ciborium's attic storey and whose horse seems to tie together the classical South and the Gothic North.

Arnolfo, although not Roman, was, with his workshop, the great sculptor of late-century Rome. His Roman work begins with the unfinished but tremendously powerful stone Charles of Anjou enthroned on the Capitol. It ends with his commissions from Boniface VIII, particularly for Boniface's sepulchral chapel at Saint Peter's in which the pope's tomb, in Gardner's words, "formed the focal point of an elaborate complex of mosaic, sculpture and wrought iron." [76] Arnolfo's decades in Rome also produced the *presepio* figures for the Christmas crib in Santa Maria Maggiore (where a precious relic of Christ's own crib is

The Physical City

preserved) and, almost surely, the Vatican bronze Saint Peter who extends his toe to pilgrims' kisses and advertises the regal enthronement of the key-holding saint. Arnolfo also carved tombs, parts of which remain to stun the viewer with their beauty and power, like the fragments of Cardinal Riccardo Annibaldi's tomb in the Lateran and, above all, the figure of Honorius IV once in Saint Peter's but now in Santa Maria in Aracoeli.

"Conspicuous in Rome are the tombs." [77] Almost as compelling as Arnolfo's own work is that by an anonymous sculptor who carved the figure of and draperies around Cardinal Ancher Pantaléon (d. 1286) at Santa Prassede. The work of Giovanni di Cosma may be artistically inferior, but it is very visible in Rome; and it presses upon the viewer the three-dimensional reality of thirteenth-century cardinals and cloth (and imagined angels) in Santa Maria sopra Minerva (Guillelmus Durandus, d. 1296), and Santa Maria Maggiore (Consalvus, Gonzalo Hinojosa, d. 1299), in Santa Maria in Aracoeli (Matthew of Aquasparta, d. 1302), and in Santa Balbina (a papal chaplain instead of a cardinal, Stefano Surdi). These serious, substantial images, with their heavy ceremonial shoes and their elegant draperies expose a view of life that could produce real solemnity. They also expose the fact that the curial cardinalate was magnetic to artists as well as to household clerks.

Certainly not all of the work of the late thirteenth century in Rome reached the level of Giotto, Cavallini, and Arnolfo. The pretty little church of Santa Maria in Cosmedin, which Boniface VIII's nephew, Cardinal Francesco Caetani, redecorated (and decorated with his family's arms), is a good example of work at a lesser level. Here the derivative Deodato built a Gothicked ciborium in a church for which as for most Roman churches (of which one could argue there were surely enough) a total Gothic rebuilding was out of the question. But even, perhaps particularly, surrounded by the second rate, by Giovanni di Cosma and Deodato, one is aware of the prodigious, heavily patronized, and significant artistic creativity of late thirteenth-century (pre-Avignon) Rome. This activity was accompanied by conventional, repetitive decorative detail, for example, the stars of its spangled ceilings (as in the chapter-house of Tre Fontane and in the cave church off the Nomentana), the frieze of its borders (as in the mosaic panels in Santa Maria Maggiore and Santa Maria in Trastevere and on the old painted ceiling above the simulated consoles at Santa Maria in Aracoeli), and the shapes of its stylized flowers and decorative pieces of building metal (of both of which good examples survive at Tre Fontane). The ecclesiastical furniture produced by late thirteenth-century creative activity joined the already characteristic aspects of twelfth- and thirteenth-century Roman churches—the constantly cosmatesque floors with their changingly shaped pieces; the archpriests' or bishops' thrones; the twisted (Jerusalem-remembering) paschal candles; the small tufa blocks; the brick campanili.

A great deal of the physical city of thirteenth-century Rome remains. Inside the miserable ring of modern suburbs which surrounds Rome, a few medieval towers still lift themselves. The Caetani fortress in the tomb of Cecilia Metella, although oddly restored, still dominates the Appian. The heart of the city is still commanded by Honorius IV's wall (at least a reworking of it). Although the once arched Corso and the Via di Papa may not, the island still looks in some part itself. The destruction, or supposed destruction, of 140 towers in the 1250s and the long series of more recent modernizations might not have been expected to leave so much more that is medieval than those familiar jumbled parts, great or small, of hundreds of churches that reach back to, or beyond, the thirteenth century. A bridge, the Nomentano, and a gate, the Porta San Paolo beside the pyramid of Gaius Cestius, remain to show us what bridges and gates looked like. The city within the Aurelian walls, with its disparate but not cleanly divided neighborhoods, once tied together by processions and interlinking property interests, the thirteenth-century city of Saint Peter's men, would seem to survive in many places, but to survive without its men.[78]

But sometimes the men themselves seem to jump out, or up, at you. Stand in the rear of Santa Maria in Aracoeli, and you see lying amidst the cosmatesque work a flat tomb. From it look up at you the hands, the feet, the face of Matteo, the household scribe of Luca Savelli, who died in February 1313. He is stylized of course, but men are stylized. He seems more there, more really present, than the cardinals of the sculpted tombs, more to be a co-heir, to have a mother, and perhaps to have held a lean-to with a vineyard in the *rione* Colonna— less is demanded of him, in keeping the world's order, and he demands less than Ancher or Surdi. He was trying less hard, or less obviously and expensively, to impress. There are other men like Matteo; there is a fragment of one (who died in 1323) now plastered in the corridor wall of a palazzo (number 5) on the Via della Dogana Vecchia (and suspiciously close to Sant'Eustachio). At Santa Prassede there is a splendid spicer, on the floor between the nave and the right aisle, with shoes like Matteo's, but dressed in the garb of a pilgrim to Compostella, with hat and shell.[79] The inscription around his watching figure says: "This is the tomb of Giovanni da Montopoli, spicer. What you are, I was. What I am, you will be. Pray for me, sinner, do penance." You pray.

CHAPTER

II

The Ideal City

WORKING IN THE FIELDS OF distant Worcestershire or Gloucestershire, the ordinary thirteenth-century peasant would never have seen the physical city of Rome. He would not even have been able to think of it in any serious sort of physical detail. He would probably have spent his whole life within twenty or thirty miles of the place where he was born.[1] But as unlikely as his having been to Rome or his having been able to imagine its detail is his not having heard of it or having some idea of what it stood for. Of all the cities of the world only Jerusalem and its sacred satellites can have competed with Rome for a place in thirteenth-century men's minds.

The peasant's neighbors, the monks in monasteries, knew more of Rome. Most of them must at least have known someone who had actually seen the city, walked in and out of its gates, and prayed above its martyrs' bones. They could be expected to appreciate in some detail, in the 1280s, the disastrous news that part of Saint Peter's had fallen "where the altar of the apostles stood, with their celebrated and far-famed statues." The two most familiar English monk historians of the twelfth and thirteenth centuries both put the city of Rome into their work. William of Malmesbury wrote, as Master Gregory would, in the great tradition, of Rome's fall from being the mistress of the world, and he decorated his lament with the verses of Hildebert. But William also, in the fourth book of his "Deeds of the Kings of England," wrote an important and somewhat knowledgeable description of the city, with its gates and its martyrs' shrines. Matthew Paris in his "Itinerary to Apulia" drew a plan of Rome, which was, as he said, "the end of the road for many." [2]

The idea of Rome which was spread through Christendom, although often not very intricately realized, was complex. It was, in various ways, the idea of many Romes combined—ancient and modern, pagan and Christian, governmental and spiritual, holy and sinister. It

The Ideal City

was this ideal complexity, magical and mysterious, which was the real Rome to most thirteenth-century men.

The magic and memories of the place made people believe that wonderful treasures were buried beneath its surface. That common medieval belief, and knowledge, that many beautiful and valuable old things were buried in barrows and hoards in the ground, with all its historical and psychological implications, concentrated itself upon "Golden Rome." The richest of treasures was thought to be the treasure of Octavian, Augustus, and that treasure was sometimes connected, as it was by William of Malmesbury, with a story of treasure, various in its location and its telling and long and deep in its appeal (as Jean Cocteau's brilliant use of it in his version of "Beauty and the Beast" shows); but, in the middle ages, the treasure was buried particularly in the soil of Rome.[3]

There was an image in the city of Rome, in the Campo Marzio some versions say, on top of the middle finger of whose extended right hand was inscribed, *Percute hic!* ("Strike here!"). But for a long time no one understood what the inscription meant, until finally it was interpreted by a certain shrewd clerk (who, according to William of Malmesbury, was the, in twelfth-century memory, magician pope Silvester II, Gerbert, from the end of the first millennium). The clerk dug away the ground where the shadow of the image's finger fell when the sun was above it, until he found steps which led him down into a noble underground palace. He entered its hall and there found a king and queen and their court, all magnificently dressed, at dinner, but no one said a word to him. And they were all of gold in William of Malmesbury, and gold knights were playing with gold dice. The clerk looked into one corner of the hall and saw a great polished stone, a rubylike carbuncle, which gave light to the whole place; and he saw in the corner across from the carbuncle a man (or, in Malmesbury, a boy) standing with a drawn bow in his hand and on his forehead written, "I am who I am. No one can escape my bow. Especially cannot that carbuncle which shines so splendidly."

The clerk wandered through the rooms of the palace, but no one spoke to him. In the stables he touched the beasts, but they felt like stone. He went back to the hall, but said in his heart, "I have seen wonderful things (*mirabilia*) today, all the heart could desire can be found here, but who will believe me?" So as a sign he took a gold cup and one of the magnificent knives from the table; but as he slipped them into his costume to carry them away, the statue in the corner shot his arrow and smashed the carbuncle to pieces and threw the whole hall into total darkness. The now contrite clerk could not find his way out. There in the palace he died his wretched death. (Avaricious Silvester's morally informed end, better known, was less dire.) Again, this richness of fable and gold, some thought, lay beneath the Campo Marzio, a place that a fourteenth-century Roman notary could remember in casual dating as *ubi erat exercitus* ("where the Roman army was").[4]

The persuasion about the ancient riches of Rome, of which the local application of the legend of the golden palace was one expression of one face, showed quite another face, one that has been seen in the documents recording the leases of Roman lands. In them, the landlord reserved for himself part of any gold, silver, or other valuable metal or stone found in the land. The land of Rome was queerly rich; the men of the Calcarario who ground up marble into chalk knew it. Both the perplexity of the expression and the combination of the mysterious and the mundane are characteristically Roman.

There was a related movement, a similar dispersal, in the various explanations of individual Roman monuments. A good example of this movement exists in Master Gregory's discussion of the equestrian statue of Marcus Aurelius then at the Lateran but which later, in the sixteenth century, was moved and became the center around which Michelangelo reconstructed the Capitol.[5] Master Gregory assures his readers that the immense figure of horse and rider was known by three names to three different groups of viewers. Pilgrims called it Theodoric. The Roman populace called it Constantine. Cardinals and clergy of the Roman curia called it either Marcus or Quintus Quirinus. Master Gregory tells something of its (mistaken) history: that it had stood on the Capitol, that Gregory the Great had used its columns for Saint John Lateran, that the people of Rome had set horse and rider up again before the papal palace at the Lateran. Master Gregory describes the position of the horseman and the bird on the horse's head. He explains that the names given it by pilgrims and Romans are based on foolish fable which he rejects, but explains rather the basis for the correct name, Marcus, which he says he has learned from the aged, from cardinals, and from learned men.

He then tells a conventional story of Marcus Crassus's defense of the Romans against the Mesi. Not only is Master Gregory's story given an additional complexity, an extra dimension of error, by his and his informants' choosing the right name for the wrong reasons, it is given an even further dimension by his misrepresentation of what must have been common belief. The oldest (essentially twelfth-century) redaction of the *Mirabilia,* the most common of Roman guidebooks, and Master

The Ideal City

Gregory's predecessor, also makes a distinction, although a less pretentious one.[6] The *Mirabilia* begins its account of the equestrian statue by saying, "At the Lateran is a certain bronze horse called Constantine's, which it isn't." The *Mirabilia* then, after saying that whoever wants to know the truth can read the story, tells a story much like Master Gregory's. Obviously the confusion about the horseman's identity was pretty broadly appreciated by the time of Master Gregory.

The guidebooks' and the medieval observers' dual (and more than dual) view of the Roman past and the relation of individual monuments to that past comes into sharpest focus when it observes the disparity between, or the combination of, the classical-pagan and Christian pasts. Sometimes this duality is handled perfectly straightforwardly (if not perfectly historically). It is, for instance, in the *Mirabilia*'s story of the origin of the feast of Saint Peter's Chains.[7] That account tells of Octavian's assuming power after Caesar's death, and of his brother-in-law Anthony's opposition to him, of Anthony's repudiation of Octavian's sister and his taking to wife of Cleopatra "queen of Egypt, most mighty (she) in gold and in silver and in precious stones and in people." The *Mirabilia* tells of the great sea battle of Epirus, of the death of Anthony on his sword, of Cleopatra's vain attempt to seduce Octavian, and then, in her husband's mausoleum, of her taking two serpents to her breasts. Octavian took great riches from his victory and triumphed in the East and in Rome, where he was received by the senators and all the people. Because the victory took place on the Kalends of August, the whole city came to have a huge festa in honor of Octavian Caesar Augustus on that day. And that festa continued, according to the *Mirabilia,* moving into a story with a cast from mixed centuries, until the time of Arcadius, the husband of Eudoxia.

This good woman, who ruled the empire for her son Theodosius after her husband's death, was divinely inspired to go to Jerusalem to visit the sepulchre of the Lord and other sanctuaries. There she procured from a Jew the chains with which Herod had bound Peter; in her joy at finding the chains she decided that they should go back to the place where the body of the blessed Peter rested in the dust. She came to Rome on August 1, and there she saw in its full rapture the ancient pagan feast in honor of Augustus, the celebration of which no pope had been able to stop. Eudoxia approached the pope, Pelagius (from the mid-sixth century, much later than any of the candidates for the "true foundress Eudoxia," who is most commonly identified as a combination of Eudoxia, wife of Valentinian III, and her mother, the wife of Theodosius II, both early to middle fifth century); she went, too, to the senate and people. She said to them all:

I see you so eager in your celebration of the festa in honor of the dead emperor Octavian, because of the victory he won over Egypt. I ask you that, for me, you give over the honor of the dead emperor Octavian for, instead, the honor of the Emperor of heaven and his apostle Peter, whose chains I have brought out of Jerusalem. Just as that other emperor freed you from Egyptian slavery, this heavenly Emperor would free you from slavery to evil spirits. And I wish to make a church to the honor of God and Saint Peter and to put the chains in it. And our apostolic lord should dedicate the church on the Kalends of August and it should be called "Sanctus Petrus ad Vincula," Saint Peter at the chains. There each year our apostolic lord should come to celebrate Mass. Just as Peter was freed by an angel, the people of Rome can go away freed from their sins with his blessing.

The people acceded to the request. The church was built. The pope dedicated it on August 1, just as Eudoxia *christianissima imperatrix* proposed. There, at San Pietro in Vincoli, Saint Paul's chains were added to Peter's, and each year on the first day of August the Roman people came and celebrated the feast.

The first day of August and the chains gather together the two great Roman traditions, put them together by showing the change from one to the other, through the agency of a woman whose very title, *christianissima imperatrix,* joins the two. The story also shows the movement from Jerusalem to Rome (rather forgetting Constantinople). It shows, in men's minds, Rome's becoming the second Jerusalem. The story is not the only medieval explanation of the chains, it cannot be historically correct, but it is historically interesting. Best of all, it illustrates, with a nicely balanced rhetoric, a common way of dealing with Rome's dual tradition, by showing the connection of one with the other, on a day, at a place, and the replacement of, or the enrichment of, one by the other.

This sort of replacement is seen again in the *Mirabilia* in its story of the Pantheon.[8] In this story the *Mirabilia* tries to recall for its readers the time "of the consuls and senators" (actually the time of Augustus) when the "prefect" Agrippa had (according to the account) subjugated the Suevi and Saxons and other Western peoples and was then asked to fight the Persians (because Rome was warned by the alarm in the statue that stood for Persia among the "multitude of statues" on the Capitol, the legendary alarm system of empire). Agrippa worried about the problem; and in his worry a woman appeared to him and told him that she would make him victorious if he would build her a temple of a sort that she would show him. Agrippa replied, *"Faciam, domina"* ("I will, madam"), or as his phrase translates with surprising modernity into the vernacular guide based on the *Mirabilia,* the *Miracole de Roma, "Madonna, volentieri."* The vision showed him how to build the temple, and he asked her who she was. She said that she was Cybele, mother of all the gods. She said that Agrippa should make a libation to Neptune, a great god, that Neptune would help him, that Agrippa should dedicate the new temple to Neptune and to Cybele, and that they would both be on his side and he would win. Agrippa told this to the senate. He then went forth and conquered the Persians and forced them to tribute. Victorious, he returned to Rome and built the temple and dedicated it to Cybele, to Neptune, and to all divine spirits. He called it the Pantheon. He placed on its roof, above its eye, a great gilt statue of Cybele, and had gilded its bronze tiles.

Then much later, in the time of the Emperor Phocas (early in the seventh century), Pope Boniface IV saw this wonderful temple dedicated to Cybele in front of which Christians had often been afflicted by spirits (*daemonibus,* the ancient gods turned devils), and he asked the emperor to give him the temple. It had in the classical past, it was thought, been dedicated to Cybele on the Kalends of November, and so by Boniface it was dedicated on the Kalends of November to Blessed Mary ever Virgin. The temple had, the *Mirabilia* says, been dedicated to the mother of the gods; the church was now dedicated to the mother of all saints. It was decreed that on the first day of every future November the pope should come to the Pantheon to say Mass, and that the people should receive the Body and Blood of Christ as on Christmas Day, and there should be a festa for the Virgin and all saints, and the dead should have, through the churches of the whole world, Masses for their souls. The pattern of prototype and change, of day and feast, the rhetorical balance, the pairing of pope and emperor are all very similar to those of the story of Saint Peter's Chains. In each case a day and a Roman place are specifically tied to the two great Roman pasts, and they to each other, and a Roman day and place are made doubly wonderful and doubly attractive to pilgrim tourists.

There existed, however, in the guidebooks as elsewhere (even in stories for sermons to be spread about the world) a more potent and imaginative, a more universal and significant combination of the two pasts, one that showed foreknowledge as well as prototype of Christianity in ancient Rome, among classical heroes themselves. The great instance of this sort of combination again centers around Augustus, often remembered as a powerfully benevolent historical figure, for various reasons, including his having ruled at the time of Christ's birth and, in fact, at the time of this story. Augustus, the model of strength and peace and good government, is at the other end of the imperial spectrum from the evil, brooding, destructive, perverse persecutor, Nero. They are counterfigures, and both are important to the Christians' imagined history of the Roman past. The *Mirabilia*'s connection of Augustus with Christ is a story with a past, with many tentacles and relations, cousins and ancestors of various lines, eastern and western, of which the finest, but far from the closest, was the Fourth Eclogue of Virgil, for the middle ages the greatest and purest, and again partly because of this Eclogue, of Roman poets.

In the *Mirabilia*'s version, the senators, having seen the beauty of the emperor and the peace and prosperity with which the whole world lay subject to him, came to him and said: "We want to adore you, because there is divinity in you. If there were not, all fortunes could not lie subject to you." Augustus initially opposed the idea and asked for time. He then called to himself the Tiburtine Sibyl and told her what the senators had said. She in turn asked for three days and then returned and told him in her verses (learned from Augustine's Sibyl) that a king would come from the heavens. And then on the spot the heavens burst open and a great light shone upon him; and he saw in the sky, standing over an altar, a beautiful virgin, holding a boy in her arms. And he heard a voice saying, "This is the altar of the Son of God." The emperor immediately fell upon the ground in adoration. Later he told the story to the senators and they were, as he had been, filled with wonder. This vision, the *Mirabilia* continues, occurred in the chamber of Octavian the emperor where now is the church of Santa Maria in Capitolio. That is why it is called "Sancta Maria Ara Caeli."

The location of the legend is significant. It is the governmental Augustus who is involved, in the place where great old government was remembered and present government was transacted. If woods were changed to miracles, Virgil's evocative line from the Fourth Eclogue could apply to the legend maker's song:

si canimus silvas, silvae sint consule dignae.[9]

In general the joining of Christianity and the classical past in the *Mirabilia* and related texts is not so pointed, so intellectualized, so carefully conceptualized, as it is in the case of Augustus's altar or of the Pantheon and San Pietro in Vincoli. In general the *Mirabilia* seems a

The Ideal City

sort of palimpsest with one civilization written over the other, or perhaps a tapestry with the two stitched together. The fascinating thing about this palimpsest is that it is sometimes hard to decide what is background, what foreground. One is repeatedly shown Christian churches so one may mark the places where there were once great pagan temples. But sometimes one seems to be shown classical landmarks so one may know what martyr acted there (as at the sewer by the palace of Septimius Severus into which Saint Sebastian was thrown). Beyond this there is much accumulated jumble of gates, bridges, measured columns, obelisks, and of Christian act—the place beyond the Porta Appia where Saint Sixtus was beheaded and where the Lord appeared to the fleeing Peter, and Peter said to Him, *"Domine, quo vadis?"* and the Lord said, "To Rome, to be crucified again." [10]

But none of the jumble of the guidebook seems frivolous because of the importance of the matter, both matters, and of the place, Rome. None of it, or little, even in the style of the *Mirabilia,* seems dead in a city where the juncture of civilizations it proclaimed was still alive and visible, where cardinals were locked up to elect a pope in the Septizonio, the palace of Septimius Severus, where one could go to see the Ara Coeli, where at the end of the century one could go to Santa Maria in Trastevere and see in Cavallini's mosaic the picture of the *Taberna Meritoria,* on the site of Santa Maria, where the oil had flowed in Rome, at the birth of Christ in Bethlehem, and had thus connected physically the promised birth of Christ with Augustus's City. Moreover, the jumble, the complexity, of ideas about Rome was less disorderly, more controllable, because much of it was actually the complexity of joined opposites.

One of the most potent of all the ideas about Rome in the thirteenth century was the idea that Rome stood for government. It stood for government because it was still remembered that in Rome, in the distant past, serious government had come closest to being realized. Both the *Mirabilia* and the related *Graphia Aureae Urbis* include an old description of offices of government in their more general description of the city. *Roma caput mundi regit orbis frena rotundi* ("Rome, the world's head, holds the round globe's reins") was, the *Graphia* makes clear, a pleasing verse and inscription in guidebooks as well as on crowns and coins. And Nicola Pisano remembered it when he built his great fountain at Perugia.[11] That strange medley, *macedonia,* of fragments of republican and imperial Roman history, which the twelfth and thirteenth centuries recalled, illustrated for them as it had for their predecessors, the surface that both hid and revealed the secret of true governance and political power. When in the eleventh century William of Poitiers had tried to make William the Conqueror sound as if his prudence was the sagacity of planned government, he spoke in the idiom of Rome; and William of Poitiers wrote in a tradition that had moved through Bede, Charlemagne, and the Ottos brokenly to himself.

The general governmental memory was of course remembered in connection with specific places within Rome. In the *Mirabilia* it is specifically connected with the Capitol, "which was the head (*caput*) of the whole world, from which consuls and senators were accustomed to rule the globe. . . . It was called the golden Capitol because in its wisdom and its splendor it overpowered all the realms of the whole world." [12] But the whole of guidebook Rome was seen as emanating an aura of past governmental greatness, resounding with the echoes of great names, not just Augustus, but Julius Caesar, Trajan, Antoninus, Nerva, Hadrian. Thirteenth-century Rome had a senate and senators, admittedly oddly realized. The twelfth century had recalled republican institutions to join distorted imperial ones. Rome saw itself, as it was seen from the distance, as growing from a great, and always a particularly governmentally great, past.

The earliest beginnings of Rome's great past found in the middle ages a rather peculiar legendary statement. It was peculiar because it needed to combine two elaborately known but disparate pasts, a fact that in context is unsurprising in a Christian city whose inhabitants bore names like Enea and Ottaviano. It could not, and could not want to, escape the great heritage of Trojan Rome, of Aeneas and Romulus, a Rome enshrined by Virgil for both classical and medieval Rome. But the history of Rome, like the histories of other Christian medieval places (like, for example, Genoa or Saxon England), had to fit into the history of the Old Testament too. By the time of the writing of the thirteenth-century manuscript of the "Graphia" this fitting together, for Rome, based theoretically on the legends of "Hescodius" (probably a distorted Hesiod) had been accomplished, although by the early fourteenth century the history of "Hescodius" was already being sharply attacked, at least in northern Italy.[13] This fitting together was not only not uniquely Roman, it was also not a very tight juncture; and it was complicated.

The story began with the building of the tower of Babel, after which Noah with his sons took ship for Italy and built a city near what later became Rome. Janus, or Giano, Noah's son, together with an-

The Ideal City

other Janus, his grandson or nephew, the son of Japhet, and Camese, a local man, built a city, Ianiculum (Gianicolo). They ruled, eventually Giano alone, after having built with Camese a palace also called Ianiculum in Trastevere, where "now" is the church of San Giovanni ad Ianiculum, by the Porta Settimiana. Giano also built and then ruled from a palace on the Palatine. In Giano's time, Nemroth-Saturn came and made a city called Saturn on the Capitoline; and Italus and Hercules and Tibris and Evander and Coribas and Glaucus and Aventinus Silvius came and built cities, and so did Romae, daughter of Aeneas, with her Trojans. Finally, in the four hundred and thirty-third year after the destruction of Troy, Romulus of Priam's royal blood, in his twenty-second year, on April 17, built a wall around all these places and called the collective place Rome; and noblemen from all the surrounding races came and lived there with their families.

Like medieval Christians, medieval Jews had to deal with the central historical importance of Rome. They, too, had legends of foundation, the miracle of the planted reed, the efforts of the sage Abba Kolon.[14] The city of Rome was alive with history, fabulous and "real." One might have thought that so much "real" history would have obviated the necessity for fabulous history, that the good history might have driven the bad history out; but in Rome it certainly did not. The whole place glowed with fable and magic. "When the Sibyl had come to Rome, there came to meet her the entire city, both great and small." [15]

Rome's wonderful governmental past was naturally attractive to medieval "Roman Emperors." The magic of past politics entranced the Hohenstaufen. Its sibyl spoke to them, particularly to Frederick II.[16] Through Frederick's long ruling life, until his death in 1250, Rome and its idea were important to him. What the nature of that importance was depends upon the sort of man Frederick was. About what sort of man he was, although he is in lots of ways very visible and he certainly has been carefully studied, there is room for considerable difference of opinion. He can seem, as he has to those who have known him best, a genius, a man of rare perception and ideals, even of great, of grandiose, political understanding, the hero of the thirteenth century. He can also seem, as he does, for example, to me, a bright, pathological, egocentric bore, a man incapable of individual personal relationships, of real human relationships, a man incapable of relating seriously to his own surrounding external "reality," and so pressed to a life of hyperbolic fantasy, rhetoric, and violence (which he could express, and we can see, because of his exalted position). The things

1. Charles of Anjou by Arnolfo,
in the Palazzo dei Conservatori
on the Capitol,
see pp. 67 and 98.

2. The two cities, crowded and open,
within the walls of Rome,
a late medieval drawing
from a manuscript in Turin:
Biblioteca Reale, Varia 12,
fo. 28, see p. 15.

3. San Tommaso in Formis: Cosmati mosaic of
John of Matha's vision, see p. 14.

4. Cardinal Ancher, and drapery, in Santa Prassede, see p. 68.

5. Boy with a thorn in his foot,
piece of antique sculpture
in the Palazzo dei Conservatori
on the Capitol, see p. 90.

6. Steps of Saint Peter's
after coronation of Boniface VIII
(January 1295) with part of the
procession moving off toward the Lateran,
from a volume of Stefaneschi's
Opus Metricum once in Boniface VIII's
library and now Vat. Lat. 4933 (fo. 7ᵛ)
in the Biblioteca Apostolica Vaticana,
see p. 189.

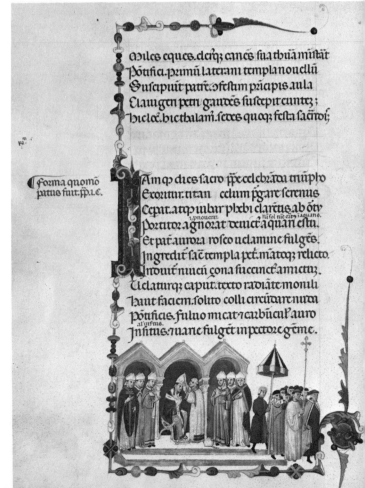

7. Rome, representing Italy, with Saint Mark, by Cimabue, in the upper church at San Francesco in Assisi, see p. 9.

8. Cloth merchants and their customers, from a fifteenth-century manuscript in Bologna, see p. 53.

9. Pewter pilgrim's badge with figures of Peter and Paul, probably thirteenth century, discovered in Dublin and preserved in the National Museum of Ireland, see p. 54.

10. Angels and Cardinal Colonna, in detail of the Coronation of the
Virgin, mosaic by Torriti in the apse of Santa Maria Maggiore, see
pp. 66 and 170.

11. Nativity, mosaic by Torriti in the apse of Santa Maria Maggiore, see pp. 66.

12. Nativity, mosaic by Cavallini in the apse of Santa Maria in Trastevere, see pp. 58 and 66.

13. Fresco of fisherman from Tre Fontane, see p. 67.

15. Mounted saint on the ciborium by Arnolfo at Santa Cecilia in Trastevere, see p. 67.

14. Thirteenth-century metal brace from Tre Fontane, see p. 67.

16. Pope Saint Silvester showing the
images of Peter and Paul
to the Emperor Constantine,
fresco in the oratory of
San Silvestro at Quattro Coronati,
see p. 84.

17. The Theater of Marcellus,
before restoration, see p. 42.

18.   Face of Christ from Last Judgment by Cavallini at Santa Cecilia in Trastevere, see p. 67.

19.   The Ponte Nomentano, see p. 69.
20.   The Three Kings by Arnolfo in Santa Maria Maggiore, see p. 67.

21. The Pantheon,
restored, see p. 77.

22. Mosaic Senatorial Donor,
Chapel of Santa Rosa
in Aracoeli, see p. 117.

23. Island in the Tiber from downstream
before the building of the
anti-flood embankments, see p. 17.

24. Ciborium of Saint Paul's outside
the Walls, hanging angel censing,
see pp. 67 and 216.

25. Angel from the tomb of
Boniface VIII, *Grotte* of Saint Peter
in the Vatican, see p. 26

26. Tomb of Matteo the Scribe, detail, floor of the Aracoeli, see p. 69.

Frederick did and said about Rome must be interpreted, even be seen, by every historian in terms of his own Frederick. But one thing about Frederick must generally be seen and admitted. For a medieval emperor, Frederick was very Italian. He came from Italy, and, in part, he lived there.

In April 1212 Frederick, a boy of seventeen, came to Rome to be received as uncrowned emperor by the people of Rome and by his distant guardian, the pope, Innocent III. In November 1220 Frederick, by now twenty-five, came to Rome, to the Vatican, for his coronation at the hands of the pope, Honorius III. But Frederick's great flirtation with the city came later in 1236 and in the years that followed. In 1236 began "Frederick's wooing of the Romans with rhythmical high-sounding phrase." [17] Frederick II tried to make the Romans of Rome seem, perhaps to the world and to the Romans themselves, and certainly to himself, the citizens who created emperors, the basis of, and the partners in, his power and his victories.

This is most apparent physically after the Battle of Cortenuova, in which Frederick defeated the Milanese in November 1237. In victory, Frederick celebrated a great triumph which, he told the Romans, was after the pattern of the triumphs of his predecessors, the old Roman emperors. He sent to the senate and people of Rome the captured Milanese *carroccio,* a chariot, the symbol of the city which bore its standard: "Our wishes would . . . be far removed from reason if we, illumined by the radiance of the Caesars, were to tolerate the Romans' being left without a share in the rejoicings over a Roman victory." [18] The chariot was installed on the Capitol. There is still, in the Capitol, a *Sala del Carroccio* with its inscription. Like Frederick's famous statue on the bridge gate at Capua, the *carroccio* was a palpable token of his self-conscious "Romanness," but in a different way. Unlike the Capuan statue, this sign was deposited within the city that Frederick tried to revive, so that it might meet his own pretensions.

Frederick's pattern touched his successors, and a great apologist of the Hohenstaufen could move himself to Roman rhetoric over Corradino in Rome in 1268, saying that Romans "remembered now that they were of the blood of Romulus." [19] The Hohenstaufen pattern of securing Roman allegiance was, at least by Frederick II, supported and the rhetorical superstructure given some underpinnings by the emperor's buying tenements from Roman nobles and regranting them to their former owners as fiefs. But in general, in spite of the dreams historians have dreamed of it—"The shades of Rome, of the Romans and their Caesars, had tasted blood: they began to stir again. . . ."—

The Ideal City

it was all dream.[20] Rome did indeed have valuable memories of its ancient past, but valuable for guidebooks, for cadging tourists' coins, for building the wealth of men and trades who grew fat from the pilgrims' coming, not for building a state (at least not a secular state). It was an economically, not a directly politically, valuable memory (and only valuable economically in conjunction with other dreams). Its direct political use by Frederick has a queer, false, insubstantial air as it also has in Dante but not, I think from Charles Davis's work, in Nicholas III or in the republican ideas of Ptolemy of Lucca. The dream was necessary to Frederick, psychologically, not politically; he had to think of himself as an old emperor in old Rome, gloriously independent of his contemporaries.

The closest thing to a secular rekindling of the ashes of old political and constitutional Rome came, in the thirteenth century, through the efforts of Brancaleone, a Bolognese. And it is in fact to Bologna, rather than to Rome the city, that one should look in the twelfth and thirteenth centuries for a serious attempt to recapture that particular ancient Rome.

Not only was the imperial-political Roman memory, in Rome, a limited one, it was also overshadowed, dominated, in most areas of thought and act by the other memory, the memory of Peter and Paul. In repeated statement, Rome was the city consecrated by the very glorious blood of the two martyrs for Christ, the two princes of the apostles, bearers of the Word, teachers, who died, wonderfully, at the hands of the same persecutor, at the same time, in the same place, this place, Rome. Upon the ground enriched by their blood, in rhetorical fancy, had grown the power and importance of new Rome. And the rhetorical fancy was matched by economic and ecclesiastical political fact.

Peter's primacy (enhanced in medieval sentiment by Paul's closeness) and his presence in Rome were the assumed basis for the presence there of pope and curia, although the real basis had more to do with past politics and Caesars than thirteenth-century Roman denizens were perhaps prepared to realize. One aspect, however, one parable, of the importance of the popes' Caesarean heritage was clearly recognized. The thirteenth century did not forget the donation of Constantine to Pope Saint Silvester I. In a city which had long known pictorial propaganda, a new cycle of the story of Silvester and Constantine was painted in mid-century at the church of Quattro Coronati. Earlier, Innocent III had preached, in his sermon for Saint Silvester's Day, of the great Constantine, cleansed from his leprosy through Silvester's baptizing him, who kept for himself Byzantium and gave Silvester the West (and gave the Church tranquillity). Later Nicholas III, in his urban constitution, spoke of this baptismal cure as one of the historical reasons for papal rule. But the story was used rather than needed by Innocent and Nicholas. It helped them, and, again, it connected the old and the new Rome, the political and the religious. Both popes surely would have regretted the public discovery of the literal falsehood of the story of the donation; but it is hard to believe that either would have been deeply, personally surprised or, in any serious way, embarrassed by the discovery. The donation was one of many "reasons" for papal power; each enhanced the others and helped to rationalize reality, or desired reality. But if any of these "reasons" was in fact necessary, it was not the donation. It was in fact subsidiary to the past presence in Rome of Peter and Paul.[21]

The presence of the active, greedy, public, pretentious papal establishment kept Rome's relics burnished and obvious. It gave Rome the jubilee. It brought masses of litigants to Roman courts, and diplomats and petitioners and penitents to the Roman curia. Papal government made Rome rich with court incomes and with the profitable demands it made upon the trades that fed, housed, clothed, doctored, and wrote for people. The court was of course not always within the city. It moved to Anagni, Viterbo, Orvieto, Perugia, even Lyons, but it remained the Roman court. (Where the court was, there was Rome.) In the thirteenth century, Rome was its most normal home; and from it Rome assumed the airs and riches of a capital city.

Peter and Paul were among the most prominent of a mass of spectacular relics. These relics were the real hoard buried in Roman earth or in Roman altars for which all the other buried treasures, gold, marble, the palace at Campo Marzio, can be made to seem only the language of symbol. It was the hoard of relics that made Rome the glowing center of the thirteenth-century world, that, with the court, made Rome rich by drawing pilgrims to it. The richness of Rome as reliquary made it a constant festa. On each day of the calendar (so at least it seems) some church was lighted to celebrate the feast of the saint of whom it held some fragment. Often there was a concelebration in several churches; the city was (and is) like one of those boards which are maps with lights which respond to the pressing of buttons marked for each of a group of categories—botanical gardens, museums, churches; Saint Panfilo, Saints Pietro and Marcellino, Saint Erasmo. It was this richness, this existence as the great and final focus for the medieval fascination with deposited relics, a richness that was made relatively richer after the plundering of its great rival Constan-

The Ideal City

tinople at the beginning of the century, that gave Rome its most powerful, positive aspect, its strongest and most distinct "idea" in the thirteenth century.

Not all of the relics were dead buried coin. The most exciting of them were connected with Christ, with His life and suffering; and Christ was alive in the thirteenth century. Although the violence of Christ's reality was dulled, as well as enhanced, by the accentuation of the Real Presence within the Eucharist, the great importance of the Eucharist (honored in pyx and under ciborium, honored by Francis) was due to the pressing reality of Christ Himself to thirteenth-century men. In approaching the Eucharist they were approaching the real Christ, as well perhaps as hiding from the painful implications of a less pious and more active response to His commanding presence.

The importance of the Veronica—Dante's *la Veronica nostra,* probably the greatest treasure at Saint Peter's, the more than iconic relic before which poor William of Derby was crushed and which Innocent III and his successors caused to be carried to Santo Spirito each year—was an importance close to the Eucharist's. On it (although not on it alone in late thirteenth-century Rome), one saw Christ. (And Santo Spirito, like other thirteenth-century hospitals, was a place where Christ's work was done.) The earth and cross of Saint Helen, a more important predecessor of Eudoxia with her chains, enriched Santa Croce in Gerusalemme. The column of Christ's flagellation (at Santa Prassede) was brought, by a Colonna, from the Holy Land. The Jerusalem that Christ had touched had been and was being moved to Rome.[22] At Rome, too, the number of churches directly dedicated to the Savior was unusually high.[23]

Although the serious Christ was the great center of the Roman reliquary, His seriousness was suspended in a more various mixture of relics and observances, perhaps including His own rather curious relics at the Lateran. Relics of new saints joined those of old ones (as, for example, Saint Francis joined Saints Cosmas and Damian at San Cosimato in Trastevere) just as new saints in mosaics joined old ones (as did Saint Francis and Saint Anthony at Santa Maria Maggiore and the Lateran under the Franciscan pope, Nicholas IV). Roman churches covered the relics and even houses of the old mistily remembered saints of the heroic era of Roman martyrs, as at Saint Cecilia and Santi Giovanni e Paolo. The treasuries of Roman churches, like those of churches everywhere, contained inscrutable bits of overly dismembered distant saints both famous and obscure; and Rome was bursting with churches (414 according to the Catalogue of Turin's final sum).[24]

In Rome, however, as elsewhere, in dedications and celebrations, if not in relics, the Virgin dominated. Her feasts segmented the year. Her churches were great and small—the Popolo, the Minerva, the Pantheon, Aracoeli, Santa Maria in Trastevere, drifting down to churches unremembered except in old lists. As she was connected in Trastevere with miraculous oil, she was connected at her great church, Maria Maggiore, with miraculous snow. There the legend grew that on a hot Nones of August snow fell, and as Cybele before her had showed Agrippa, she with the aid of dreams and the Roman patriciate showed Pope Liberius where to build her church. There is clear evidence that this old legend was increasingly popular in the thirteenth century. Thirteenth-century popes repeatedly granted indulgences to those who visited the church on the anniversary of the miracle. In about 1222 or 1223 Pope Honorius III, who had been a canon of Santa Maria Maggiore, ordered the papal court to celebrate the feast. By 1269 the old local feast was acknowledged by the international Franciscans. By the early fourteenth century the mosaics of the snow had been placed on the façade of Maria Maggiore (and the snow still falls there on this feast of the Virgin).[25]

Mary, in fact, epitomizes the confusion of the thirteenth-century view of sanctity and religion. She above all was involved with the living Christ. It was she, at Cana, who provoked Christ to human charity (in that part of the gospel that was the text for Innocent III's and subsequent popes' celebration of the carrying of the Veronica to the hospital of Santo Spirito). In connection with her the pious man was almost forced to see Christ as the man who did and advocated certain specific acts in Palestine. But the figure of Mary was also particularly susceptible to the dilution of vapid sentimentality. Her figure lent itself to a vague unfocused religiosity. She could most easily absorb the extravagances of Cybele and Minerva. And her figure, perhaps even more than Peter's and Paul's, contributed its tone to the general mysterious miasma of religion that hung like a cloud over the relics and ceremonies of thirteenth-century Rome and attracted to itself the fascinated hordes of pilgrims.

Having assembled the classical and Christian memories of Rome, and having put together again its governmental and its religious aura, any historian might reasonably presume that he had gathered the essential components of its complex thirteenth-century image. He would think perhaps that he had tasted at least briefly all the flavors, seen at least in passing all the hues, of the "idea" of Rome.

He would be grossly in error. He would have excluded the major

The Ideal City

paradox within the idea. Against all the white magic and wonder of
Rome (with certain contrapuntal darknesses like the figure of Nero or
Simon Magus on the Via in Selcis, defeated figures), against all this
splendid complexity, stood the evil Rome, the Rome of black magic,
greed, poison, and fever. Against, or with, the new Jerusalem stood the
new Babylon. Against the lady of the world of the pilgrims' hymn, rose
with the red blood of martyrs, white with virgin lilies, stood the nox-
ious sump of the satirists' poems; against the flower of all cities, the
greedy gullet or, in another image, the leech. An anonymous poet
wrote:

> What shall I do in Rome?
> I don't know how to lie.[26]

The Rome which connoted greed to the world was essentially the
Rome of the papal curia. "My Lady Money" transacted "all business in
the Curia." [27] To frightened and harassed provincials, money seemed
the key to everything there. Letters to and from Rome are full of the
problems of purchase and bribery. Sometimes it seemed wise to pretend
to have more money than one really had, sometimes less. The horrid
sophistication of the place had to be met, and in meeting it innocence
was of necessity lost. As a representative of Canterbury reflected in
1188, after having gained experience at the curia: "How sweetly innocent are the days of youth, the child playing at his games, . . . how
blessed they seem from the hard age of man." In bitterness, not admiration, the poet wrote, "I have seen, I have seen, the *caput mundi."* His
was not, or was only ironically and complexly, the golden capitol. This
poet was Walter of Chatillon, and his verse, at its best, caught most
fully the glancing, sharp, mixed metallic beauty which satire on Rome
was capable of producing—the keys, the coin, the colors, the grammar, the wept-for ruins, the broken marriages, the mute flock, the
crying law (*Clamabat decalogus*), jangling coin, jangling meter, jangling rhyme, woven perfectly together. This purest, or least pure, statement had perfected the tone of thirteenth-century satirical statement
before the century had begun. (A heavier, more solemn, less exhilarat-
ing, but sounding and scripturally intricate form of denunciation of the
curia, but more especially of the people of Rome, had also been by
then long written and established by the great Bernard of Clairvaux,
not only but particularly in the fourth book of his *de consideratione,* to
which Saba Malaspina need only have looked to find rationalization for
the local ambitions of his late thirteenth-century popes.) [28]

The Rome where one lost money, and jewels and presents, was

essentially curial Rome, the Rome of expensive lawsuits and disguised simony, but it was also by implication and extension the entire city. The *Miracole* says of the Porta Taurina (the Porta San Lorenzo or Tiburtina) that it has sculpted on it a double-headed ox (*bove*), with a dry, thin face on the outside and a fresh, green, fat face on the inside. And according to the *Miracole* the double head stood for those people who came to Rome poor and left it rich.[29] People could get rich as well as poor in Rome; it was the same process in reverse.

And the butchers, the poulterers, the vintners, the money-lenders, the taverners into whose hands the foreigner fell were not narrowly curial. Debts to them delayed him when he would go home. Moreover, Rome, even in its decay, still seemed urban, as Italy did, to much of rustic Europe. In its anonymous, transient urbanity, it projected a frightening sense of untrustworthy sophistication—a variant of the medieval Constantinople theme.[30]

The evil city was particularly connected with poison and fever. In its damp heat, mysterious malaria brooded, the shadow, in disease, of the spirit of the place. Hugh of Evesham, the English cardinal, the "Phoenix," master physician, cardinal priest of San Lorenzo in Lucina from 1281 to 1287, was thought to have been made a cardinal so that he might stay close to Rome and protect Pope Martin IV from the fever. In the year of Hugh's elevation, his old associate, William Wickwane, archbishop of York, wrote to him to warn him to keep dangerous, poisonous concoctions away from his house. After Hugh's death it was said both that he had died from poison and from fever.[31] In writing of the year 1240, Matthew Paris, the English monastic chronicler, wrote of other English deaths in Rome. "In Rome four monks, exhausted by the wiles and delays of the curia, laid themselves down and died—in the words of Solomon, 'a broken spirit drieth the bones' —and thus perished a significant part of their convent. But the exact cause of their deaths is not known, whether they were from disease or despair, plague or poison." [32] The treasure hoard and memories of Rome were, like Saint Guthlac in his fen, surrounded by feverish, diabolical, poisonous mists and damp. The gold was guarded by fever; the cross was surrounded by corruption.

So the thirteenth-century idea of this "politically negligible, and universally powerful," "more than one city," rich with its legend-swathed peculiar monuments—Castel Sant'Angelo, the pyramid "of Remus's tomb," the marble horses—with all its dappled qualities seems as various and centrifugal as a local chestnut shell.[33] But actually the diversity is rhetorically balanced and controlled. It makes it-

The Ideal City

self up into surprisingly neat patterns of contrast and paradox and repeated image and argument.

Of these patterns, three related major ones seem dominant. First of all, Rome, as one sees her in later drawings, weeps among her ruins and sings the song of Master Gregory and William of Malmesbury, the song of the distance between past greatness and present failure. In this paradox of distance between past and present, between different conditions of the same place, the idea is sometimes a relatively frivolous celebration of classical engineering or just classical ingenuity as it is, at its frailest, in Master Gregory's visual machine of the Priapus (the boy with the thorn in his foot now at the Capitol). Sometimes, at its strongest, it reaches a melancholy grandeur, as it does in the conventional but final passage of the *Mirabilia:* "These and many other temples and palaces of emperors, consuls, senators, and prefects existed in pagan times in this Roman city, as we have read in ancient histories and have seen with our own eyes and have heard from the old ones. We have tried to write down, as well as we were able, what great beauty and splendor there were, of gold and silver, of bronze and ivory and precious stones, to preserve their knowledge in future memory." [34]

Against this melancholy stands the pattern of Christian joy, of a Rome once barrenly pagan with its pathetic Cybele now enriched with martyrs' blood and Christian understanding. And against both of these, in their different ways, serious patterns stands a third, mockingly but not lightly ironic, exposing the distance between pompous pretensions and actual corruption. All three patterns swathed themselves in the mystery of a magic place, a place unlike all other places. This complex but balanced, "mobile," idea of Rome prevailed in the century before Dante, Cola, and Petrarch applied their minds to the meaning of the idea of the city.

CHAPTER

III

Who Ruled Rome?

THE QUESTION OF WHO RULES IN any relationship is a terribly difficult and complicated one. This is immediately apparent to anyone with the slightest relevant experience. In the earliest and most obvious relationships, in the family, between lovers, it is always, in the end, after the most imaginative and penetrating definition, impossible to say who controls, who dominates. It is only less complicated (because the evidence is less complete), not less difficult, when the relationship is historical. It also seems less complicated when the relationship is communal, partly because seemingly repetitive multiplicity disguises complexity, because in large groups men are more easily seen to be acting in stereotyped patterns, to fall into categories and classes, to be relatively blunt and obvious objects with gross needs and fears.

Historians have sometimes made the power relationships between medieval leaders and followers, landlords and tenants seem simple. (With a wink, they say that of course we know where the power lay.) But this simplicity can be defended, if at all, only in extreme cases—when the labor market was noticeably unbalanced; when defense depended upon the control of a weapon, a place, a specific hoard. In most medieval communities the pattern of rule was complex. The pattern must be stylized in order even to talk of ruler and ruled. Admitting that stylization, the ruled must cooperate for the relationship to exist; they must in some negative way at least consent, must find some relative advantage in being ruled. This is not at all to suggest any disguised democracy, at least not of any attractive and recognizable sort, in medieval government. It is to suggest that it was particularly difficult for medieval rulers, especially distant ones, to implement their rule, impossible for them to do it, really, without obviously effective threats, bribes, or advantage to the subject.

This general uncertainty clearly applies to the city of Rome in the thirteenth century. But thirteenth-century Rome had its own unusual difficulties and simplicities. There are thirteenth-century places of

Who Ruled Rome?

which one can say, this is an independent commune, or this is a part of the kingdom of England or France, or this is a community whose lord is the duke of Normandy—always remembering the difficulties in defining those statements. But this direct simplicity, again probably always misleading, is not available for Rome. Was it subject to the pope, to the emperor, to a local noble oligarchy, to a communal government? The emphasis of one's answer would certainly change from period to period within the century, but it would generally have to be curiously uncertain and mixed. And this Roman peculiarity is matched by, complemented with, another. Rome, "unable to become a stable commune, yet the capital of empire and church," seems to have been queerly slack in its governmental and communal organization.[1] In some lights it seems almost not to have existed, to have been only an ideal answer to the constant mental necessity for a city: "Arthur had to have a Camelot," and Peter's men had to have Rome.[2] The structure of gild, citizenship, nobility, party, electorate is startlingly nebulous in thirteenth-century Rome.

All this uncertainty is unnerving. But it has advantages. The absence of an easily identified ruler and of a neatly observable governmental organization encourages the viewer to ask more fundamental questions, to look at a slightly deeper level into the organization of this community, to try to find out what did hold it together. There is in thirteenth-century Rome peculiarly little superstructure to conceal the real, perhaps the real economic, nature of the town. (Unfortunately the reality which is not obscured is, because of inadequate evidence, not blindingly compelling.)

Although one cannot give the insubstantiality of Rome an easy governmental name, it did, as a governmental unit, exist. It was a community with citizens and denizens, with nobles, courts, and officials, and its constantly repaired wall. It had a government. It had a seal. It had its own money, the *denari provisini senatus,* of the type of Champagne (which had been used in Rome particularly between 1154 and 1184), issued by the senate after 1184, with *Roma caput mundi* inscribed on its obverse and *Senatus P. Q. R.* on its reverse.[3] Most significantly perhaps, Rome had an army, a militia.[4]

Much of the shape and form that the community really found for itself seem to have been found when the army was active against the armies of neighboring competing towns, against Tivoli in the twelfth century and against Viterbo in the thirteenth. The Roman army was a mixed body, not very clearly visible, at least as feudal and mercenary as it was municipal. Still, in mustering its army, Rome was aware of itself and its common interests, interests distinct from those of Tivoli, Viterbo, and the pope. In its growth in the school of neighborhood Latin wars, this in many ways communally backward medieval Italian city echoed oddly the growth of great Rome. "Viterbo," Gregorovious wrote, "was the Veii of the Middle Ages to the Romans; they hated the town with a hatred bordering on frenzy." [5]

The focus of Roman government throughout the thirteenth century (and, in fact, at least from the relatively spectacular communal revolutions of the mid-twelfth century, consolidated in Clement III's peace of 1188) was the senate with its senators. It was the focus in the sense that whoever controlled the senate had already gained, in some very signal way, control of Rome. It was also the focus in the sense that the authority of virtually all municipal offices and commissions stemmed from, or passed through, the senate. In a way then, one can say that thirteenth-century Rome was ruled by, or at least through, the senate, and that that sacred name, the "S" of SPQR, was the mold and mirror of Roman governance, the repository for ideas of communal order, a sort of governmental Neoplatonic λόγος. The senate summoned the army. The senate made treaties. The senate established weights and measures—like the "palm of the senate." The senate pronounced legal decisions.

But the substance and the position of the senate were very strange. It lay between the mixed and changing controlling powers above it and the slack, ill-organized municipal government beneath it. It itself was of a pliable substance. It suffered quick, stunning changes of size and shape. Within a few years during the pontificate of Innocent III (1198–1216) its size varied from one to fifty-six and back again. This indecision is classically preserved in a very formal document, the form of the oath which Angelo Malabranca, the then senator, swore in 1235, that he and his successors would make their successors swear to abide by his treaty with Gregory IX. "Each senator," Malabranca swore, "would compel his successor, or successors, if there should be more than one. . . ." [6]—in 1235 the Roman senate's size was so unfixed that the senator could not predict whether it would have one member or more than one member.

Nothing of the senate's position or of the actual government of thirteenth-century Rome can be understood without rehearsing, at least briefly, a little outline of thirteenth-century Roman constitutional-political history (although it has been told very often before).[7] The outline itself will remain unintelligible, at the level of meaningless words, if some of its points are not probed a little more deeply.

Who Ruled Rome?

The treaty of 1188 between Clement III (a Roman, Paolino Scolari of the *rione* Pigna) and the city of Rome established a lasting, but not clearly detailed, pattern for the relationship between the pope and the city. The pope recognized the senate's existence as an approved institution; the senate recognized the fealty it owed to the pope; the pope promised, besides one hundred *lire* a year for the city's walls and one-third of the city's "money," to give to the senators and the other municipal officials their accustomed emoluments. Thus, at the accession of Innocent III the senate seemed a clearly established body. The senate of Innocent's immediate predecessor Celestine III (1191–1198) seems, like his own, to have suffered grave fluctuations in size. It varied between fifty-six "elected" members and the one active ruling senator, Benedetto Carushomo, the bridge restorer. Giovanni Capocci and Giovanni Pierleone seem to have been among Celestine's other single senators.[8] In the difficult but interesting years between 1198 and 1205, to which one must return for a slightly fuller examination, there was sometimes a senate of one, sometimes of two, and sometimes of fifty-six. From 1205 until 1238, through the rest of the pontificate of Innocent III, through that of Honorius III (1216–1227), and most of that of Gregory IX (1227–1241), there was regularly one senator. During the remaining years of Gregory's pontificate there were two senators, until in his last year Gregory made as his only senator Matteo Rosso Orsini, who after the pope's death ruled Rome as its "Guelf" dictator until he was joined as senator by Giovanni Poli (a former senator and a member of the Conti family). Under Innocent IV (1243–1254), who succeeded the short-lived Celestine IV (October-November 1241), there seem sometimes to have been one, sometimes two senators until the senatorship of Brancaleone degli Andalò, who came to power in 1252. His dictatorship would seem to mark a change in the nature and significance of the office.

For this whole period, the half-century from 1198 to 1252, the essentially annual procedure of senatorial elections is extremely obscure. The pope seems sometimes merely to have appointed the senator or senators, although even in these cases there was almost surely always some sort of popular acclamation that represented election. It is clear that in some cases the pope did not appoint; the elected senator could not have been a man whom he would have chosen. But even in these cases there can hardly have been the sort of popular election that narrative sources have been thought to imply because there is no observable workable electoral organization. There is certainly in this period some experimentation with election through medians and some use of *rioni* for electoral purposes. There was talk in Arnold of Brescia's time of two thousand popular electors.[9] Still, all the evidence is very vague and unsatisfactory. Although there seems to have been some real sense in thirteenth-century Rome of the difference in meaning between the words "citizen" (*civis*) and "denizen" (*habitator*), and although there were certainly fourteenth-century regulations concerning the way in which foreign merchants might establish citizenship, there really seems to have been no clearly defined body of citizens in thirteenth-century Rome. (Although one may of course argue that the evidence for such a body is lost, the remaining evidence seems to me to point in the opposite direction.) What the sources tend to talk about in Rome (beyond governing councils) is an undefined *populus* shouting in the piazza of the Capitol. This sort of body cannot effectively elect. It can only approve.

But, although it is extremely difficult to say how a senator was elected, it is not at all difficult to say what sort of man was elected senator. The senator was a member of a powerful Roman family and faction. He was Pandulfo de Suburra, Pandulfo de Giudice, Annibaldo degli Annibaldi, Giovanni Poli, Luca Savelli, Angelo Malabranca, Oddone Colonna, or Matteo Rosso Orsini. He was clearly the visible peak of a force dominant in Rome, papal, antipapal, half of a compromise couple, or something less blatantly political and more complexly familial. But he was, however chosen, the voice of real and dominant power in Rome.

In the second half-century, from 1252 to the death of Boniface VIII in 1303, things were much more obvious and direct. After Brancaleone's experiments with setting up a serious communal state, which must be looked at separately, there was an aristocratic, particularly an Annibaldi, reaction. Brancaleone died in 1258. He was succeeded for a short time by his uncle Castellano. In 1259 he was succeeded by two senators favored by the pope, Alexander IV (1254–1261); they were an Annibaldi and an Orsini, and they in turn were succeeded by another Annibaldi and a Savelli, then by a Conti and an Orsini, then a Colonna and a Poli (in each case one senator was a papal relative).

This renewed aristocratic-papal domination was troubled and insecure. It coincided with renewed attempts to hand over the senatorial chair to foreigners (if Brancaleone and Emanuele Maggio, who interrupted his rule, are remembered to have been foreigners), but foreigners of different type. The two great candidates of 1261 were the Hohenstaufen claimant Manfred and the English Richard of Cornwall, king of the Germans and brother of Henry III. Richard of Cornwall

Who Ruled Rome?

was not merely the "Guelf" candidate but also the candidate of a pair of cardinals with English connections: John of Toledo, cardinal priest of San Lorenzo in Lucina, and Ottobuono Fieschi, cardinal deacon of Sant'Adriano. Their maneuvers make it clear that in 1261 there was enough of an electoral process to suggest bribes and propaganda— but then of course so would the collecting of a party even without election.[10] Still it is ironic that this most tantalizingly half-revealed election, at least after the pontificate of Innocent III, occurred as the electoral senate (whatever it had been) was dying. It was the Annibaldi cardinal, the ranking member of this potent mid-century papally related (to the Conti), local family, who supported the senatorial candidacy of Charles of Anjou. And it was Charles who not only took the Regno from Richard, or his nephew Edmund, and Manfred, but the senate from the aristocratic families of Rome—or at least he changed their role in it. From the time of the negotiations of 1263 until the Sicilian Vespers of 1282 (and their Roman echo in 1284), Charles of Anjou dominated the Roman senate, as he did all central and southern Italy, either by controlling it directly or indirectly or by shaping its reaction against himself.

Charles is remembered as a man who disliked sleep and avoided laughter. He seems to have been singularly unlovable, impenetrable, even for a medieval king. But he was in many ways a very efficient ruler. Arnolfo's imaginatively plain, strong statue of him on the Capitol is great in the way he was great. It is full of powerful authority. Charles himself seems, interestingly, to have been lacking in all the qualities that made and make his brother Louis IX attractive. He was direct but opaque. If he wanted a city like Rome, he threatened it with his army and his fleet in the river. If he wanted to rid himself of a royal rival, Corradino, he executed him. When he came to Rome and accepted the senatorial office he established himself in the most obvious palazzo, the pope's Lateran, and antagonized Clement IV who might otherwise have been even more of a royal puppet. (Charles later moved to Quattro Coronati.) Still, in connection with all this Charles, with his external force, seems in some ways to have ruled Rome remarkably well, to have given it much the sort of government it required. He took his responsibilities seriously, and he interpreted government broadly. When Innocent V lay dead *aput Urbem, ubi habemus regimen,* Charles thought that it was his job, one of the duties that his really ruling the city entailed, to see that the dead pope had a fitting sepulcher. "Victorious he climbed the Tarpeian rock and the stones of the Capitol." [11] But like everything else that he did, he did it in a way eventually unbearable to human sensibilities.

The French popes, Urban IV (1261–1264) and Clement IV (1265–1268), were both supporters and dependents of Charles of Anjou, but they wanted some measure of independence from him (although Clement IV complained that Charles was too economical, plain, and spare in the men whom he first sent to govern Rome). The popes did not want merely to replace a gigantic Hohenstaufen ogre with an equally gigantic and threatening Angevin ogre. They wanted to stand between Charles and the secular forces of central Italy. Urban had at first insisted that Charles, if he accepted the throne of Sicily, not accept the senatorial chair at Rome or that of any podestà. But Urban found alternatives to Charles more repulsive, and he withdrew his original stipulation; he then forbade Charles to accept the senate for life, but rather, regardless of his oath to other Roman supporters, to hold it at the will of the pope and for a limited period. This sort of nervous negotiation dominated much of the history of Charles's relationship with the city. It has left its traces in double letters and corrected drafts. It makes odd contrast with the direct quality of Charles's actual rule which began in 1264 with his sending a prosenator and Provençal knights to take over the Capitol and the city.[12] He himself arrived and was invested in June 1265.

In 1266 after the defeat of Manfred, the pope demanded that the reluctant Charles resign the senate. In May he did. Disputes between the pope and the newly elected senators, Luca Savelli and Corrado Beltrami Monaldeschi (of Orvieto), particularly over ecclesiastical debts to Roman merchants, began immediately. In 1267 a rising of "the people" installed a "popular" government under Angelo Capocci, captain of the people and member of an extremely interesting Roman political family. Capocci chose as senator Arrigo of Castille, a rich, romantic, military cousin of Charles of Anjou. Arrigo's relations with the pope were strained even before he took Rome (or followed it) to the camp of Corradino, the new Hohenstaufen contender against Charles. Arrigo's fall followed the fall of Corradino in 1268 (although Arrigo's prosenator had not been willing to give the fleeing Corradino the protection of the Capitol), and Charles of Anjou returned to full power, through his ruling prosenators in Rome. He ruled for ten years.

Then in 1278, with the election of the Orsini pope, Nicholas III, the Roman nobility took the senate to itself again. Nicholas gained recognition of the papal states from the new emperor, Rudolf of Habs-

Who Ruled Rome?

burg, and in his famous senatorial constitution he forbade the election of future senators who were foreign potentates or the close relatives of foreign potentates, with the specific proviso that he did not mean to exclude the Roman nobility. As a pope and as an Orsini, and in him the two are often indistinguishable, Nicholas created local cardinals (Orsini, Malabranca, Colonna), and, having thoroughly informed himself about the position of the senate with documents whose transcriptions are preserved, he recalled local power to local hands. To an Orsini senator succeeded a Colonna and a Savelli, and to them a Conti and an Orsini.[13]

Nicholas III died in August 1280. He was succeeded the next February, in another reaction, by the French pope, Martin IV (1282–1285). Nicholas's last senators, again a Conti and a Savelli, calling themselves "senators and electors ordained by the magnificent Roman people," gave the rule of the senate for life to Martin in March, not as pope but as an individual man.[14] In April Martin granted the senate to Charles for the duration of Martin's life. But early in 1284 the Romans, led by an Orsini faction, made bold by Angevin difficulties in the south, stormed the Capitol, killed the French garrison, imprisoned the French prosenator, and recognized Giovanni Cencio Malabranca, an Orsini relative, brother of a cardinal, as captain of the people.

After a short period of renewed constitutional experimentation, at least with names, and the reconciliation of Annibaldi and Orsini (whose fight lay at the very center of the 1284 revolt), within the year 1284, the pope presented the city with two senators, an Annibaldi and a Savelli. Martin IV was succeeded by Honorius IV (1285–1287). Honorius was a Savelli of Rome, the son and brother of senators. Granted the senatorship, he gave it to his brother Pandulfo, one of Martin's senators.

The Savelli ruled Rome. Their localness, their wisdom, their popularity, and perhaps their relative weakness compared with Annibaldi, Orsini, or Colonna made their rule a successful and easy one. But the pattern for the remaining years of the century is in them clearly established. The papal *signor,* using the concentration of power first seized by Brancaleone and then made conventional by Charles of Anjou and transferred to papal hands by Nicholas III, ruled Rome through the senate and through a senator or senators who represented his family or a combination of family interests (as in 1292) that would make the greatest possible interests of his own family bearable to a preponderance of Roman power-bearing families. The solidity of the pattern is clear under Nicholas IV (1288–1292) because he was a client rather than a member of the family he represented, the Colonna. Even under the most sordid pope, Boniface VIII (1292–1303), this need not be a sordid system. Boniface's first senator was again Pandulfo Savelli. It was to Boniface's advantage to have Rome ruled peacefully (if that peace aided rather than interfered with Caetani ambitions).

Before Boniface's election two interesting interruptions had occurred in the papal-senatorial pattern. During the disturbed papal vacancy after the death of Nicholas IV, the city of Rome seems to have fallen into a violent, murderous, economically destructive (and pilgrimage-destructive) chaos. For a time there was no senator. Then in the autumn of 1293 two interesting senators were chosen, a Stefaneschi and a Sant'Eustachio, members of two eminently respectable and not highly controversial Roman noble families of the (very high) second order. The vacancy itself ended with the short peculiar papacy of Celestine V (July–December 1294). Celestine was, among other things, a tool of Charles II, the son of Charles of Anjou. His papacy threatened Rome with the return of the Angevins. It consolidated momentarily the reacting Roman families, at least long enough for them to accept a Caetani.

Now surely the glaring point about all this senatorial history (even admitting the significant change to a sort of controlling signory in about 1252) is that the principal governmental organ of thirteenth-century Rome, the senate, was a shapeless, formless, variable thing. It responded to local and foreign popes, to popular uprisings, to Hohenstaufen and Angevin emperors and kings, and repeatedly to the tough, propertied noble families of Rome. (When the senate is seen as having a classic form at all in the thirteenth century, an odd vision, it is the form of a pair of ruling nobles—an Orsini and an Annibaldi.) This is all another way of saying that the thirteenth-century senate and Roman government in general were unobstructive and flexible and responsive. That was their genius.

If one dips within the narrative of senatorial history at a few points, examines a little more closely something of its texture, one moves a step closer to understanding the forces to which Roman government responded. One of the most revealing periods in the century and at the same time, and connectedly, one of the most difficult to understand is the first half of the pontificate of Innocent III, particularly the years from 1202 to 1205. Innocent III began his rule of Rome strongly. The exact events of his coming to power in Rome are somewhat confused. But he seems successfully to have refused to give a cus-

Who Ruled Rome?

tomary money gift to the Romans and to have controlled, through a subject median, the selection of the senator. In the autumn of 1201 he wrote of Rome, "by the grace of God we hold it in our power."[15]

There had already been some signs of difficulty. The pontificate was born, like many others through the century—it was a state of Roman nature—with as its enemies the nepotistically favored relatives of the late pope, in Innocent's case the Orsini. And, equally a state of Roman nature, it was burdened with the hungry relatives of the new pope, in this case particularly Innocent's brother, the notorious Riccardo Conti. There were, further, two vocal leaders of the opposition to Innocent's control of the city, and countryside, two former senators with famous family names, Giovanni Capocci and Giovanni Pierleone. They seem to have been men with some demagogic powers, and so doubly dangerous because to family faction they could attach a temporarily spirited mob. Innocent was also forced to face early in his pontificate a recurring, dangerous external problem, the intermittent Roman war with Viterbo. That Innocent supported Rome in the first encounter between Rome and Viterbo has been interpreted as an indication of his sensing his weakness within the city in spite of good signs and brave talk.

The elements which threatened Innocent's peaceful rule exploded in 1202 after, but not perhaps as a very direct result of, a second Viterbo war in which Innocent was suspected of having tempered the treaty so that it favored the Romans less. The explosion was a complicated one, bursting in its confusion in various directions, with various centers, at various levels. It was both a civil war (if the pettiness of thirteenth-century Roman disturbances deserves that pretentious name) and a constitutional crisis (if the word "constitution" is not too pompous in this primitive setting). Like Roman fireworks, it was loud, quick, spectacular, diverse, and led to nothing.

But in spite of its chaotic quality, a repetitive image dominates any description of the affair. It was conducted around, within, next to, the towers of Roman noble families. Towers were built; towers were destroyed; towers were threatened not only with siege machines but with other towers. The most tantalizing of towers was the one of Pandulfo of the Suburra's which was called *fagiolum,* the "bean" (or perhaps the "beech" or the "cloth"—curtain); the great central symbol of the whole affair was Riccardo Conti's Tor de' Conti.[16] It threatened and dominated; it was hated and opposed. It was said to have been built with illicit funds taken from the church by Innocent III and given to his brother. Its bulky base still stands, mammoth even after its partial destruction in a mid-fourteenth-century earthquake; it controls the path from the Lateran to the Vatican and the Capitol to the Colosseum; it looms over the Forum of Nerva.

Towers stood for family and faction. At the center of the affair were the Conti and the Orsini. But they were surrounded by other families, the Frangipane and the Pierleoni, the great declining families of the twelfth century, the Capocci, quickly upwardly mobile, the Annibaldi, rising to increased prominence through their marriage connection with the Conti, and the family of the Suburra, loyal to Innocent. The hatred between Conti and Orsini, natural as the immediate economic circumstances made it, had deeper roots. There was, the *Gesta Innocentii III* says, old rivalry between the family of Celestine III's father, the Boboni, and the family of Innocent III's mother, the Romani de Scotta, a Roman city family.[17]

The dispute burst out under Innocent when, in his absence, the Orsini attacked the city holdings of his relatives. Pandulfo of the Suburra, as Innocent's senator, tried to prevent further trouble by exiling troublemakers of the two parties, one to the area of Saint Peter's, the other to Saint Paul's. An Orsini cousin (Teobaldo di Benedetto di Oddone), who visited them often, was murdered on the road between Saint Paul's and the city. The Orsini tried to rouse the city "to excite the furor of the people" against Innocent and Riccardo, particularly by conducting, or trying to conduct, an elaborate funeral for the murdered man, publicly mourning and lamenting before a papal palace and before Riccardo's house.[18] The never very serene surface of conventional Roman aristocratic behavior was broken by the breaking of houses and towers.

The second provocative incident or set of incidents which led in fact to greater violence was in a way less central to the most smarting problem of Innocent's early pontificate (that is, the Conti-Orsini rivalry), but it was also in a way more generally significant and indicative of problems less confined to the relationship between specific families, although it itself grew out of a relationship between two specific families. It pointed more clearly to the disturbances which naturally surrounded the fall of some fortunes and the rise of others—or more exactly the changing of fortune from one family to another. It was also more flamboyant.

It grew out of the fact that the Poli family, who held of the papacy large tenements in the *campagna* but whose fortunes had fallen disastrously, had borrowed money to sustain themselves from Riccardo Conti. They could not repay him. Riccardo planned to recover his

Who Ruled Rome?

losses by having a Poli heir marry his daughter—presumably as a step toward the Conti absorption of Poli holdings. The Poli agreed and then reneged. Innocent and Riccardo moved toward the confiscation of Poli holdings. The Poli reacted histrionically (in order, the *Gesta* says, to arouse the people). They howled their poverty and dressed in rags. They scurried naked through the streets in distorted penitential processions. They made a foray on Saint Peter's at Eastertime and disturbed the divine offices with their clamor and what the *Gesta* calls "blasphemies." They threatened and insulted Innocent as he proceeded crowned through the city.[19] And they handed their property, like ancient client kings, over to the senate and people of Rome. The Poli as they fell provoked the furor they intended. They provoked a tower war. They did not in the end injure the Conti in any clearly noticeable way, nor topple them from power (although they must have made them more cautious). Within a few years the Poli inheritance was safely in Riccardo's possession. In the next generation the Conti of the Poli estates had taken the name Poli, and seem, ironically, to have identified themselves with the interests of their estates, to have become Poli rather than Conti, able to oppose Conti popes.[20] In this the Poli had won, or had a Poli prototype of Marxist theory.

These outbursts were accompanied and, essentially, followed by temporary adjustments in the senatorial constitution. The existence of a single senator disturbed the dissidents who sought a return to the fifty-six, whom it was believed would less directly represent papal interests. Innocent sought in his usual placating manner a usual technique, a resort to medians. But the opposing parties broke into two groups. The pro-papal party led by the ex-senator Pandulfo, held the Capitol; the antipapal party dominated by Giovanni Capocci moved to the area around the church of Santa Maria *domne Rose* (in approximately the present location of Santa Caterina dei Funari), in the church and in the house or dome or tower of their partisan Giovanni Staccio. They captured medians and pressed for the selection of at least two senators favorable to them and their rebellion.

In the spring of 1204, Innocent III, having returned from illness and to Rome, tried to pacify the conflicting parties and named a new median, Giovanni Pierleone, who named a distant relative as senator. But the dispute broke out again in late spring, and the dissidents, meeting again at Santa Maria *domne Rose,* chose a senatelike group of their own, the *boni homines de communi.* Giovanni Capocci's power increased, and the war of the towers raged, particularly around the Capocci center near San Pietro in Vincoli (where the Capocci towers now stand) and around the Suburra towers on the Magnanapoli. Although the war went well for the dissidents during the summer, it then turned against Capocci, because, it is thought, his overbearing leadership annoyed his compatriots or frightened them. With the turn of fortune, the rebels, as well as the papalists, agreed to a treaty. Arbiters were selected for arranging a pattern of senatorial selection. Eventually it was agreed (in an essentially pro-papal, 1188-type decision) that fifty-six senators should be elected (with Innocent agreeing to accept this part of the decision but suggesting that it was unwise and unworkable). By the spring of 1205 the old system had returned, and Innocent was able to choose directly a single senator. He chose Pandulfo.

Some of the cries of those involved in Innocent's battles were the last cries of the condemned; some, like the Orsini's, were the cries of those trying to keep their heads above water until the more affluent times of a more favorable pope or a more propitious alliance. They were the cries of family faction. In this they predict the end of the century as well as announce its beginning. In 1303, the dying Boniface VIII, in the Vatican, perhaps half-mad, the victim of his own grandiose Caetani plans, brutally beaten by the Colonna and the French, saw the city as a complex of family places. He was caught, a prisoner in the Orsini Vatican; he longed for escape to the Lateran, where the Annibaldi, the enemies of both Colonna and Orsini, ruled. The turbulence of 1303 like the turbulence of 1203 was family war—and at least two of the dominant disputing families were the same in both years.

There were also significant differences. The event of 1203 proved both the sagacity and strength of Innocent III. In 1303 the pope was truly beaten, and the year can be seen not unreasonably as marking the death of the old Roman papacy—although not (with one of those repeating ironies which grace the history of the city) the death of Caetani family fortunes. (The importance of propertied senatorial Roman families, after all, preceded Gregory I's turning of his own family properties to the use of a newly imagined Roman bishopric.) But another significant difference between 1203 and 1303 points up the significance of the century's history, the century of Brancaleone, Charles I, and Nicholas III. In 1203 there was constant dramatic appeal to an admittedly poorly defined populace and there was constant manipulation of the senate which was inextricably involved with, and was the natural respondent to, political change. Neither populace nor senate was completely ignored in 1303—the senate was the Orsini place. But they were the senate and populace of a relatively secure, although not securely held, signory.

Who Ruled Rome?

Boniface VIII's pontificate had contributed to this constitutional tightening and simplification, as well as to the continued inflammation of family disputes. The senators of Boniface VIII's great year, 1300, were Riccardo Annibaldi of the Colosseum and Gentile Orsini, successful men in war and peace, efficient rulers in a difficult although wildly profitable domestic year, men from great conventional senatorial families. At least two of their inscriptions, both impressive, can now be found on the Capitol. But it seems to me that their inscriptions make clear that they were both Boniface's men in what was in that year Boniface's Rome.

Although one could argue that the most interesting facet of the relationship between the pope on the one hand and the city and senate on the other was its developing clarity, one could argue equally forcefully that in spite of that development the crucial point about the relationship was its persistent lack of clarity. Boniface VIII himself, or his chancery, exposed this lack of clarity in a document so peculiarly revealing that it seems almost to have been invented for the purpose (and, in a contemporary political sense, it in a way obviously was). On October 23, 1298, from Rieti, Boniface VIII appointed two citizens of Rome and "noble men," Fortebraccio Orsini and Riccardo Annibaldi, senators of Rome for the six months following the approaching feast of All Saints (November 1). He explains his power of appointing first in extremely high and theocratic style, the apostolic dignity, the ordering of the universe, the pope's responsibility for all the faithful living anywhere in the world, and his special responsibility in ruling the Roman people and his special affection for them, a people to whom he was not only vicar of Christ and highest pontiff as to all the others, but their own bishop. These Roman people also feeling a special affection for him, Boniface continues, in order specially to honor him, have according to the will of all bestowed on him for life the rule of the city to exercise himself or through others, *prout in instrumento publico super hoc confecto et communis Urbis sigillo munito . . . continetur* ("just as it says in the public instrument drawn up about this matter and sealed with the city's communal seal"). God's universal plan, the special relationship of the pope with the city, and the sealed public instrument together conspired to grant Boniface the authority to appoint the men whom he chose as senators, two nobles from potent families.[21]

Although the change (and, in spite of Boniface's words, one cannot deny the real change) from Innocent to Boniface is traceable through the intervening century, it was not simple, obvious, or devoid of revealingly complex incident. The first step away from Innocent would seem to have been a step in the opposite direction, toward a more freely elected (whatever that may have meant) senate. The relationship between Honorius III (1216–1227) and the city seems, like that of Honorius IV, his great-nephew, to have been relatively peaceful and easy. The Savelli were an ideal family, clearly local and urban, aristocratic but not overly powerful, rich and of course nepotistic but not abusively so, connected with the senate and increasingly with the people. Honorius III's existence probably encouraged the growth of relatively independent local government. The single serious urban disturbance of his pontificate (1225–1226) was provoked in part by the animosity of Riccardo Conti, Innocent's relict brother, and involved the senator Parentius, a man who was singularly free from papal influence and seemingly dependent for his support on civic sources.

The pontificate of Gregory IX, the second Conti pope, before 1235 also pressed Rome toward the realization of civic independence. Gregory's complete absorption in his war with Frederick II made him incapable of dealing savagely with Rome. In 1228 a "Ghibelline" rising in the city, heralded by an attack on the pope at the altar of Saint Peter's on Easter Sunday, drove the raving pope from the city. (He was summoned again after the terrible flood of February 1, 1230; and he further warmed the then frightened Roman hearts with gold.) Although Annibaldo Annibaldi, in 1231, acting against the heretics of Rome by senatorial edict, seems to have been completely in Gregory's control, or in complete sympathy with him, Rome showed real senatorial vigor and independence in the years from 1232 to 1234. Rome showed its independence under the leadership of two senators with unusually interesting backgrounds, Giovanni Poli, Riccardo Conti's pro-imperial son, and after him Luca Savelli, Honorius III's nephew. Luca's year (1234–1235) was one in which the city fought to control its senate and the *campagna* and planted again on its possessions boundary-stones marked "SPQR." The *Vita Gregorii IX,* hostile to the urban claims, talks with shock of the effort of the Romans (the people of "Rome, the gift of Constantine") to control the election of the senator, the minting of money, ovens for making bread, the fees for pasturing animals.[22]

In 1235, under Senator Angelo Malabranca, the city returned to peace with, and subservience to, the pope; but it was a city that could see itself and its government more coherently and independently. With the dictatorship of Matteo Rosso Orsini, in the vacancy after Gregory's death, city government took another step. This in many ways repulsive, domineering "Guelf" ruled the city and formed alliances with neigh-

Who Ruled Rome?

boring towns. He locked up the cardinals in the Septizonium and made them elect. Their imprisonment is one of the most brutal and sordid stories of thirteenth-century Rome; cardinals were dead and dying in the putrid enclosure. The English Robert of Somercote's body could not be disposed of properly. Guards stood above the cardinals' cubicles, a victim wrote, and showered them with urine.[23] Matteo's rule was very clearly independent of an ecclesiastical superior. He was Rome's governor. This relatively coherent Roman unit, seen and in some ways understood, constantly honed on wars with Viterbo, was then prepared to be given into the hands of Brancaleone degli Andalò.

Brancaleone was a Bolognese knight who had fought for Frederick II, a man of good family and already of considerable experience when he was summoned by "the Romans." [24] To them he brought the governmental idiom of northern Italy. Whatever the real source of Brancaleone's initial power in Rome, whatever the group who summoned him, it cannot have been the pope or, in the usual sense, the Roman nobles, because Brancaleone was so extremely independent of them both when in fact he was not hostile to them. He clearly tried to build up a Roman state independent of the controlling papacy (at times even condescending to, and protective of, the pope). This is shown both in Brancaleone's coins and in his attitude toward surrounding towns, Tivoli, for example, which he took for the city against the pope's will. Brancaleone's title, captain of the people, is also thought to have been significant, a proclamation of the basis of his power, a statement of those in whose interest he ruled. His supervision of the reorganization and ordering of the gilds was probably in good part an effort to build a workable non-noble basis for Roman government. His reputation and his acts alike argue his enmity to the Roman nobility. He is most famous as a destroyer of towers; according to Matthew Paris he had about 140 of them torn down.[25] His putting to the gallows of two Annibaldi in his second captaincy (when his teeth had been set on edge by his imprisonment and maltreatment after his first three years of office) caused a shudder of amazement as far away as Matthew's Saint Alban's. After Brancaleone's death, the news of which was brought to King Henry III while he was staying at Saint Alban's, Matthew wrote: "He was truly the hammer and rooter of the haughty magnates and evildoers of the city, and the protector and defender of the people, and the representative and lover of truth and justice." [26]

When, after his first three years, Brancaleone was replaced by Emanuele de Maggio, Emanuele, in contrast with Brancaleone, was said by Matthew to have cared only about pleasing the nobles, and particularly the Annibaldi. Matthew's Brancaleone was so loved by all the people that when he died in 1258 his head, according to Matthew (and Matthew thought that this was excessive), was put in a precious vase on top of a column and honored like a religious relic (and one recalls the orthodox fears about the dead Arnold of Brescia). When, between Brancaleone's two captaincies, the rule of Emanuele had proved too interested, a league of the popular party (*confederatis . . . popularibus*), counseled by an English master-baker, a co-citizen of the rioting Romans (*concivis eorum*), Matthew of Belvoir (a cell of Saint Alban's, it has been pointed out), rose and freed Brancaleone.[27] When the pope excommunicated Brancaleone after he had maltreated the Annibaldi and their allies, he seems to have claimed the privilege of the Roman commune's officials.

None of this would make sense, go together, be understandable, if Brancaleone's rule had not been in some sense, as it has always been claimed to be, popular. But it is very easy to go too far with Brancaleone. Gregorovius in one of his most peculiar passages certainly did: "Were definite information concerning his government forthcoming, we should find that under him the democracy rose to greater power, and that the guilds attained a more secure position." [28] About Brancaleone, definite information does not come forth easily, and disconcertingly much of it comes from Saint Alban's. It seems impossible to say who elected him, who supported him (except that it was certainly not the Annibaldi). The evidence suggests more his creation of gilds than their creation of him. He was pleasing because he brought order. He was able to produce order probably because he demanded and got a three-year office and noble hostages, and because he brought with him, like podestás elsewhere, his own court. He was a popular dictator. He had perfected, insofar as it was in his century to be perfected, the use of the powers of what has been called the popular party; he succeeded and surpassed Giovanni Capocci and Luca Savelli. What the party really was, we do not know. In Brancaleone the power of the "people" would seem to have reached both its apex and its end. He had in fact with his unified commune, and the taste for order that his successes encouraged, prepared the signory for Charles of Anjou's use.

One must constantly return to the enigma of the electorate, and put together the little pieces of evidence, and wonder what they mean. When Brancaleone died of a fever that came upon him as he was laying siege to Corneto, he was succeeded by his uncle Castellano, whom the citizens of Rome (*cives Romani*) elected, it was said, because Brancaleone had told them to. This was done without the pope's assent, and

Who Ruled Rome?

he, according to Matthew Paris, complained that even as a simple Roman citizen (*simplex civis Romanus*) he ought to have been called to a senatorial election.[29] Against this implication of real election ought to be placed the election of Martin IV as senator for life, an act performed by the outgoing senators as senators and electors "chosen by the magnificent Roman people." Again against this should be put the bribes for Richard of Cornwall's election. Gregory IX seems to have held the nomination after 1235; Nicholas III (in his Roman policy attacking the pretensions of Charles of Anjou) claimed it for the Romans. Clement IV, playing his coy games with Charles in 1266, said that the Roman people (*Romanus populus*) was in possession, and had been for a long time, of the ordering (or "electing," *ordinandi*) of the senate. Richard of Cornwall's Roman informants wrote to him in May 1261 that he had been elected "in the customary manner," by a large multitude of the people of Rome congregated in the Capitol (*in Capitolio . . . Romani populi multitudine congregata*) and that he had been elected unanimously and harmoniously, except for the dissent of some sons of perdition. And the documents around his election suggest, at least, the importance of *rioni*. From the end of the century (1294) Cardinal Pietro Colonna wrote to Jayme II of Aragon to explain the facts of senatorial election, that in fact a popular election could not come to much without the support of the Orsini or the Colonna.[30] Motivation in these incidents is not hard to understand. It is easy enough to see why these people said and did what they did, but it is virtually impossible to see what they were talking about or acting upon.

Again, beneath the pressure of Hohenstaufen and Angevin, the enriching power of the pope, the propertied (with, particularly, country seignorial and feudal property) power of the families like Orsini, Colonna, and Annibaldi who blatantly shared control of the city at times like the vacancy after the death of Nicholas IV, beneath all these, was there a coherent body of citizens, a privileged group within the general populace, in whom resided the sort of power and control that electors have? Was there an institutional, coherent, rational connection between the superficially dominant power groups and a broader popular group beneath them? The most frequent answer to these questions would seem to be, "There must have been." Rome, this theory goes, must have been like Siena, but in Rome the documents are lost. It is, on the basis of local evidence not lost, a queer answer and a queer theory. The local evidence seems to me to argue real institutional incoherence, a Roman populace (repeatedly but not always called *populus, populi*) that could be gathered in a piazza to shout assent, like the

English country electorate before property qualifications. The old answers have been given, it seems to me, by historians who could not imagine a city in which the control by and of the populace was not urban institutional.

The word "citizen" (*civis*) was certainly used in thirteenth-century Rome, as it was, for example, in 1281 of the witness Tommasio, called "once of Cera [Ceri?] and now Roman citizen (*cive Romano*)" or when, on his dated (1299) tomb of Cardinal Consalvo in Santa Maria Maggiore, Giovanni di Cosma calls himself "*civis Romanus.*"[31] It means something in a document to find a man called a citizen of Rome and an inhabitant of Trastevere. But what? When, in about 1208, the Roman senator wrote to the consuls of Terracina telling them that it was insupportable that Roman citizens should be imprisoned there, in what way was he distinguishing from other men the specific men of whom he spoke? Nicholas III in his senatorial ordinance seems to define citizens of Rome as those who were born there, but it is an obliquely used definition.[32] When, in a document of 1218, a *habitator* of Gallese sells days in a mill to two "*habitatoribus alme urbis,*" surely the document is telling us where these men lived, not anything about their citizenship status.[33] Clearly, however, the mid-fourteenth-century collection of Roman statutes means something very specific when they establish which foreign merchants shall be understood to be Roman citizens—those who have in the city the greater part of their goods, movable and immovable, and a house in which they live continuously with their household—and they do again when they say that no one shall be considered a Roman citizen who does not have a house or vineyard, the house in the city, the vineyard within three miles of the city, and unless he has lived for three years in the city with his family or household.[34] One may ask, even remembering the lateness of this evidence, what would be the purpose of this sort of definition if it were not for identifying a politically responsible group. The answer is crashingly obvious. The purpose of this sort of definition is to describe the men (specifically merchants) within the city who should be permitted to enjoy the commercial advantages (the advantages of relatively toll-free commerce) of citizenship.

Still, even in the thirteenth century, even among the shouting crowd on the Capitol, there must have been some definition. It is impossible to believe that women and children went to shout. But even that qualification is less absurdly obvious than one might expect. Who were children? Men under the age of complete majority, twenty-five? Although women would not have "voted" on the Capitol, was their po-

Who Ruled Rome?

sition in Rome and the countryside not closer to men's than one might expect? Their property was, it would seem, carefully protected by law.[35] They were active as guardians in a familiar, although not necessary, pattern: if the father died the child's guardian was his mother; if the mother was dead, the guardian was his grandfather; if the grandfather, his grandmother. The consenting witnesses in an act at a Roman castello could be described as "all the people of the castello, men and women (*homines et mulieres*)." [36]

The *populi Romani* were summoned by trumpets and bell, by bell and trumpets, by bell and voice of herald, or by bell and trumpets and voice of herald to the piazza of the Capitol to be made into a *parlamentum* to respond to the senator, to vote unanimously to support his fight against the Colonna, to approve a treaty, to confirm the privileges of the canons of Saint Peter's.[37] They were to perform a political act, to approve, and perhaps (at least of a fallen senator) condemn.

In 1255 Brancaleone summoned them to the Capitol and made them form a parliament in the accustomed way at the sound of bell and trumpets and voice of herald (*fecisset in Campitolio ad sonum campane et bucinarum et voce preconis parlamentum more solito congregatum*).[38] They were brought to the Capitol to "express the will of the Roman people," to say whether Oddone Colonna should be attacked with an army formed of a fifth part of the men or with a general army. But there were those in that parliament, according to the account sealed by Lord Pietro the chancellor, who wished to disturb the stability and peace of the city. And in the parliament, on the Capitol, they made noise, and they threw rocks. And after this incident, which caused grave confusion, the Lord Pietro put the question as to whether or not the senator Brancaleone should be permitted to act as he would, with full power, against Oddone and his accomplices and against those who had thrown rocks (because Brancaleone had said that this was too criminal to be overlooked); and no one whom Pietro could hear answered no. Even in this unlikely parliament the official verdict was unanimous. It is not really clear that a parliament could answer no, when the proper answer was yes. To be effective, it would seem, men of the ilk of these parliamentary dissidents would have seriously to riot or revolt. Then they could, or could help to, change government, as they (*Romani plebei populum*) did in 1237 when they overthrew Giovanni Poli and replaced him as senator with Giovanni Cenci.[39]

In his necessary quest for public opinion in Rome, the senator was of course not restricted to the crude instrument of the *parlamentum*.

Most obviously he had a council or councils. These councils could advise him so that he would not bring the *parlamentum* to a stage of actual revolt when he chose to call it. These councils obviously became more necessary as the senate itself changed from a large representative body of fifty-six or fifty to a body of one or two selected nobles, and even more when that body was converted to a single foreigner and his foreign prosenator—more, that is, to the extent that local counsel was needed at all, because a strong foreign army, when it was effective, lessened that need.[40]

The phrase *"consilium generale et speciale"* has been thought to describe two representative institutions of different size, the *generale* large and the *speciale* small, which met to advise the senator in the church of the Aracoeli. This may sometimes have been the case, but it would seem likely that the phrase came to be a generally descriptive one for a full council and that although there were councils at the Aracoeli of very various sizes they were not very rigorously institutionalized.[41] The senator summoned, it would seem, the citizens most appropriate to the occasion. In this his action would not vary much from that of many contemporary rulers. In this he would behave very much as he did when he found decision in legal disputes; he then summoned the appropriate judges, took their opinion, and proclaimed sentence (except that by the judges he was, conventionally, much more bound). But, however institutional, there is clear, and again unsurprising, use of council in thirteenth-century senatorial government even under Charles of Anjou, as in 1281 when the royal vicar Philippe de Lavène took counsel of the *anziani* and heads of gilds.[42]

In an interesting example, Matteo Rosso Orsini, *"Dei gratia alme urbis illustris"* senator (the appointee of Gregory IX acting in 1242 during the vacancy after Gregory's death) proclaimed a treaty with Perugia and Narni. He did this with the council of the city gathered in the Aracoeli where it was customarily gathered (*congregato Urbis consilio in domo Sancte Marie de Capitolio, ubi consuetum est more solito congregari*), at the mandate of the senator and at the petiton of the notary who was the syndic of the commune of Perugia. There were eighty-five of these councilors or counselors. They are not on the whole a surprising group. They include three judges, four notaries or scribes, and a lot of familiar names, four Annibaldi, a Capocci, a Malabranca, a Frangipane, an Astalli, a Sant'Eustachio, a Boboni, Cenci, Crescenzi, Surdi, Grassi, Rossi, a Capudzucca, Giovanni Poli (the only man given his usual, formal title *comes* before his name), a man identified as the

Who Ruled Rome?

lord (or Dom) Bonaventura of the cardinal bishop of Porto. These men more than counsel, presumably, gave substance, the assurance of real local backing, to Matteo Rosso's "Guelf" treaty.[43]

The council is not always described as responding unanimously. In 1241 the senators Annibaldo and Oddone Colonna, at the insistence of pope and cardinals, promised to return to the stipulations of the treaty between the pope and Angelo Malabranca and to protect ecclesiastical liberties.[44] This action was taken at the Aracoeli, before "the councilors of the city fully assembled." After a full discussion (*deliberatione plenaria*) the greater and wiser part gave their assent and approval (*de assensu et voluntate maioris et sanioris partis ipsorum*). This very conventional medieval electing and approving formula is different from, but no more compelling than, "unanimously."

Rather more compelling in some ways is the description of another set of councils, in the year 1257, concerned with provisions of the treaty with Tivoli, and summoned by the foreign senator Emanuele Maggio.[45] On Tuesday March 6 the general and special council of the city was assembled by bell and herald at the mandate of the senator, in the *palazzo vecchio* of the Capitol, to consider the enactments of previous special and general counselors (the phrase is turned around) from the time of Brancaleone. On Saturday March 10 another council was called, this time in the *palazzo nuovo*. The description of this council is suggestive. Its members were summoned by sound of bell and summons of summoner, a selective sort of summons. It was, the document says, the special council of the commune of Rome, to which were called the *anziani* and certain other provident men. The document refers back to the Tuesday meeting as a meeting of the general council, and it refers to that meeting's specification of the further special meeting of the senator "with special councilors and others" described this second time as "the senator with the special council and others who are not of the council (*qui non essent de consilio*) if the senator wishes to call them." The next meeting, on Wednesday March 14, was of a general and special council of the commune of the city, again in the *palazzo vecchio* of the city, again at sound of bell and voice of herald. The "greater part" of the Saturday special council had approved the stipulations of the treaty; all of the Wednesday general council "with the exception of about two" approved. The plan was submitted to a parliament of the whole people by the senator because, he said, he wished in this to follow fully their voted will; they approved the submitted plan unanimously.

This 1257 document with its cluster of councils seems to be describing real and different things. While it does not distinguish a special and general council from a general council, it does distinguish both of these from a special council. In its talk both of *anziani* and of men who may be assembled although they are not of the council, it strongly implies a specific and known membership of the special council. Its distinction between the summons of the summoner for the expanded special council and the cry of the herald for the general council points up the fact that a bell and a cry assume that the auditors will know who is summoned; the members of the general council must have known they were members of it. Even the movement out of the Aracoeli and into the *palazzo vecchio* for the general and the *palazzo nuovo* for the special council gives the document a believable air—although the fact that the scribe did not know the correct date for his initial meeting may make one a bit cautious about too heavy a dependence upon his phraseology. The fact that the treaty plan was too intricate, practically, for a large body to do much with it but accept it or reject it makes if anything the distinction between councils seem more serious.

The whole mechanism of councils and parliament here does not of course suggest democracy. It operated during a very peculiar senatorship, that of the antipopular Emanuele between two senates of Brancaleone. The voting, with its everybody but about two, suggests nothing of serious division and counting, any more than does Emanuele's pious phrase to the parliament. What the document and the councils do suggest is more institutionalization than does most of the evidence from thirteenth-century Rome. It argues caution in denying as well as in accepting the existence of institutions. At least in 1257 the councils had some recognizable form, even if, in themselves, they had no independence—they came only at the senator's summons, and they voted yes.

In the list of the counselors of 1242 some men are called "lord" and some are not. This does not seem a completely random selection: all the Annibaldi are "lords," as is Giovanni Poli, the Boboni, the Frangipane, the Malabranca; Giovanni Capocci is not. The lords are not a convincing roster of nobility, but the notary presumably had something of the sort in mind. In fact, in spite of the blatant and overwhelming importance of the Roman nobility, its definition, like so much else in the Roman constitution, seems to have been vague, unless one limits the term to men with titles from external sources, and in that case the group loses its force in the context of the city.

There were consuls and proconsuls and later the honorific title "Magnificence." [46] Early fourteenth-century (1305) antimagnate legis-

Who Ruled Rome?

lation lists the names of Roman noble families. In the mid-fourteenth-century collection of laws there are specific legal restrictions on barons and baronesses. In the same collection, men elected to play in the Testaccio games and then not playing were forbidden to hold office for five years unless they could plead old age (were forty or over), illness, or other real incapacity. This suggests something of a game-playing, Testaccio-crunching, officeholding elite, as does Saba Malaspina's talk of nobles playing games on horses to celebrate Charles of Anjou's entry into Rome. These suggestions are confirmed by the collection of statutes' definition of *cavallarocti* for the purposes, as it says, of its own laws. They are men who have already held office as *cavallarocti* (knightly office?) within the city or who have played in the games of the Testaccio and Agone. It reinforces an impression gained from a rhetorical phrase, used for example by the senators Gentile Orsini and Stefano Colonna in 1306, talking of everyone coming into the Capitol, "Roman citizens, both noble and foot (*cives Romani, tam nobiles quam pedites*)." Nobles fought as well as played, it is assumed, on horses.[47] But in spite of the extreme vitality of the thirteenth-century Roman nobility, and in cases of some families its seeming to last forever, the caste seems to have been covered with little controlling, defining superstructure. Who was a noble is almost as hard a question as who was a citizen. The nobility can certainly not be considered a crisp, edged governing class—but it must be admitted that the title is often used in contemporary documents with no sense of internal confusion.

From nobility one moves naturally to party. Were Romans clearly Guelf or Ghibelline? The answer must be a distinct *no*. Saba Malaspina might well lead one to the opposite conclusion with his talk of Arrigo of Castille's catching all the unsuspecting leaders of the "Guelf" nobility, like fish in a net, on the Capitol, and his talk of Giacomo di Napoleone Orsini as the *dux* and *magister* of the "Ghibellines." But the fact that Napoleone and Matteo Orsini and their connection, Angelo Malabranca, dominate the "Guelf" list suggests a difficulty even within Saba's text. The Orsini, through an early fourteenth-century Orsini cardinal, if report and assumption are correct, provided the most striking denial of Guelf-Ghibelline connection from the whole period. An Aragonese correspondent writing to Jayme II from Avignon on February 7, 1324, heard the cardinal reply to the teasing John XXII who had called him a Ghibelline, "Truly, Holy Father, I am neither Ghibelline nor Guelf, nor do I understand what the words mean. . . ." And the listening Aragonese heard further: "Truly,

Romans have many enmities and friendships, and they help their friends, whether they are Guelf or Ghibelline. They help and love their friends, but you will not find that any true Roman is either Guelf or Ghibelline." This, by itself, also goes too far, at least for the thirteenth century. It would be unreasonable not to note in Roman nobles political persuasions that could understandably be described by the two "party" names. Although these persuasions seem seldom to have been binding or consistent or to have spread to whole families, there were certainly men and even families who have a noticeable Guelf or Ghibelline aura. But, in general, political grouping and regrouping in thirteenth-century Rome was surprisingly facile, independent of ideology, and dependent upon momentary interest. Party is certainly not that which provides the seemingly missing firmness to the institutional structure of medieval Rome.[48]

To pursue the meaning of government in thirteenth-century Rome one must return to the senator and his officials. As Halphen wrote, "La pièce essentielle de cette constitution était le sénat." [49] By the thirteenth century, the senator's only rival, the prefect, had become ceremonial and hereditary, the decoration of a potent landed family (the Vico of Anguillara—the family of the Trastevere tower). One can look at a thirteenth-century senator. In a mosaic in the chapel of Santa Rosa in Aracoeli, Saint Francis, exposing his stigmata, places his hand on the shoulder of a small, kneeling, donor senator, before the enthroned Virgin and Child and a standing John the Baptist. In the Palazzo Colonna, in a mosaic from Aracoeli, Saint Francis, again with stigmata, between a Colonna column and Saint John the Evangelist, presents Senator Giovanni Colonna to the Virgin and Child. Both senators are dressed as senators with distinguishing decorations at their necks and on their hats, with red or purple robes and white collars.

The senators are also visible in repeated administrative act— repeated but far from continuous. Their most common recorded activity, as one would expect of medieval governors, is purveying justice. The sort of job the senator did can be seen in the litigation of a family in the *rione* Pigna.[50] In March 1252 a woman named Luciana, the daughter of Bartolomeo di Giovanni Nasigrassi ("Fat Noses") and the widow of Giovanni, son of Niccolò Ciapi of the Pigna, was trying to get her portion of her dower lands and money back from her in-laws, the heirs of Niccolò Ciapi. On March 14 the palatine judge Tommaso de Oderisciis and the *primicerius* of judges Consolino di Giovanni Scannaiudei (who was also advocate in this case) gave their counsel to

Who Ruled Rome?

117

the senator, Giacomo Frangipane. They advised him of the justice of Luciana's claim and they recommended his putting her in possession of her properties. On March 19 Frangipane ordered that the decision be sealed and that two named mandatories put Luciana in possession of specified properties in the Tuscolan highlands, houses in the Pigna (on two sides of which were the properties of a shieldmaker's heirs), and a vineyard outside the Lateran gate (on one side of which was the property of the Frangipane and on another the lane that ran to the mill of the Camera—Giacomo was dealing with neighbors, at least of his extended family).

On December 4, 1253, the same judge palatine, acting this time with the judge Stefano di Benedetto, confirmed the previous counsel and advised immediate execution. The two judges issued a formal confirmation on December 12 (in which Stefano was described as counselor in this case). On January 15, 1254, the then senator Brancaleone ordered that the counsel be sealed with the senatorial seal and that a named mandatory (not one of the earlier ones) invest Luciana with the same properties that she had theoretically been given two years before by Giacomo Frangipane.

This is the way senatorial justice worked. Selected judges heard the case. They proceeded, we can feel reasonably sure from the description of other cases broken in process, according to quite standard procedures established by Roman civil law.[51] Having come to the point of sentencing, the judges instead wrote an advisory council. The sentence was issued and sealed by the senator who thus remained the official judge. This tight restriction could cause difficulties when the senator and his vicar were absent from Rome, so that in 1271 the practical governor, Charles of Anjou, made it possible for a judge to give sentence when the vicar was outside the city. Although this problem must have been considered earlier (as when in May 1254, Brancaleone, who used a vicar too, was governing with his notary in the senatorial pavilion in the Roman camp above Tivoli), it was particularly severe under Charles.[52] The counsels presented to the senator or his vicar, in the thirteenth century, were regularly submitted by a palatine judge and one other judge.

The case of Luciana's dower points up the slowness of effective justice. Appeals and various sorts of recalcitrance could postpone the enforcement of a judgment. (We do not know that Luciana actually got the properties even in 1254.) But it is equally noteworthy that the fall of the old senatorial government and the accession of Brancaleone did not either halt or change the verdict in Luciana's case. (The sense of continuity that one felt in Emanuele Maggio's handling of Branca-leone's treatment of the Tivoli treaty is confirmed.) In fact, the same judge palatine gave the same opinion to the successive senators. This impression of judicial continuity is confirmed by noting a counsel of 1250 given to the senators Pietro Annibaldi and Napoleone di Matteo Rosso Orsini.[53] The 1250 counsel was given by the judge palatine Pietro, the son of the lord Consolino the *primicerius*, and his colleague was Tommaso de Oderisciis, advocate.

Judges were among the most important officials of the city of Rome. In the thirteenth century there were two main sorts of judges, palatine judges and judges called *dativi* or, by the end of the century, simply judges of Rome. The latter were judges drawn from a fairly large pool and commissioned for specific cases. The office of palatine judge is thought to have grown out of that of the old judge ordinary. There were seven judges ordinary, from seven offices in the papal household. It has been argued that the seven were replaced by one pal-atine judge—but this argument is made against surviving evidence which lists palatine judges in the plural. Although there seems to be only one to take the oath of 1235, there seem to be four of them in a witness list of 1257. By the end of the thirteenth century, particularly perhaps under Charles of Anjou, they had multiplied and become spe-cialized; there was a judge palatine over appeals and other extraordi-nary cases (by 1328 his court sat by custom "in [on?] the *podia* of the Capitol"), a judge for widows and orphans and the poor, a judge of the camera of the city.

The judges were selected by the senator. In an interesting docu-ment of 1255, Brancaleone delegated to Arriverio "de Carbonensibus" the job of finding judges and notaries for the following year. In 1283 Charles of Anjou had southerners from the Regno among his eight judges as well as his other named officials.[54] But by the end of the cen-tury, in 1299, the papal camera of Boniface VIII was still paying sti-pends to Roman judges. The pope paid thirty *soldi*, nine *denari* to the *primicerius* of the judges of the city and twenty *soldi*, six *denari* to each of forty-five judges. (He also paid a sum to twenty notaries and to the college of notaries.) [55] When the pope had seized the signory he still at another level was involved in judicial administration in the antique way—perhaps nothing shows the lack of clarity in Roman adminis-tration more clearly, or more clearly reflects the persisting indecision about sources of authority of the sort seen in Boniface's appointment of senators.

Around the judges in Roman senatorial witness lists one finds

Who Ruled Rome?

scribes and notaries.[56] In 1202 Adamo d'Isola, scribe of the senate, and Cencio d'Isola, chancellor of the city (and elsewhere of the senate), appear together.[57] In 1257 Emanuele's parliamentary process was recorded by Andrea Romanutii, scribe (*scriba*) of the senate. It was witnessed by Invectardo, the notary (*notarius*) of Emanuele the senator, and three other *scriniarii*, or local authorized notarial scribes. A 1233 Giovanni Poli ratification of the peace between Rome and Viterbo was recorded by Romano, scribe of the senate. The business of the pact between Rome and Tivoli in 1259 was recorded by Giovanni di Pietro Gualterii, *scriniarius* of the Roman church and "now" palatine notary. In 1255 in Brancaleone's disturbed *parlamentum* it was Pietro, chancellor of the city, who recorded the votes. When in May 1235 the senator Angelo Malabranca and his urban officials took their oath over keeping the peace with Gregory IX, at least five *scriniarii palatini* took the oath, although it was redacted by a scribe not identified as an urban servant.

In 1283 Charles of Salerno (writing for his father, Charles of Anjou) talked of twelve official urban notaries, four from the city and eight from the Regno. He placed them: two in camera; six in criminals; three in civils; one in appeals. In almost every way visible, Charles's urban administration seems slightly more adequate than other senators', but, in general, his twelve and Angelo's five give a sense of the size of the official Capitoline writing office which could easily be supplemented in a literate town full of scribes and notaries. The writing office also probably gives a fair idea of the general size of the official staff. Perhaps a better one comes from the information that Charles of Salerno wrote to his father's urban chamberlain in 1283, that sixty officials should eat from his, the chamberlain's, table.[58]

There were other necessary officials of a wide range of dignity and importance from the bell-ringer and the two trumpet-blowers and the banner-bearers and the porters and doorkeepers on the one hand to the important chamberlain and the treasury-wardrobe keepers, of whom there were two in 1231, or the relatively dignified physician of 1283 on the other. There were summoners, guards, a senatorial seneschal.[59] In 1306 the two senators appointed a new doorkeeper for life. He, Paolo de Nigris, is described as a Roman citizen. The document of appointment, written by Giovanni de Fuscis de Berta, notary and scribe of the senate, and intended to be sealed with the senatorial seal, says that Paolo is to receive all of the customary emoluments of the office from the camera and the senate. It explains why a new doorkeeper was needed; even in the lifetime of the deceased Egidocio di Nicola di Pietro, keeper of the second door, there were terrible deficiencies in its keeping and also that of the third door, through which one gained access to the great hall of the Capitoline palace; it was notorious, through everyone who came to the Capitoline courts, that the doors were kept so poorly that shame rather than honor accrued for the senators and Capitoline officials. To regain order and decorum Paolo was appointed.[60]

In 1235 twenty-eight men called *iustitiarii*, men involved in the carrying out of justice and the execution of sentences, took the oath.[61] Some of them are identified as *iustitiarii* of a place, two of the Lateran, two *de Porticu*; some are themselves designated as of a place like Ponte or Trastevere; one was a Caffarelli, one a Curtabracha, one a Lombardi (Jacobus Henrici Lombardi), one Bartolomeo was called Bibolus, and one Paolo di Giovanni "Nasi Crassi" was almost surely the uncle of Luciana of the Pigna-Lateran-Tuscolan dowery and so almost surely himself a man of reasonable property.[62]

These *iustitiarii* lead to the constant, pressing problem that must occupy any historian, the question of how the peace was kept in thirteenth-century Rome, insofar, as, of course, it was kept. The most helpful answer comes again from Charles of Salerno in 1283. He wrote to the chamberlain that he, the vicar, and the marshal (the focal official in these affairs) should decide if thirty mounted police guards were sufficient in keeping the city, and that, if they thought not, they might raise the number to fifty. Charles had troops who could do this job, and he had external sources of revenue which could pay them. Nothing is a better key than this fact to understanding Charles's success in governing the city. He did not need to rely on the urban budget scraped together from the rents and rights of the city, the money from salt, coinage, port tolls, rents from Jews, and judicial fines; thus, too, although they were controlled by an autocrat and were perhaps careless of local sensibilities, his officials presumably need not depend so heavily on tips and bribes.[63]

One sees then that from the senator, the chief governmental official, radiates an administrative bureaucracy. This one might say, and I think not unconventionally, was how Rome was governed. But the senator's was a very feeble radiance, and the slightest serious thought makes one realize that this cannot have been an important community's government, if one means by government anything more than a simple administrative outline. The shape of the senatorial office was too indistinct, the control over the senator too undefined and variable, the hand of the senator in the community too short-fingered, for all this

Who Ruled Rome?

to be called the real government of Rome. One asks hard, in this context, who ruled Rome, and finds always (except for brittle, limited periods) an unsatisfactory answer. No one ruled Rome. No group ruled Rome. Every possible resolution crumbles. The clarity of every answer becomes muddy upon examination. Rome in the thirteenth century, except for brief intervals, would seem not to have been ruled, not to have had rule. But we know that that is not true, because, whether or not there were sixty men at the chamberlain's table, the community functioned. It found for itself a coherent, profitable existence. This combination of frail government and continued existence should be made to have some value by the historian.

One must explore Roman government another way. That is not to say that one should not have taken the conventional path first. It is really helpful to know of Rome that official governmental thought was focused in a senator who echoed the ancient name (revived in the Romanesque twelfth century), who could be subservient to pope or emperor-king or "people," and who can in certain lights, at certain times, look very much like an ordinary Italian podestà. It is valuable to see that both city and senate are distinct enough for each to have a seal, to see that salt, coins, and pope all pay, that the Capitol was really used, that judges spoke through the senator. At the very least, official government is one face of, one way of expressing, real government. But having seen the seals and the doorkeepers, the traditional governmental outline, one can look again for interesting existents, see where arresting positive evidence remains, and then, even if the material is sparse, try to say what, in terms of this total community, this evidence means.

There is one group of governmental officials in thirteenth-century Rome that catches the eye immediately, because they existed, because of the intricate importance of what they did. They are the Masters, and Submasters and other officials, "of the Buildings of the City." It is extremely, and helpfully, curious to find in such a lightly governed community, this group supervising the way people actually lived, and built their houses, and threw away their garbage. These Masters, three in the surviving documents, supervised, in response of course to complaint, the examination of each offending area. They inspected documents and looked at the physical place. They, and in late century, by 1279, their judge, made "arbitration" or passed sentence, at the Capitol, at the bench in the portico where justice was returned, before witnesses, sometimes identified by *rione,* summoned by an identified curial summoner, listed in a document redacted by their own scribe or notary. If their *arbitrium* was unacceptable, one of the parties could appeal to the senator (through whom from "the people" their power came), as the proctor of the nuns of San Ciriaco in Via Lata did to the senator Pandulfo Savelli in 1286 about building over an old wall of the nunnery.[64]

A repeated problem, and a significant one, with which the Masters dealt was the encumbering of the accesses to Saint Peter's. In 1233 two of the Masters, one a Grassi and the other a dative judge, Pietro Malpilii, acting for themselves and with the power delegated by the third judge, a Boboni, acted in a dispute between the chapter of Saint Peter's and citizens of the Leonine city, over houses and other structures built out into the street that led to Saint Peter's, particularly in the piazza before the church and on the Ruga Francigene.[65] The Masters went to the Leonine city and inspected the buildings with their own eyes. They looked at various pertinent instruments and at Leo IX's privilege to the basilica; they consulted the wise. They ordered that porch structures, porticos and loggias (*proforula et porticalia*) which stuck out before the houses into the street by more than seven *palmi* had to be cut off and removed, and they ordered the Submasters to see that this was done.

In 1279 the Masters were again faced with encumbrances around Saint Peter's, particularly, this time, around the piazza itself, called the Cortina di San Pietro. The more active Masters were the *scriniarius* Deodato, and Giovanni Staccio, the son of Angelo the son of Giovanni Staccio—presumably the man whose house held the rebels against Innocent III. The Masters with their fellow went to the piazza and looked at the houses, one of which belonged to the heir of a miller, and another to the heir of a canon of Saint Peter's. After listening to advocates on both sides and after careful deliberation, particularly with the judge, Don Angelo di Pietro di Matteo, who was assigned to the Masters' office, they ordered that houses not project farther into the piazza than a line drawn between two houses, one of which belonged to Santo Spirito in Sassia. It was further ordered that projections be removed under penalty of a fine. But the fine was not to be applied immediately. The houses need not be torn down immediately, while "in these days" the papal court was staying at Saint Peter's, "because in these days the houses in question have as their tenants those who follow the papal curia." So execution was postponed until the departure of the pope. In 1306 the Masters were again concerned with the neighborhood. This time the Masters ordered their Submasters, one a notary, one a mason, to investigate the garbage-filled vacant lot across from Santo Spirito. It

Who Ruled Rome?

was they who found the marble markers that identified Santo Spirito as the negligent owner.[66]

Although the Masters' concern with Saint Peter's was significant, they were not concerned with it alone. They worried about the dye coming from the Jews' house near Santa Maria *domne Berte,* about the tower of Master Enrico the doctor (actually near Saint Peter's), about an obstructive portico belonging to an ex-Pavian in the *rione* Trevi, about the boundaries of San Sisto, about lands in the Isola di Porto. What is so very interesting about the office of the Masters is that it shows the ability of the Roman community, in important matters, to function in such a powerful and sophisticated governmental way. That this ability was the expression, at least sometimes, of the deliberate choice of the community, or part of it, to govern itself shrewdly in important matters is strongly suggested by the choice of the Masters as future arbiters over problems of construction and waste by consenting parties in private contracts.[67] One finds, moreover, the prior and counselors of the saltpans, proceeding in a manner not unlike the Masters, in inspecting, acting, giving counsel, on the site of the salt workings.[68] More blatantly impressive, one can see that in times of crisis the community could provide those *boni homines,* however elected, who met to adjust the "constitution," to save the city from chaos or tyranny.

Rome was clearly a city that, however lightly governed, could respond governmentally in time and area of real need. Salt counselors and *boni homines* together recall another interesting set of officials active in thirteenth-century Rome, the *grasciari* or *grascierii* (provisioners) who helped supply the city with its necessities—its *grascia* of wheat, barley, and pork. Significantly, the city was not completely responsible for this work, even insofar as it was a governmental function. In time of need, a pope like Martin IV could still feel and take the responsibility for provisioning the city, for buying grain in Sicily to be sold at a proper price in Rome, the semicharitable business to be in this case arranged by a canon of Saint Peter's and a brother of the hospital of Santo Spirito in Sassia.[69] It is perhaps precisely in its areas of greatest governmental activity that a governmental monopoly is most noticeably absent. It is here that one sees most sharply the action of the pope, of the consuls of the two great gilds (even in a city with weak gilds). Again, in this weakly gilded city, it should always be remembered that gild statutes (at least surviving ones) predate urban statutes. In the middle of the century one already sees a judge of the merchants of the city (an officer who presumably linked gild and urban government) actively giving counsel—and giving counsel to the Masters of

Buildings. In 1255 the judge Pallone with his counselor, the now familiar Tommaso de Oderisciis, gave counsel to the Masters that they should prohibit the destruction, by a man named Romano "Citadini," of the equipment of a Santa Maria in Trastevere mill on the Trastevere side of the river. The action was provoked by a canon and proctor of Santa Maria, whose petition included a description of the course the counsel should take. The participation of the governed is firmly felt.[70]

This participation can be seen in another way by moving through the confusion of papal with communal government in thirteenth-century Rome. Papal government offered Romans an alternative set of courts. A contemporary historian has written recently, "In the eyes of the commune of Macerata papal authority is nothing more than a potential judicial loophole."[71] It was a loophole, and also something else, perhaps, more important for the men of Rome. The central Tuscan patrimony around Rome offered the layman an alternative. Only here, theoretically, could a case concerning violation of property come before papal ecclesiastical courts if both parties were laymen, and only if both were inhabitants of the Tuscan patrimony.[72] This Tuscan privilege points up the potentialities of confusion between courts and laws in Rome. And, at least for the historian, the facts of cases can be very confusing. It is sometimes very difficult to tell why a case was tried by a court of papal delegates, a papal auditor, or a court responsible to the papal vicar of spiritualities in Rome and not by a senatorial court; and it is often equally hard, perhaps harder, to tell why a case was tried by a senatorial rather than a papal court. One is pressed toward thinking that in the eyes of the governed, of the contesting parties, the source of the court's authority need not be important. This, if it is true, and I think it is, is very revealing.

First one should glance at some of the ways in which papal courts offered themselves to thirteenth-century Romans. Early in the thirteenth-century collection of San Cosimato's documents, one finds a little letter of justice from Innocent III in Ferentino to his vicar, Guido Papareschi, cardinal priest of Santa Maria in Trastevere (Guido's family's neighborhood church), and to Hugh cardinal priest of San Martino.[73] Innocent delegates to the two cardinals, as he had to the cardinal vicar orally, a property dispute of common form between the Roman Benedictine monastery of San Cosimato and the Roman church, also in Trastevere, of San Salvatore de Pede Pontis—a perfectly ordinary affair.

In a better explained case, from November 1212, Innocent is again active, this time proffering a decision of arbitration.[74] A man

Who Ruled Rome?

named Oddone Benencasa had sold half of a house commonly called
the Benencasa Arch (Arcus Benencase) to the nobleman S. di Rainerio
di Stefano (probably Stefaneschi). As part of the deal in the contract of
sale, Oddone had promised not to sell or give away the other half of
the house without S.'s permission on pain of no less than six hundred
lire. But since Oddone had failed to secure S.'s permission when he
alienated the house, S. demanded the money from the monastery of San
Cosimato as Oddone's heir, because Oddone had entered the monastery
and taken with him all his goods. It was opposed that because Oddone
had brought his goods to the monastery and transferred them alive
(inter vivos), the monastery was thus in possession of Oddone's goods
by right of gift not by right of inheritance, and so the monastery did
not have the obligation that inheritance might bring, particularly because no action was taken at the time of the death of the donor, and
because the caution of the civil law regulated to S.'s present disadvan-
tage the collection of penalties stipulated in contracts. Innocent, having
considered the matter, offered that the 250 *lire* deposited with the senator in this case be returned to the monastery, but that for the sake of
peace the monastery grant 150 *lire* to the noble S.

Neither of these cases gives one much sense of the presence of a
physical court. Quite the opposite is true of a dispute between San Cosimato and a man named Giovanni Obicionis Calloboccoris (or Callo-
boccionis), in 1247.[75] The case was heard by Beraldo, priest or archpriest of Santa Maria in Monterone, who acted under a commission
from the papal vicar, Stefano, cardinal priest of Santa Maria in Trastevere. The proctor of the convent claimed victory in the case particularly on the grounds of their opponent's contumacy. The recorded
hearing at the beginning of March has a peculiar physical presence because it was held in front of the church of Santa Maria in Monterone
before a group of witnesses which included, besides a "Buckabella"
and (more famous name) a "Buckamazii," a shoemaker, a blacksmith,
and a water-drawer (or men who had chosen these as surnames). "In
front of the church of Santa Maria in Monterone" is still a nostalgi-
cally medieval, although bustling, place (and very urban rather than
papal), with the church's primitive, although seventeenth-century,
façade facing the timeless dark Arco Sinibaldi surrounded by working
artisans.

This place with this case shows the pope in the city. So in a less
salubrious way does the working of the inquisition at Aracoeli, as in
May 1300 from his cell there the inquisitor of the Roman province,
Fra Simone de Tarquinio, made a Roman, Angelo di Pietro di Matteo,

his judge delegate in dealing with the disposition of certain Colonna properties confiscated by the inquisition because of the Colonna's political heresy.[76] This place with this case shows the Caetani pope's hand moving the government of local property, actually moving on the senatorial hill, in the convent of the senatorial Franciscan church.

In general, however, the papal government in the city of Rome offered alternate courts to the normal urban ones for making official and reasonably permanent the transfer and division of property. The papacy also offered alternate seals for authentication—the necessary dressing of government—like the seal of the papal chamberlain and general auditor of causes of the papal camera with which he could validate an *inspeximus*.[77]

With this in mind one ought to look back to the senate. What does the senate as court seem to be doing? Often it seems to be supervising, lending a seal of validity to the transfer of property. So the monastery of San Cosimato petitions, and on the basis of the petition, the palatine judge and his colleague the advocate give counsel to the senator to invest.[78] If the investiture were to be a successful one, the negotiations which would make it generally palatable had almost surely already occurred. In an apparently slightly more elaborate case, Giovanni Colonna as senator wrote to the palatine judge Corrado da Offida ordering him to conduct an inquest over a San Cosimato property dispute in Marino.[79] A dispute over San Cosimato's property boundaries had arisen earlier and then been settled by elected arbiters. The settlement had been officially preserved in a public instrument. But the established boundaries had, according to the convent, been violated and the markers disturbed. The judge, if he found the complaint true, was to return the situation to that which the arbiters had settled, after having reported his findings to the senatorial curia. Here the senator attempts to be the re-creator not the preserver of arbitration. If in property cases a senator found for one side, he was technically able to put the victor in possession of property, as in a 1300 case in which the urban mandatory, acting under the commission of a palatine judge, acting under powers of letters sealed with the seal of the senate, put the proctor of the hospital of Santo Spirito in Sassia in corporal possession of property.[80]

The two pressing concerns of judicial government, property and violent crime, were of course not always separate. They are clearly connected, and connected with Roman "colonial" government, through the Tivoli properties of a murderer named Oddone di Giovanni Benencasa.[81] (The murderer's Christian name and surname are identical with

Who Ruled Rome?

127

those of the Oddone of the Arch, and properties from both arrived eventually in the hands of San Cosimato in whose archives both are preserved, but the early Oddone, who went into San Cosimato, died too early for them to be the same man. The two Oddones may have been relatives; their existence may be a coincidence, even if a dizzying one, the sort that produces archival headaches.) Oddone killed Bartolomeo di Benedetto di Bartolomeo. In punishment his property was confiscated and divided into quarters which were awarded "by the council of Tivoli according to the statute of the same city for murder."

We know about this because of the resale of awarded quarters. In November 1232 one part was being sold by the "consul of the Romans and count of the city of Tivoli" (*Romanorum consul et Tyburtine civitatis comes*), Matteo Rosso Orsini. (Consul was a title and office that connected Roman nobility with subject places; the Orsini had held the consulship in Tivoli, at least intermittently, for some time; and they would continue to be involved with San Cosimato's Tivoli, actually Cassano, property.) Matteo resold his Benencasa quarter, for sixty *lire,* to four men, one of whom at least was a Benencasa, reserving any dower rights that Oddone's wife or daugher-in-law could show. In October, the senator, Giovanni Poli, and the justiciars of Tivoli constituted by the city by authority of the senate had sold for sixty *lire* the Benencasa quarter that had gone to the city (Rome). But surprisingly this presumably absent, if not dead, homicide himself appears in December—to receive property. In that month a citizen of Rome (Giovanni di Gregorio de Rufino) restored to Oddone and his heirs "all you have given to me" and also the cancelled and cut act of donation. Pretty clearly Oddone had, in a process not otherwise unknown in the thirteenth century, gotten rid of some of his property during (or before) confiscation and reclaimed it afterward. In June 1237 Oddone and his son Giovanni (whose name forms part of a familiar alternating pattern) with the consent of their wives Tyburtina and Maria sold property in Cassano in the territory of Tivoli to the monastery of San Cosimato for one hundred *lire* (*provisini senatus*). As late as 1276 the monastery was still acquiring property from the family; it then got a house in Tivoli from Donna Agnese the daughter of Giovanni di Oddone Benencasa, Oddone's granddaughter.

Murder did not make the Benencasa destitute, nor did it go unpunished by Tivoli's Roman masters. Whatever its cause, it resulted in an intricate rearrangement, but not a general dispersal of Benencasa property. Murder is a central and serious concern in the earliest Roman statutes, surviving as a collection from the middle of the fourteenth century. Nothing is clearer in these statutes than the problem of defining what exactly murder was, when it had been committed, in mid-fourteenth-century Rome. In a sporadically violent society, one in which there was close identification between the government and powerful noble families and particularly one that was still very dependent upon the principle of self-help (even when government was used in civil cases) and war, killings of various sorts occurred. It cannot always have seemed wise to repay them with other killings; confiscation, even partially evaded, must often have seemed a shrewder deterrent to violence.

Still by the mid-fourteenth century, when the act of murder had been properly established, it was considered a capital crime. It joined a group of other crimes which the statutes, echoing their society as well as other communes and tradition, one must presume, viewed with particular horror: arson, incest, sodomy, rape, or kidnapping of a child.[82] The edge of murder that provoked most horror was still murder within the family. "Abominable to God and man" was man's shedding his own family's blood, though not that, if immediate, of a patently faithless wife.[83] The statutes, although capable of horror, were generally shrewd in their efforts to prevent violence, particularly widespread violence, and to protect property. It was a capital crime (in a statute attributed to specific senators and probably from 1312) to lead twelve armed men by night to the house of one's enemy, and all of the executed criminal's goods were to be confiscated. Similarly leading a band of between six and eleven men was not a capital crime; and the fine for it although very heavy was far from total confiscation—at least for the rich. For similarly leading a group of between two and five men the fine was very much reduced. Not only were punishments cannily graduated, but also collected fines were sensibly distributed. Sometimes part of the fine went to the maintenance of the city walls; sometimes half of it was to go to the camera of the city and the other half to the injured party. It was specifically provided that a murderer's house should not be destroyed (as a heretic's was after trial by inquisition), but rather that its value be divided, half to go to the city, half to the heirs of the victim. The statutes forbade, except in the case of specified major crimes, the use of torture.[84]

In general the statutes use a pattern familiar in collections of medieval laws in trying to prevent repetition of crime and growth of violence. Repetition of the crimes of robbery and thievery and slow restitution are punished by heightened penalties: from fines, to cutting off an ear (as in the case of a second thievery), to cutting off a foot (in the third), to hanging (in the fourth). There is also a concern with age in

Who Ruled Rome?

protecting the child under ten and one-half and giving lesser sentences, except for homicide, to children between ten and one-half and puberty (fourteen).

The statutes also show an interest in class that is connected without doubt with the general effort to keep over-mighty nobles from destroying the peace. Crimes were sometimes to be punished more heavily as the criminal ascended the class ladder. The purpose of this class-graduated legislation is most apparent in cases like the punishment for maintaining private prisons (least for an ordinary man who imprisoned someone for less than an hour, most for a magnate or his bastard who imprisoned for over two hours and did not pay his fine within ten days). In a tariff of offenses recalling the Anglo-Saxons, but in this changing them, a commoner who in a fracas caused another to lose his eye had to pay a fine of five hundred *lire,* but a knight had to pay three thousand; if the loss were of finger or toe, a commoner had to pay one hundred *lire,* a knight two hundred, and a more powerful magnate six hundred. False testimony also was graded, and, most interestingly, the production of a false instrument by a notary. (One might not have expected so many classes of notary.) A foot paid five hundred *lire,* a knight one thousand, and a baron or magnate four thousand; if he did not pay within his ten-day term he was to lose his right hand, and after conviction he could never be a notary again.[85]

This mid-fourteenth-century Rome was a violent society. (It provoked a law that forbade, with a fine of one hundred *soldi,* shooting a bow or throwing a stone in the Aracoeli or any other church with glass windows.) The suppression of violence is always a government's job. (Of Charles of Salerno's notaries, six it will be recalled were used in criminal cases and three in civil cases.) Petty violence as well as grand is readily apparent in the Roman statutes. For "making a fig" at someone the fine was twenty *soldi;* for throwing anyone to the ground on purpose and with malintent, one hundred *soldi;* for throwing anyone into a well, one hundred *lire;* for saying insulting words, twenty *soldi;* for putting one's hand to a knife threateningly, forty *soldi;* for throwing stones, three *lire;* graduated fines, from fifty to one thousand *lire,* for striking someone in the face, depending both on the nature of the scar and the class of the criminal (with the face defined, ear to ear, top of forehead to end of beard); for breaking a tooth, ten *lire* (half to the camera, half to the toothless); for putting filth or dung or anything shameful in someone's mouth, twenty-five *lire;* for making fortifications for disturbing the peace of the city, one thousand *lire.*[86]

Again, these statutes survive from the fourteenth not the thirteenth century. They are, as a collection, later than the statutes of Roman merchants and later even than the statutes of quite minor neighboring towns.[87] But both crime and remedy must in many cases represent thirteenth-century custom. Some of the things we are told, the severe rights of punishment of a father over his family, for instance, must have been at least as true in the thirteenth century when extra-familial supporting institutions were in general less well developed. Animosity to dice-playing and its dangers (as opposed to the favored chess), sabbatarian laws for shops, things of that sort are less clear. Would the concern over having taverns close at the proper hour have increased from the thirteenth to the fourteenth century, or the protection of citizens' rights to fish, or the regulation of the sale of food? The fourteenth-century statutes enjoin anyone who keeps a pig in the city to keep it penned up so that it does not wander through the city. Because, another statute says, of the harm to beasts and men done by wolves and other ferocious beasts in the city, anyone who kills a wolf within the city will be paid ten *soldi,* outside of the city, five.[88]

It is possible that the wolves and the pigs came, I suppose, only after the popes had gone to Avignon, possible too that the statutes are merely copies of those of other cities (but it would be odd to copy them in the absence of wolves and pigs). In general they suggest that a hundred years after the death of Innocent IV the city of Rome was a wild, messy, irregular place, in which wolves joined demagogues and the Colonna in their assault on public order; but they also suggest that Rome had a government which, effective or not, was accustomed to face these problems with some sagacity. It was a sagacity that faced even the problems of sex and marriage. The time and the accusers were limited in cases of forced seduction and adultery; while in cases of desertion a series of fines were established (and a reward for informers) to keep the husband at home, and particularly to keep him from living under the same roof as his concubine. Counterfeit husband and counterfeit coin were both opposed by the statutes, but, of course, with very different punishments; the mistress who made the one paid at worst ten *lire;* the coiner who made the other paid his life.[89]

So when one asks what Roman government was, and one answers (as I will) on the basis of property, one must always remember the basic surging violence which Brancaleone, Charles of Anjou, and these statutes tried to answer. It was a violence which threatened property, and that threat in itself made violence undesirable. But it was also violence involved with property in many more ways than simply threatening it. It was violence, on its grandest scale, often supported by heavy

Who Ruled Rome?

feudal and seignorial property in the *contado*. It was violence which fought to control office and property-granting power, the urban government, the papacy.

A major job of Roman government was, obviously, to act as an arbiter between conflicting powers, to help define for them their own general self-interest. This sort of governmental purpose and its slow realization by the greedy governed was made brilliantly clear a generation ago by Sir Frank Stenton's exposition of the English magnates' retreat from chaos under Stephen.[90] Stenton's model is generally applicable, but its application to the Roman situation is particularly clear; one can watch the greedy governed more easily because there is so little (and such dispersed) government to hide them.

What was government in Rome? One can answer with cases and edicts. A dispute in 1239 between the canons of Santa Maria Nova and the monks of San Sebastiano shows what real government was.[91] The two religious houses both had water mills in the country at a place called Marmorea or Mont'Albino. The problems that arose from their using the same watercourse in the same neighborhood had come to a head by May 1239, because of San Sebastiano's repairing and building of mills. The two houses were unable to arrive at a settlement of their conflicting claims over expenses and damages and over the watercourse and its direction, so the case was brought to litigation before Rainerio, the powerful cardinal deacon of Santa Maria in Cosmedin.[92] Its particular problems had to do with the manner and expenses of constructing works to guide, divide, and rejoin the watercourse so that it could serve the mills of both efficiently without doing damage to either, and also with arranging a compromise partition of interest in a mill under construction. The two parties accepted their cardinal judge as arbiter in the case, and they bound themselves to accept his arbitration under pain of loss of two hundred *lire*. The cardinal set out general guidelines for a compromise which was to be defined further after their evaluations by two commonly elected master masons. The cardinal pronounced his compromise in the Church of San Clemente in Rome, in the presence of two proctor canons from Santa Maria Nova, two proctor monks from San Sebastiano, and a little court of which some members appear as witnesses: two proctors of bishops (Siena and Volterra), a monk of Saint Paul's, a member of the cardinal's household, the son of the painter Giunta of Pisa, and a member of Giunta's household.

Now, again, the happenings in this ad hoc court of Cardinal Rainerio, surrounded by this medley of witnesses, in a pretty little Roman church, seem to me to say what Roman government was all about, to tell very specifically of the problems with which Roman government had to deal. It had to help its propertied subjects, individual or corporate, come to arrangements about the division, transfer, distribution, and use of their properties, to their own best mutual advantage. It had to give an air of legitimacy (a necessary not a frivolous air) and some assurance of permanency and record to the peaceful expression of power based on wealth. It had to present a forum or fora for the non-destructive settlement of dispute—settlement based essentially upon the realistic appraisal by the parties of their own interests and power. It had, and in this it shows its sophistication, to suggest experts, like the master masons, to help in definition. To sum up, it had to offer arbiters and a place for arbitration where compromise could sensibly, realistically, and respectably be arrived at.

The presence of arbiters and arbitration is everywhere apparent (and beside them are counselors giving counsel). But arbiters arbitrating are not, I think, except in form, so very different from judges judging. They are perhaps more legally flexible and more direct, but both record a peaceful definition of realized actual existents (a realization most convincingly established, perhaps, by the clause in individual contracts between consenting parties which agrees to renounce specific statute). The action of the cardinal for the two religious houses is not really very different from the case, for example, in which in April 1274, the *iudex forensis Tuscie et Colline* and a colleague judge give counsel to the urban vicar in a dispute between the monastery of San Silvestro in Capite and three men of Gallese all named John (and called in repetition "the three Johns") over a tenement in a place called Pomaro.[93] San Silvestro wanted, and thought it could get, a clear decision, a victory, and sought judgment not arbitration—either because its right and power permitted this extravagance, or because the business of compromise went on outside the purview of court and documents. But there is nothing really to suggest that an urban vicar, even with the backing of Charles of Anjou, could enforce this sort of settlement if it did not basically correspond to the local fact of possession and power. It could give the settlement formality, memory, the splendor of the senatorial seal, summoners and mandatories, but not real force. Force depended upon acceptance. Neither the cardinal's act nor the vicar's seem very different from the ordinary recorded sort of arbitration like that of an arbiter chosen in 1215 by San Cosimato in Trastevere and Albascia, widow of Pietro di Angelo and guardian of his son.[94] Their arbiter, in deciding the disposition of a Trastevere oven and garden, is performing the typical governmental act.

Who Ruled Rome?

133

That there was a complexity of authorities (cardinals and vicars) for supervising these acts of government must generally have been a boon to the governed. Thus, the superficial confusion and complexity of authority that would seem to have denied effective government in Rome may have in fact offered a loophole to the Romans as papal government did to Macerata. They were not too firmly bound by one authority; they had room to maneuver. There is no reason to suppose that Romans were unduly confused by the various authorities "governing" them in the thirteenth century or that in choosing a court (cardinal's or vicar's) parties needed too much to worry themselves. They, in the end, had to find, with power and reason, the settlement to their own disputes. Any reasonable guiding and recording authority would have sufficed to make it formal.

This idea of the governed governing is reinforced by the fourteenth-century statutes. They stated convention, and they guided the Romans in governing themselves (although they may have done more besides). The statutes protected the landlords who leased their vineyards for a yearly rent of one-quarter of their new wine from fraudulent and depriving release, echoing the words of more than a century of contracts; the statutes thus stated, affirmed, offered governmental formality to existing convention. The statutes fought filth in public streets and piazzas: no one was to throw air-infecting garbage in public places, streets, or piazzas (with the exception of the river) under pain of paying a twenty *soldi* fine; no butcher was to throw offal in public places, except the river, on pain of a forty *soldi* fine. Informers, in both cases, were to get half the fine. The law was a little machine for self-government and self-policing.[95]

Much of this is true of course of any medieval city or place. But in some ways Rome was peculiar, and its peculiarity was directly connected with its major sources of income. To understand this, it seems to me, one should return repeatedly to Angelo Malabranca's 1235 senatorial edict in protection of pilgrims.[96] In it the senator is a herald's voice which points out community self-interest, the self-interest of a community whose prosperity depended upon income from pilgrims (indirectly from their donations and directly from their rents and purchases) and income from the papal court (directly from its rents and purchases, its largesse, and its distribution of office, and indirectly from its attracting to itself foreign diplomats, litigants, and petitioners). Malabranca proclaimed that his regulations were issued not only for the peace and tranquility of the inhabitants "of this happy city," but also for the comfort and benefit of all those coming to it, particularly those who in their devotion sought to visit the doorstep of the basilica of the prince of the apostles.

To protect the pilgrims, the Romeseekers (*peregrinos, Romipetas*), he issued restrictions, the violation of which would bring the wrath of senate and people and a fine of one gold pound (half for the city walls, half for the basilica). The senator said that he understood that in the area around Saint Peter's lodging-keepers forced pilgrims into their lodgings, even making raids upon those already settled elsewhere and forcing them to come along. There should be no more of this violent hospitality. Pilgrims should lodge where they wanted to, buy what they needed to, where they wanted to, and do it under the protection of the senate.

This, too, this act of Malabranca's was what Roman government meant. It meant the help of a wise ruler who could implement the definition of Roman palate for foreign money. It meant the controlling of greed so that greed could best be satisfied. But because the income which fed the greed was from such a curious source, not from the manufacture of cloth or the production of wool or oil, it did not particularly demand a rigorously institutional gild and commune sort of government. It could perhaps better respond to its flexible income flexibly.

Again and again the areas of firmness in Roman government and organization, good men producing order, the masters of the streets, consuls and judges of merchants, the control of garbage, the concern with housing and provisioning (including the excellent provisioning in 1300, a stunning job), show awareness of, and response to, the city's real economic needs. Its economic needs in terms of local land transactions, the control and use of powerful noble factions, and the protection of pilgrims, were suitably met by a "fluid system of political relationships." (And Roman government might be more readily understandable if it were seen against the pattern established by the study from which this phrase came, the pattern of medieval Muslim cities, whose organization, like Rome's, was not based on hard institutions. And yet with all this flexibility, admittedly, one sometimes seems aware of a sort of crusty underhive—as in the physical city—of institutional activity, only briefly visible, as some random document exposes it— of rectors of judges and of clergy, judges of Saint Martin or Martina, Inquisitorial registers.) [97]

That Romans understood clearly the value of the presence of the pope is clear from their insistence (although not constant insistence) that he be present in the city. Brancaleone called Innocent IV to Rome and protected him there. That Romans were correct in their under-

Who Ruled Rome?

standing is made clear, or seems to be, by, for example, the pathetic fall in rents of Hospitaller houses when the pope had gone to Avignon.

The shadow of the pope hangs over the whole century in Rome. It is a dark shadow under Boniface VIII, at the very end of the century, dark in a number of ways. One of its indications is the title used by the senator, Guido de Piglio, in 1303. He repeatedly called himself, *Nos, Guido de Pileo, domini pape nepos, Dei gratia alme urbis senator illustris*—pope's nephew first, senator afterward. But the overwhelming position of the pope, even the *prepotente* Boniface VIII, seems to have a different quality when the pope is seen to be, as he was, one of the chief financial assets of city and citizens—like a factory or a mine or a beach. What power he had over the city was in large part due to the financial advantage he brought it. (And, as a hostile observer noted, he was one man and could die.) [98]

Who ruled Rome? One must answer, a complex of forces and properties and ideas expressed by urban government, by the papacy, by the local nobility. But if one thing ruled Rome more than another, it was surely money. Rome's rulers might fairly be said to have been the little *denari provisini* of the senate with their mocking caption: *Roma caput mundi.* Mockery recalls the verse of Walter of Chatillon, who, in a mocking refrain, had written royal *laudes* to "King Coin," *laudes* appropriate to the ruler of Rome:

Nummus vincit, nummus regnat, nummus cunctis imperat.[99]

CHAPTER
IV

The Popes

IN THE LONG AND OFTEN UGLY history of the popes the thirteenth century is a brilliant passage. The eighteen men who, in the century, sat in Peter's chair were all (even perhaps Martin IV) in their different ways distinguished men. None was cheaply shoddy or pettily immoral. In fact none, except Boniface VIII, was notorious for any sort of personal immorality; and at least three, Gregory X, Innocent V, and Celestine V, were renowned for their saintliness (although none was a Gregory I or a John XXIII). None, except Celestine V, was uneducated; and he had been to school. Most were even learned, well-educated, often in the best schools and universities (Paris and Bologna) in Europe; they were teachers because of their knowledge not just because of their position. Some, Innocent III, Innocent IV, Innocent V, were in different ways great scholars. Three of them, Innocent V, Nicholas IV, and Celestine V, were men from the new, or relatively new, religious orders. Had this succession of men, this monarchy of merit, sat on any other throne in Europe— England, France, Naples—the world would have been, and still would be, dazzled by it (even though no single pope may have come within sight of Louis IX's variety of distinctions). They are history's philosopher-kings.

They were the elected monarchs of the Western world's church, the elected guardians of its intellectual institutions, its high priests and medicine men, the leaders of its whole literate community. These men are too big and too important and too well known to be exposed in any serious way, here, in a brief description of the city of Rome. Every thirteenth-century literary gossip wrote what he knew of them, and more, in his provincial chronicle. The popes were watched by all the dazzled, envious, imaginative, disapproving eyes of Christendom. They themselves wrote great books and issued great laws (which have little specifically to do, often enough, with Rome itself). They had an extremely efficient chancery, and their carefully written letters are preserved in archives all over Europe. Their chancery, moreover, produced for them

The Popes

139

and saved for us a series of registers which record, in outgoing letters, many of their variously important official acts.[1]

Although each of the thirteenth-century popes hides part of himself and his life (often an aggravatingly important part), each can be known almost as well, if not quite in the same way, as one knows one's neighbors in one's own village. These popes are, in different degrees, fully observable as people and, those who ruled, rulers; and they have all been studied by modern historians. In sharp contrast with most Roman senators, of whom one knows too little to talk of them really helpfully, one knows too much of Roman popes, bursting still with life, to talk justly of them here. No sensible reader would expect their complexity (or the complexity of the scholarship about them) to be caught in one chapter in a small book. Still they cannot, just because of their pressing presence, be ignored. Whether or not they ruled Rome, each was in his turn, to borrow Matthew Paris's Alexander IV's words, clearly at the very least, its most important "simple citizen." One cannot simply overlook the gem within the ring.

One is tempted to think of these thirteenth-century popes as a group, as if they strolled, as popes, through the rooms of the Lateran together. But of course they did not. No pope knew who his successor would be—unless, if the perhaps unlikely story is true, when Boniface VIII, as a cardinal, piped through a tube to Celestine V Celestine's divine inspiration to resign, Celestine in a rare moment of acumen recognized both the voice and its purpose. Even then Celestine could not have been certain of the outcome of the ensuing election, nor could any suspicious or hopeful dying pope. Still, although the popes must be seen as discrete rulers, many of them did know each other, just not as, both, popes. Most of the popes had as cardinals known as colleagues both their predecessors and their successors. Only three of the thirteenth-century popes had not been cardinals before their elections. The cardinal-electors elected from among themselves.

It was a queer business, and irregularly circular. The cardinals elected the pope; the pope appointed the cardinals. But once elected or appointed, with rare and dubious exceptions (Celestine V, the Colonna cardinals under Boniface VIII), they remained. Each electing college of cardinals was a composite group representing the tastes, politics, and compromises of a series of dead popes, although sometimes, in a day, a single pope could change the majority of the composition and so presumably the whole complexion of the college. Nicholas III did, on March 12, 1278, the day of an earthquake, by adding nine new members to the college, more than doubling it.[2]

Boniface VIII, on the other hand, if he meant what he said (one can never be sure with Boniface), was tempted to resist the whole business of appointment.[3] When the cardinal bishop of Sabina died it was gossiped about in high circles that Boniface would name some new cardinals; but when the rumor came to him, Boniface said that he thought it was time rather for deposing than creating. In spite of his creating fifteen cardinals in five promotions, Boniface was a deposer, and he was also, perhaps, in the intricate web of his various planned promotions, wary even of his own prelatical creations. A much later Genoese chronicler, writing of the Franciscan Porcheto Spinola whom Boniface selected for the archiespiscopal see of Genoa, says that during his distribution of ashes to the prelates in the curia in Rome on Ash Wednesday 1300 Boniface threw the ashes in Porcheto's face and said, "Ghibelline thou art and with the Ghibellines shall return to dust." (The "unbridled tongue" of this "passionate man" makes the history of Rome around the year 1300 crackle with his sharp, half-mad wit— no thirteenth-century historian can be ungrateful for him and it.) [4]

Papal elections were often painful and long drawn out businesses. Much depended upon each decision—unless the cardinals chose the easy, postponing way out by electing a dying man, an unsure expedient at best, though, because the touch of Peter's keys repeatedly proved disturbingly reviving. Into a new pope's hands went, by late century, the control of the now relatively formed signory. In choosing Nicholas III, the cardinals chose for Rome an Orsini signory, in choosing Nicholas IV, a Colonna signory, in choosing Boniface VIII, a Caetani signory. That the signory did not fall permanently into the hands of a single family, as one might have expected, had to do with the, in other ways, great seriousness of the papal election. The keys that bound and loosed could not be stealthily buried within the Cecilia Metella—although they certainly stayed a long time in Conti towers. (Besides, and this removed some of the tension from actual elections, the most hostile families, like the Orsini and Colonna, were bound together by similarity and marriage; and, furthermore, they knew that by pushing too strongly they might weaken themselves, an awareness indicated by the case of the paired senators of 1293, when the death of the Orsini provoked the retirement of the Colonna.)

Beyond the direct control of the signory and beyond political allegiance and alliance, and probably more important than either, the election bestowed on one man (and the advisers and powers that formed his inclinations) a vast, sweeping, enriching patronage. His taste in summer resorts, for example, could enrich or impoverish towns like

The Popes

Rieti and Tivoli. His attitude toward Rome, the city, the length of time he spent there, could alter a huge list of rents, prices, and incomes. In electing a pope the cardinals were electing the employer (although he might be a reemployer) of cooks and poulterers, of warriors and castellans, of clerks and confessors, the selector of cardinals, the favorer of religious orders, of nationalities.

The electoral conclave which followed the death of Gregory IX, the conclave which elected Celestine IV (who was pope for seventeen days in October and November 1241), was conducted in extremely painful circumstances as the sick cardinals clogged in the under-latrine of the Septizonium testify. Thirty-five years later, the conclave (acting under Gregory X's restrictive decree) sealed by Charles of Anjou within the Lateran palace in the summer of 1276 was less garishly, but quite evidently, awful. The election which followed the death of Nicholas IV in April 1292 was less physically painful, but it was very difficult. When Nicholas died, Latino Malabranca, a Dominican, a Malabranca, and an Orsini nephew, was cardinal bishop of Ostia. He, as dean of the college, summoned the cardinals to elect in the Dominican church of Santa Maria sopra Minerva. There were twelve cardinals, and six of them were Romans, seven if one includes Benedetto Caetani (a man from the *campagna,* from Anagni, whose own papacy as Boniface VIII would make his family effectively Roman). Two were French. Three more (Matthew of Acquasparta, Gerard of Parma, and Peregrosso of Milan) were non-Roman Italians. The conclave was divided into Colonna and Orsini parties; and there were in fact two Colonna and two Orsini cardinals, three Orsini if one counts Latino. The conclave at the Minerva could come to no compromise. The non-Romans left Rome. But the Romans—the Orsini, the Colonna, Malabranca, and Boccamazzi (a Savelli nephew)—remained in the city in summer, as the Trasteverine patrician Jacopo Stefaneschi (later created a cardinal by Boniface VIII) wrote:

> Joined indeed in their blood, but in wills all union disclaiming.

The new pope was not elected until July 1294.[5]

Some of the problems of this long vacancy, with cardinals meeting in clusters, were observed and recorded by a group of people from Canterbury, the archbishop-elect Robert Winchelsey and his household. Winchelsey, one will recall, had come to the curia to have his election confirmed by the pope. His party had left England on April 1, 1293; they got back on January 1, 1295. By the time they reached

Rome there was still no pope, and only the Colonna and Boccamazzi cardinals were still in the city. Caetani was at Viterbo. The others had gone to Rieti. The archbishop-elect's party, pressed for money, went about visiting cardinals and hoping for an election. In August 1293 they wrote back to Canterbury: "There is hope that we shall soon have a pope; to-day there are eight cardinals at Rieti, two-thirds of the college and they are all of the party of the Orsini." [6] But it was a false hope. The cardinals moved to Perugia, and the wrangle continued. Finally a convergence of circumstances—the long delay, a joke of Benedetto Caetani's, the probable connivance of King Charles of Naples, the religious enthusiasm of some of the cardinals like Giacomo (or Jacopo) Colonna—caused the conclave to react with favor, after a night's thought, to Latino Malabranca's suggestion that they choose as pope a holy man, Peter of Morrone, ascetic, hermit, and enthusiast (*O sancto Petro d'Aquila pastore*). Years later Giacomo (or Jacopo) Colonna wrote of Peter, factionally, "the fame of his sanctity was so great that it moved the cardinals to make him pope." [7]

The cardinals had at last, perhaps, responded to the idea of Francis (although Celestine's predecessor, Nicholas IV, had in fact been a rather relaxed Franciscan). A holy man was chosen to prop up the church, as in Innocent III's famous and probable dream Francis had held up the falling Lateran. But it was too late. The electors' motives were too mixed. Their creature was too weak—he was no Francis. It did not work. Peter was kind to the Spiritual Franciscans; he maintained his sort of holiness. But he was a miserably inefficient and pliant pope. Ptolemy of Lucca, a slightly later contemporary, for instance, repeated the rumor that Peter's chancery issued blank bulled letters, which the recipient could fill in later. It may well have been true.[8]

Not all elections were long and difficult. On March 28, 1285, Martin IV died at Perugia. Of the eighteen cardinals (six cardinal bishops, six cardinal priests, and six cardinal deacons), three were away from the curia acting as papal legates. The fifteen cardinals in Perugia met for Mass and electoral preparation on April 1; they elected Giacomo Savelli, cardinal deacon of Santa Maria in Cosmedin, pope on April 2. The quick election avoided a fight between a pro-French and Angevin party (to which Martin IV himself had belonged) and its enemies, probably an essentially Roman party recalling Nicholas III. It made peaceful compromise in the selection of this old Roman from a pro-Angevin family, chosen perhaps largely because at about seventy-five, and disabled, he seemed near death. As Honorius IV, he proved an impressive if rather short-lived pope. He ruled well, settled old arguments, made his brother senator. Saba Malaspina wrote of the senator's zeal for peace and tranquility, of his being a man of steady heart and mind. Ptolemy of Lucca (with perhaps something of Dominican prejudice) wrote of the pope:

He was a Roman of the house of Savelli, an ancient family in the City. He was a wise man who hurt no one, although he advanced his own people. He had a very prudent brother, a humane man, although both brothers were badly crippled. Honorius was crippled in both his hands and his feet so that he was unable to celebrate Mass without a machine to help him. His brother Lord Pandulfo, was so crippled that he had to be carried about from place to place. But when he became senator he was so powerful that thieves and malefactors fled from his presence. If he caught them he hanged them without mercy. Every man felt safe in his own house. The streets were as they had been in ancient days. As soon as Honorius was elected he moved to the City and built a great palace on the Aventine next to Santa Sabina, and he made that place the papal seat; and he renewed with buildings that whole hill.[9]

Admittedly it was possible—perhaps particularly for Franciscans —to view the Savelli handicaps less sympathetically. Salimbene was sharply hostile. And a northern Franciscan wrote of a nasty little pasquinadé put on Honorius's door. It said:

> Ponitur in Petri monstrum miserabile sede
> Mancus utraque manu, truncus utroque pede.
> [A wretched monster is set in Peter's seat.
> Maimed in both hands, lamed in both feet] [10]

It is worth stopping to consider the Savelli. They are such an interesting family. Although they may have been descended from the Crescenzi, they rose to prominence at the end of the twelfth century with the careful bookkeeping of the papal chamberlain Cencius under Clement III and Celestine III. Some of Cencius's written work in protection of the income of the church of Rome (to preserve a memory of it so it would not slip away) survives in the Liber censuum, a book which, in part at least, Cencius made quite personal: ". . . I, Cencius . . . raised from the cradle by the Roman church, educated and brought up by it. . . ." [11] Cencius's pontificate as Honorius III (1216–1227) (remembered in building and mosaic at San Lorenzo fuori le mura) insured his family's continued importance and probably its wealth. Honorius IV was Honorius III's great nephew. He was the son of the distinguished senator Luca Savelli and of a woman from an important family in the *campagna*, Vanna (or Joanna) Aldobrandesca of the Counts of Santa Fiora, on whose tomb in Santa Maria in Aracoeli his effigy now lies. Honorius IV's own generation was connected through marriage to both Colonna and Orsini. By Honorius IV's time (1285–1287) the family had collected property both in Rome and in the *campagna*, although it was not yet powerful in the countryside in the way that the Orsini and Colonna were, and the Savelli's "moderate baronial position" may have made it difficult for Honorius IV to control the papal state.[12] The Savelli seem to have been consistent in their reasonable support of Angevin interests, and it was the Frenchman Urban IV (1261–1264) who raised Giacomo to the cardinalate. Honorius IV had been educated at Paris—he remembered its sweetness in his old age.[13] The contrast between his relatively formal education and Honorius III's relatively informal (or so it would seem) education is characteristic of a general change in the education of important prelates throughout the Western church (the institutionalization of university education) from early to late century. (That the change was not universal should be clear, in a second, from a comparison of the very formally educated Innocent III with the much less formally educated Boniface VIII or the uneducated Celestine V.)

The first of two really impressive Savelli attributes is the close connection of the family with the city of Rome and Roman government. The family, in contemporary records, exudes Romanness. Its senators, especially Luca, but Pandulfo too, are particularly connected with the Roman people: Luca in revolt against Gregory IX; Pandulfo in the ancient order of the streets. The second, and in Roman eyes connected, attribute of these people is their distinction in virtue: the crusading sincerity of Honorius III; the order of Honorius IV. They were not untainted with nepotism—the Boccamazzi seem suddenly to swarm in Rome. The popes Honorius were not proclaimed saints. But they were remarkably responsible, serious men of high personal and public morality.

That the thirteenth-century Roman nobility (a body that does not often provoke much admiration) could produce such a family is flattering to it, and a striking phenomenon. The Savelli chapel in Aracoeli, the Franciscan senatorial church, where some Savelli (nephews but good governors) lie beneath their tombs with their echoing *stemme*, is a shrine of real distinction because of them. It would be hard to find a thirteenth-century London or Paris family or a family from the English countryside who produced such a succession of men—not the Bohuns or the Warennes or even the Clares or perhaps even the Giffards.

The Popes

The connection of the in some ways rather scruffy Roman nobility with an elective monarchy which remembered Saint Peter and aspired to real education could be morally elevating; or the needs of that monarchy could make high moral and intellectual qualities within a family a cause for its promotion.

The Savelli are not, however, just a fascinating, isolated phenomenon. They are, in their rather muted style, symptomatic of a way of behaving characteristic of men in office during a period flamboyantly marked by the dates 1170 and 1303. Thomas Becket, as he stood by the altar in Canterbury, and Boniface VIII, as he sat in his palace in Anagni, were magnificently conscious of role; so were lesser men like Geoffrey Plantagenet at Dover in 1191. And Geoffrey definitely and Boniface probably were conscious of playing the role according to the pattern which Becket had set and which had been broadcast through Europe immediately after his death. Becket's life was a series of roles. Its exaggerations help expose the fainter lines of paler lives. When his hand, like Turpin's, held the crozier, he was fully and exuberantly the archbishop. This concept of living fully within an office explains in great part why thirteenth-century bishops, kings, and particularly popes extended themselves so far beyond the limitations which their backgrounds would seem to have predicted. It helps explain the mysterious magnificence of Louis IX and Edward I and, in their different ways, a whole series of European kings. When Peter the Chanter instructed Philip Augustus in the duties of kingship he was helping to form role; and when Philip retorted, "If you ever make a king, make him as you describe him," he was joking about something that was almost conceivable.[14] The role of pope, at least by the end of the thirteenth century, was the most magnificent official role on the whole European stage. In playing it the Conti and Orsini and Savelli did not forget their familial pasts, but they at some times and in some ways, even within the notorious Roman sump, managed to transcend them.

In general what sort of men were these whom thirteenth-century cardinals elected pope? In the nature of things they were generally, but not always, quite old men. None was a layman when he was elected, but Hadrian V never became a priest.[15] He had been a cardinal deacon, and he died just a little more than a month after his election. Clement IV had risen to fame as a layman and as a friend and counselor of Louis IX and his brother Alphonse of Poitiers; and Clement as pope remained the father of two living daughters.[16] With the exception of three Frenchmen, a Savoyard, and a Portuguese, the popes were Italian; but, again in the nature of things, most of them were, particularly from their time as cardinals, experienced diplomats and aware of a wider world. Of the Italians, seven were from Rome and the *campagna* (and these seven included no pope who died immediately after his election—probably a coincidence, possibly because they could survive in Rome's climate, surely not because they understood its poisons).

Although there were so many popes (eighteen), the many did not at all divide the century evenly among them. From 1198 to 1241 only three popes reigned. From 1198 to 1261 only six popes reigned, and one of these for only seventeen days. In the longer period, from 1198 to 1261, popes of one family, the Conti di Segni of Anagni—Innocent III, Gregory IX, Alexander IV—ruled for thirty-eight and one-half years (although in fact it was Alexander's mother not his father who was a Conti). Through the century, a little coterie of Roman and campagnole families controlled the see for considerably longer. Besides the Conti, the Savelli ruled for twelve and one-half years; the Orsini (Nicholas III) for almost three years; the Caetani (Boniface VIII) for almost nine; and indirectly through a man who was almost their chaplain (Nicholas IV from Ascoli), the Colonna for over four years. Thus this group of Roman families controlled the papacy for sixty-seven of the years between 1198 and 1303—and a short movement back into the twelfth century would extend their time.

One non-Roman family (besides the patron house of Anjou) was important. It was that of the Fieschi Conti di Lavagna of Genoa, a family with distinguished Savoyard connections. It provided two popes, Innocent IV and Hadrian V. But its importance was different from that of the Roman families. Both of its papal members were personally brilliant; and the first raised the second to the cardinalate. Their relative lack of connection with Rome meant that they could not (or could not want to) really plant a dynasty there (unlike their foreign Renaissance successors). Fieschi did, however, plant the names of Roman cardinals' churches—Sant'Adriano, Santa Maria in Via Lata—in the Ligurian countryside. They were quite noticeably men with attachments and memories. Ottobuono Fieschi (Hadrian V), ex-legate to England, brought both devotion to Thomas Becket and prayers for Henry III to Liguria. Fieschi brought treasure and duty home. When in 1254 Innocent IV lay dying in Naples in Pietro della Vigna's palazzo, according to the distant, hostile English chronicler Matthew Paris, he said to a surrounding group of weeping relatives, "Why do you weep, wretched creatures? Have I not made you rich enough?" When in 1276, Hadrian V was made pope, according to a Genoese chronicler and prelate, he, sick after the conclave and preserving the family tradition,

The Popes

147

asked his relatives, "Why are you glad? A live cardinal could do more for you than a dead pope." [17]

One may reasonably feel that both popes missed the point of their relatives' greed in grief and hope; they wanted to be Romans. Fieschi papal power and wealth could not be preserved as Conti, Orsini, Savelli power were, because the Fieschi did not have their base in Rome and the *campagna,* and in fact an unprecedented amount of Ottobuono's heavy prebendal income came from distant parts of Europe. The anecdotes about the two excellent Fieschi illustrate starkly, because of their non-Romanness, and ironically the notorious general shortcoming, if it can be called that, of thirteenth-century popes. They were nepotistic. They wanted wealth and power for family and faction. An almost contemporary chronicler said that Nicholas III would not have had his equal in the world if he had been without relatives, whom he too much favored.[18] Dante has made the world aware that this is not a celestial quality.

Most thirteenth-century popes belonged to families whose rank in society is suggested by the title-names of some of them—Conti, Visconti. They belonged to moderately elevated (or locally exalted) noble families. The basis of their families' wealth, besides office, was an amalgam of urban and rural-seignorial income. Their families had feudal dependents. They were not merchants (except insofar as almost all Roman nobles were), and they were not peasants. They themselves were of course clerks, and Nicholas IV was a scribe's son. Celestine V, as always, is an exception; he was, essentially, a peasant. So is Urban IV an exception. Urban came from a family "honest and poor," makers of footwear in Troyes.[19] His brightness led him to Paris, and his vigor led him with Louis IX on crusade. He was, when elected pope, patriarch of Jerusalem.

The most glaringly visible of all thirteenth-century popes, the most brilliantly apparent, is the first of them, Innocent III. Very few thirteenth-century faces are pressed as close as his to the glass of history. This closeness has to do with Innocent's own quality and style. He was constantly on some sort of verge or edge, pressing at the borders of the known and knowable, of existence. He was a man curiously turned out, fully deploying in act the potential of his mind. He thought, and essentially correctly, that he could do everything. He was bursting with self-confidence and a sort of optimism. He believed that it was possible for the world to be set right, for obscurities to be cleared up, for things to be done. He was anxious to do as much of all this as he could himself. It would be inexact to say that he was unreflective, but he was active rather than contemplative, a lawyer. He was, moreover, anxious to convince his world that action was good: "Do not think that Martha chose a bad part." [20] He was for crusade.

Innocent III was young and full of vigor. When he was elected in January 1198 he was only thirty-seven. His age, we are told, was the only thing that caused his electors to hesitate as they chose him ("on account of his moral distinction and his knowledge of letters"—and presumably his strength for struggle). According to Innocent's biographer, a miracle, or at least a potent omen, confirmed the cardinals in their choice. As Innocent (then still Lothario of the Conti di Segni and cardinal deacon of Santi Sergio e Bacco) sat as a nominee apart from the others in the Septizonium, three doves flew into the cardinals' presence; of them the most brilliantly white flew to Innocent and sat on his right side. Later, Innocent's triumphal coronation procession through Rome, "the whole city crowned," fully exposed him to the people and them to him. He was also exposed to them in his tough but, in the end, materially generous reshaping and controlling of the conventional money gift given by a new pope to the Roman people.[21]

Innocent's strength was connected with his self-control. The *Gesta* describes him as the raging Poli attacked him when in procession he went crowned through the streets of Rome: "He himself, with a calm expression on his face, proceeded, undaunted, showing no sign at all of fear or emotion." [22] This controlled calm was further connected with the clarity of Innocent's conception of his job as pope, his duty, the right order of things. He was untroubled by clouding doubt. He knew what to do. He found that there was a table in the passage near the kitchen cistern at the Lateran at which each day money changers changed and sold money. Innocent, roused by the sight and remembering gospel phrases with what seems an almost Franciscan literalness, had the table and its changers thrown right out of the palace.[23] Clarity is the quality that characterizes much of Innocent's thought and behavior; it is perhaps the central note of the enactments of his great Fourth Lateran Council of 1215—clarity and the complete lack of disillusionment. There can seldom have been a less disillusioned ruler.

That Innocent's attitude toward the relations of ecclesiastical and secular powers has provoked very considerable disagreement among scholars might seem to argue, in a crucial area, lack of clarity in Innocent's thought. Innocent would seem to have stated a strong position of theocratic supremacy, but to have acted, in general, and in his most important secular interventions, much more cautiously than his stated position would imply, in fact to have found within the specific case

The Popes

conventional reasons for papal intervention in secular affairs. This could argue incoherence. It can also argue a highly empirical intelligence of a sort that seems typically Innocent's. He stated clearly and from the start the most extreme possible position to which he might in the end go, and which he should have been able theoretically to defend, but in practice he did not tease the world by using the position unnecessarily. He did not weaken his position by needless provocation ("a time to keep silent, a time to speak"). Instead of being obscure, he was clear on two levels.[24]

Innocent's search for clarity can be seen in quite another quarter in the beautifully composed ending of a sermon for the feast of the Conversion of Saint Paul—and in its execution, his success. In it, the image of Paul's conversion (a blinding conversion to true light) is shown to Innocent's hearers as something also shown them by God not just to make them wonder but "to draw us out of profound confusion" and into clarity. Innocent's quest for order is beautifully exposed in the conclusion of another, a Christmas, sermon, which concludes with a discussion of "peace" and its curious reiteration through the New Testament, and finally in the mouths of the heavenly host, paradoxically the heavenly army, "Peace on earth . . . the peace of time and the peace of eternity, to which He leads us Who is God. . . ."[25]

The order which Innocent sought was not merely, nor always, mechanical. In his treatise *De sacro altaris mysterio* ("On the sacred mystery of the altar"), the superficial order is rather peculiar (rather, at least in that, like the order of the rule for Santo Spirito, which of course may not be his); it moves in commentary through the Mass and then in its fourth book delves into a really serious consideration of the Eucharist, in which (according to a scholar who was in a position to know, to have spoken wisely) Innocent collected and put in order the thought, the study, of the principal theologians of the twelfth century. Little could have been more important. The seriousness which one finds in Innocent's study of the Eucharist, in his treatise on marriage, and in his homily at Santo Spirito, is seen again in two little works, the *Encomium charitatis* and the *Libellus de eleemosyna,* in which Innocent talks discriminatingly and practically of the love of God and man, placing in the first work the love of God, of course, before all else, emphasizing in the second the importance of alms, even above prayer and fasting. Seen together these works press upon one the conviction that Innocent was, at least in important areas of his mind and in some aspects of his behavior, a really serious and thoughtful and religious man, not just a facile Martha, a worthy one, who tried (or thought he tried) to maintain the virtues of Mary too.[26]

A symbol, or an emblem, of this Martha's style occurs early in the *Gesta* when its author talks of Innocent's work as cardinal in his church of Saints Sergius and Bacchus. Finding it in ruinous disorder he renewed its roof and its walls and gave it a new altar. Innocent was the church's practical physician, and in what is at least accidentally another, later symbol, he bought a house within the enclosure of the Vatican palace for a medical doctor to live in.[27] The author of the *Gesta* seems anxious to establish the fact that Innocent's buildings were practical and grand, but not sumptuous or sybaritic. But neither the *Gesta* nor its Innocent is drab. The color becomes almost intoxicating in the *Gesta*'s rich list of Innocent's gifts to Roman and other churches: red cloth with gold peacocks on it for Santa Maria Maggiore, San Lorenzo in Damaso, and San Lorenzo fuori le mura; for San Marcello, for the altar, a black cloth with gold birds on it; for San Martino ai Monti, for the altar, a red cloth with gold stripes; for San Marco, a silver candlestick worth five marks; for Saint Peter's, a cross of gold with silver gilt feet, two gospel books, very precious, very beautiful, in gold and glaze, pearls and gems; for Saint John Lateran, imaged richness; (and for the Patriarch of the Bulgars, a great ring with five topazes in it). But, besides these, there were more practical presents, roofs and restored apses, and for San Tommaso in Formis, twenty *lire* for getting back its lands.[28]

The brilliance of the *Gesta*'s gifts is matched by another brilliance, Innocent's delight in words, in mocking internal rhyme, particularly in his rather savage early treatise, the *De miseria humane conditionis* or *De contemptu mundi*. The *De contemptu*, a set piece on the vanity of earthly pretensions and of man's life "from the heat of love to the meat of worms," has sometimes been treated as if it were grim and dour, a reminder of the dark side of the mind of this bright man; but (even if it should be borrowed) its language is too high, fantastic, lyrical, funny, and crude to allow this interpretation. In speaking of the horrors of old age ("the face wrinkles and the posture curves, the eyes mist and the joints shake") Innocent reveals his joy in his own famous (if he must have presumed transitory) youth, and equally in the easy, echoing sounds:

> *nares effluunt, et crines defluunt,*
> *tremit tactus, et deperit actus,*
> *dentes putrescunt, et aures surdescunt.* . . .

[Noses run and hair falls out,
Touch trembles and motion falters,
Teeth rot and ears go deaf. . . .] [29]

In the longer run of this tract Horace and Ovid join the game. In his
sermon for the feast of the Nativity of the Virgin, Innocent, playing
with Psalm 27, wrote of Mary, . . . *unde flos iste non defloruit nec ef-
floruit, sed floruit et refloruit,* and toyed with the forms of the verb *flo-
rescere* (to begin to bloom, to blow).[30] Innocent had a poetic ear perhaps not unlike Macaulay's.

Whether one can detect that ear in Innocent's official letters
sufficiently clearly to identify which of his letters were really written by
him is no easy problem. It (not quite in these terms) has been considered in this generation by a scholar who knows the letters very well,
and who is willing at least to point out recurring Innocentian qualities
like "parallels between words of similar sound" and "transposition of
epithets" (*affectus-effectus/fidelitas devota et fidelis devotio*); he and
scholars with whom he agrees have found Innocent's style studied, full
of antitheses, full of echoes of the liturgy and Scripture, and decorated
with classical authors like Horace and Ovid, "a richly figured style,"
crammed with scriptural texts, "studded" with prophets and apostles,
with favorite figures like Melchizedek, and favorite texts like Ecclesiastes 3:7 ("a time to keep silent and a time to speak"). Whether or
not one can tell which letters Innocent wrote, the characteristics of an
Innocentian style are clear enough. His favorites, moreover, Melchizedek and Ecclesiastes, are so appropriate to Innocent that their directness, and thus the directness which one senses in him, is almost breath-
taking; and, perhaps paradoxically, it is a directness never distantly
separated from playfulness. But in all this, it would be stupid to assume that Innocent never changed, however fixed (molded in a role)
he may seem in our vision, however the opposite of a feckless wan-
derer. Innocent's style and thought may have changed, as his face
seems to have when he grew a moustache.[31]

Although Innocent's not really unconventional playfulness is
probably apparent even in his use of quotation and symbol, he was not
frivolous. His work was concerned with matters which Innocent
rightly thought to be at the center of the church's proper concern and
his own. There is no reason to think that Innocent valued his own
scholarship less than did the author of the *Gesta* who thought it the
best of his time. Innocent's self-confidence and assumed preeminence
allow him his lightness. He was the product for which Paris and Bo-

logna had formed themselves, and he remained their grateful patron and defender. Of professors he made cardinals, not, like some Alfred or Charlemagne, to gather around himself an alien sort of learning, but rather to reenforce and express his own. He was as sanguine about his abilities in theology, it would seem, as in politics.

The *Gesta* speaks particularly of Innocent's law.[32] Three times a week he held public hearings in solemn consistory, and (like a harsher prototype of Louis IX under his tree) the "Third Solomon" spoke judgment. Learned lawyers came to hear him, "and they learned more in these consistories than they learned in the schools." One can watch Innocent acting as a judge in a detailed, extended description, written by Thomas of Marlborough, an English monk from Evesham, a student of Stephen Langton's, a bright young man.[33] (Marlborough before Innocent reminds one of another bright monk, Samson of Bury, before Henry II of England—it is definitely a scene with parallels.) Marlborough was an excellent audience for Innocent. He relished Innocent's bald wit; he paid attention to what Innocent had to say. In fact, at Innocent's suggestion (and the suggestion of the future Gregory IX), Marlborough went off to Bologna for six months in 1205 to learn some more law before waging his case against the bishop of Worcester. Marlborough's opponent, Robert Clipston, the proctor for Worcester, did poorly with Innocent. Innocent laughed at Clipston when he talked of a shortage of advocates at the curia. Clipston bored Innocent by talking too much. Clipston supported his position by telling Innocent what he had learned in the schools and what his masters had taught of prescription and episcopal right; and to this, Innocent, quickly calling to mind the appropriate cliché, said (and Marlborough recorded), "Then you and your masters had been drinking too much of your English beer when you were learning." Clipston, confused, repeated what he had said, and Innocent what he had replied.

The sharp, humorous, self-confident coarseness with which the still quite young Innocent (still in his early forties) dealt with Clipston is an aspect of that boldness and clarity with which he dealt with the church's, the world's, and Rome's problems. He had a genius for selecting and collecting old commonplaces of behavior as well as expression and endowing them with a seeming freshness which made them attractive and productive. He was an intelligent, assertive, learned middlebrow who felt no real competition pressing around him and at the same time no respectable field of learning or action alien to him. He did not isolate himself. He had his Parisians like Stephen Langton; and he had his family like his brother Riccardo. He avoided scandal (al-

The Popes

though one does not feel that he had to struggle to avoid it) except in the damning papal business of nepotism and family favoritism—and what had brought him to power, this chain of uncles and nephews, he could hardly consider bad. In this favoritism, Innocent's results, whatever his intentions, were magnificent. He can be said to have founded three Roman dynasties (or to have moved them to importance or their importance to Rome): the Conti, the new Poli, the Annibaldi. (If the use of modern analogy were not such a repulsive historical technique it would be hard to resist comparison between Innocent and his fellow youth John Kennedy; not only do they seem, on the surface, rather alike, but they might well have attracted and repelled the same observers.)

The activities of the "Third Solomon" were by fortunate chance described by a peculiar observer in a curious letter written near Subiaco in August or September 1202.[34] The letter is from one member of the curia, in the country with the vacationing papal party, to another, presumably more important member who had stayed at home. It is intimate, literary, allusive, and very sophisticated. It is informed by the "gay" tone, the lightly handled appearance, if not the actual fact, of homosexuality, which seems to color some late twelfth- and early thirteenth-century satire, as it does for instance the writings of Walter Map, although the Subiaco letter is not coarse and heavy as are the more turgid parts of Walter Map's *Trifles*. The letter moves as lightly as a butterfly, just at the edge of intelligibility, in its graceful descriptions of the lake and its surrounding greenery, of the chaplains playing like fish in the water, of the dreadful, tiring climb back up the hill to the papal camp. The writer, urban as an old Roman, admires rural beauty but deplores rural inconveniences—latrines, tents, smoke, insects, the lack of privacy, the clamor of rustics. When the letter tells after the loud day of the cicada's breaking the silence of the night, it includes within its prose a line appropriately from Ovid's *Epistulae ex Ponto* (IV, 3, line 49): "divine power toyed with human affairs." This was a line repeatedly used by historians, and sometimes, with great and skillful solemnity, as when William of Malmesbury talked of Henry I's death.[35] Its mocking casual use in the Subiaco letter illustrates perfectly that letter's tone.

In the Subiaco letter Innocent III is of course very evident, very important. He dominates the scene. He is also slightly ludicrous, pretentious, and, as one would guess from the contrast between Innocent's and the letter writer's styles, square. (The quality of the Innocent of the letter depends upon one's interpretations of the letter's tone; like one's actual assessment of many thirteenth-century popes it must be personal.) In concluding the letter the writer says that there is a gracious remedy for his rural exhaustion and exasperation, for whenever he is overcome he can run to the fountain of living water, the vicar of Christ, to sit at his feet, like Mary, and be thrice refreshed, from the treasure of eloquence and wisdom of this third Solomon. It may be very hard to believe that these phrases could be said in jest, they are, even with their heavily played trey, so solemn; but used in concluding, or helping to conclude, this particular letter, with its repeated, slightly askew, repetition of the epithet "third Solomon," they almost must be seen at least to be deliberately ambiguous and lightly ironic.

A little earlier in the letter, its writer, in telling of his desire to see his correspondent, writes, "So that you would not have to come to be punished in this place, I have humbly and devotedly beseeched [mimicking common form] the third Solomon, the Abraham of our faith, the most worthy father, the successor of the prince of the apostles, the vicar of Jesus Christ, that he would send me to you." Earlier in the letter the writer describes the four outlooks of his horsey hospice: to the south, the smoky, clamorous headquarters of the cook; to the east, placed so that he might examine urine in the morning sun, the druggist pounding endlessly in his mortar; to the north, the place where the local peasants congregated; and finally on the fourth side, "the little *tabernaculum* of our most holy father Abraham." Still earlier in the letter, in describing the classical delights and sounds of the fresh water under the hill, where Innocent could put "his sacred hands" to wash them and at the same time could put cool water in his mouth (with a play on double natures), the writer explains that on this account the "third Solomon" loved the place. It is just possible to interpret these Abrahams and third Solomons as pious unambiguous tributes (and the letter's having been saved helps make it possible)—but it is not possible for me.[36]

To turn from Innocent III, particularly the Innocent encased in the glaze of this Subiaco letter, to Boniface VIII, the last pope of the century, is to watch a startling transformation. To see this transformation with dazzling (and unfair) clarity one should look first at the Boniface who emerges from the testimony about him at his posthumous trial. For this trial, in the years 1310 and 1311, witnesses were gathered and their evidence recorded before a panel of delegates. It was evidence that might establish Boniface's moral turpitude, his having held heretical beliefs, and having made heretical statements. The trial was part of the turbid aftermath of Boniface's war with the Colonna and particularly with Philip IV of France and his advisers, and more specifically the ricocheted aftermath of the violent attack on Boniface by Nogaret and the Colonna at Anagni in September 1303. Benedict XI, Boniface's northern Italian successor, died a few hours before he was to proclaim his great formal condemnation of Nogaret at Perugia; and Clement V, his Gascon successor, pressed to the opposing trial, hesitated, hoping, presumably, that the whole affair would fade away without too much damaging the dead, the living, or the dignity of the papacy.[37] The trial did fade away. But the evidence, at least part of it, remains.

Quite naturally historians find this evidence difficult. It is suspect because of its extreme nature, because of the purpose for which it was collected, and because the scandal of its testimony is oddly repeated and reused within this case, and because it resembles too closely contemporary libellous attacks on heretics and especially those on the Templars. The evidence is, however, strangely compelling. It is very detailed both in its description of the things Boniface did and said and of the circumstances, the place, the hour, the weather, in which he did and said them. The figure who emerges from the evidence, its Boniface, is bizarre but believable, a dramatic success, and consistent in style with the otherwise known Boniface at the extreme edges of his behavior. Either very considerable artists or the man himself built this character.[38]

To go from Innocent at Subiaco to the Boniface of the trial is to go inside. From the fresh air of the camp one moves to the hung and furnished rooms of palazzos and particularly the Lateran. The trial also takes the dubious sexuality of some part of Innocent's curia to the pope himself. A shoemaker named Lello from the diocese of Spoleto testified that he was in Perugia to sell shoes at the time of the vacancy after the death of Nicholas IV and that he was called by someone from the household of Boniface (then Benedict) and told that he should take the cardinal some pairs of shoes.[39] The shoemaker went with the shoes to the cardinal, who was staying in a Perugian's house, and shoed him. Then the cardinal ordered a member of his household who was present to leave the room, and he took Lello into another inner room. There he began kissing him and saying to him, "I want you to do what I want"; he kept kissing him and coaxing him saying, "I want to lie with you; and I will do a lot for you." But Lello replied, "Lord, you ought not to do this, because it is a great sin, and today is Saturday, the fast of the Virgin Mary." Boniface answered him, "It is no more sin than to rub your hands together; and as for the Virgin Mary for whom you fast, she is no more virgin than my mother who had many children." Then Lello began to scream. Master Pietro of Acquasparta, who was outside, called in. The cardinal let Lello go. Lello escaped, ran quickly from the room, and did not even stop to get paid for the shoes. And Lello said that he was then about fourteen or seventeen years old or thereabouts.

Related stories of Boniface's misbehavior with the members of the family of his familiar, Giacomo "de Pisis," recur in the testimony. A monk of San Gregorio in Rome said that he had seen Boniface holding Giacomo's son Giacanello between his thighs and that it was notorious that he abused him as he had abused his father before him. Notto di Buoncorso "de Pisis," who used to stay with Giacomo "de Pisis" when the pope was at Saint Peter's, had seen Boniface in bed with the Lady Cola, Giacomo's wife; and he said Boniface also lay with Giacomo's daughter, Gartamicia. (For Boniface these Pisans were a garden of delight.) Witnesses had seen Boniface play at dice with the Lady Cola, with dice pricked with gold, and the lady had said, "A pope ought not play." And always on all counts the pope answered with his terrible, sharp, ridiculing tongue. Most unpleasant of the Pisan scenes is a bickering argument between Giacomo and another familiar, Guglielmo de Santa Floria, in Anagni, with one saying, "You are Pope Boniface's whore," and the other replying, "No, you are his whore." Giacomo said to Guglielmo, "You were his whore before I was, because at the time when he was still a cardinal I found you in his chamber doing that business with him." Guglielmo answered, "If I am his earlier whore, you are only his whore, because everything which you possess he gave you because you are his whore." [40]

In another story a notary from the diocese of Todi swore that he heard a discussion in the summertime between a physician recently from Paris and Boniface (then Benedict). The physician told the cardinal of the scandalous teaching of some Parisian masters concerning the life of the soul and concerning sodomy which they called no sin. Boniface (not one to lend himself to medical anti-intellectualism) replied that there was no life but the present and that sodomy was a moderate sin, rather like rubbing one hand against another. The physician has allowed the Boniface of the trial to make two of his favorite, repeated, and most shocking observations. The repetition makes Boniface's statements both less and more believable. Similarly a knight from Lucca, a man of about forty years, remembered related talk (from 1300) between Boniface and a knight from Bologna, in which Boniface had said that there was no other life than this life, and that it was no sin for a man to do what pleased him, and particularly to lie with women. Boniface said it was just as much a sin to sleep with women and boys as to rub one hand with the other. The Bolognese then shrugged his shoulders and said, "So everybody ought to try to enjoy himself." This happened in the Lateran (in the second or third room, the witness could not quite remember which). In the room there was a great bed covered with a red covering. Boniface sat in a chair and wore a white shirt and a mantle of red scarlet; and around him was a court with Bolognese, Florentine, and Luccan ambassadors on bended knees. "Heaven and hell are here," Boniface said, in Italian, at the same terrible meeting.[41]

The described chambers of the Lateran (for example, the first chamber after the place where consistories were held where Boniface wiped his mouth with a little towel) and the people multiply; the geography extends to Santa Balbina, the cardinal's vacation villa on the way to Rieti, to Todi where he was a young cleric. There is a prevalence of butchers in the stories. Boniface was (or his accusers were) very fond of meat. One day in Lent when his over-cook, Pietro da Veroli, had brought him six different meat courses, Boniface ordered him to bring another. Then Pietro brought on four fish courses in succession. This enraged Boniface. "Why," he asked, "don't you bring meat?" And the over-cook replied, "Lord, you have already had six meat courses." Boniface responded, "You ought to be cursed for wanting to be mean with the goods of the Roman church." Master Pietro replied, "No, Holy Father, I am not trying to cut expenses but rather to avoid scandal since it is Lent and I buy so much meat that it causes grave scandal." Boniface was wild with anger. He ordered that the cook's goods be confiscated, that he be deprived of all office, and that under pain of imprisonment he never enter the curia again during Boniface's lifetime.[42]

Earthly pleasure and mortality and a more dreadful third are the themes. Three monks of San Gregorio in Rome came to Boniface to denounce their abbot. They were brought to his presence by his chamberlain. They said that their abbot did not believe in the resurrection of the body, or that the Eucharist was really the body and blood of Christ, or that carnal sins compelled by nature ought to be sins; they said that he held many heretical opinions. To one of them Boniface said, "Do you have a father?" The monk replied, "No, he is dead." Boniface asked, "When has your father or any dead man come back?" The monk was still. Boniface then said further, "Go and believe what your abbot believes, because he believes better than you believe and speaks better than you speak." To monks of Saint Paul's, who similarly came to denounce their abbot and who said that their abbot did not believe in the resurrection of the body and did believe that the soul died with the body and who further said that the inquisitor Simone de Tarquinio had found their abbot a heretic, Boniface said, "You are idiots. You know nothing. Your abbot is learned and knows better than you." To his household at Santa Balbina, when he was still a cardinal, he is remembered as saying, early one morning in May as he sat in a white tunic on his bed, "Fools, fools, do you believe your bones will reassemble?" and "Idiots, idiots, what Paradise do you want? Do you believe that there is another Paradise beyond this life and that man shall rise again?" [43]

Boniface's dreadful third theme is the Eucharist. (His blasphemies oddly recall Frederick II, and one sees, either in libellous observers or in scandalous tongues, the same imagination developing the antibeliefs of the thirteenth century.) A witness named Floriano Ubertini of Bologna, a denizen of Orvieto, remembered that forty years before when he had been a butcher in Todi he had gone to the great church there to see a friend. When he got there he found a group of scholars in Benedict's room. (Benedict was then a canon of Todi.) One of the scholars called out to Floriano to ask what he wanted, and he said, "I am looking for a friend." Then the bell rang for the elevation of the Host, and Benedict said to the others around him who wanted to go to see the Body of Christ, "Why do you want to go? Do you believe, you fools, that what is held up is the Son of God?" (And this butcher, who may not have liked the scholars who, like him and other providers, clustered around the curia, further said that in Rome, a city he knew and had lived in, it was notorious that Boniface was a sodomite.)

Similarly an ancient and infirm priest, Niccolò da Sulmona ("from Sulmona" in the heart of Celestine V country), the primicerius of the church of San Giovanni Maggiore in Naples, remembered one day in the reign of Celestine V going to the papal chapel, and there he saw the cardinal turn his face away at the Elevation. Then later, in the hall, Benedict had talked of there not being a soul, of there being no difference between Saracens and Christians, and he had said that if it were not for fear of the temporal court many would hold and believe what he now said. The old priest remembered that Benedict was at the head of the room, but he could not remember on what he had sat. He thought it had been on a day in September, but he could not remember well either the day or the month. Asked if it had been cloudy or clear, he replied that he thought it had been cloudy, but that he did not remember well. Asked if the cardinal had been joking, the old priest said that he had not. Asked how he could tell, the old priest replied that the cardinal, as he said these things, had not smiled or joked but looked serious.[44]

Did Boniface joke? The interrogators appreciated the problem. He said all so sharply. Once when John the Monk said to him that what he did was not to seek counsel of his cardinals as Roman pontiffs ought but to exact unwilling consent, he replied, *Picharde, Picharde, tu habes caput Pichardicum, sed per Deum ego piccabo te,* which means alliterative, antiprovincial wrath and something like "Picard, Picard, you have the head of a Picard, but by God I'll peg you on a pike." But beyond that joke lay another: Boniface said that things would only go well in the world when there were no cardinals but only a pope. Boniface no doubt found John the Monk and cardinals like him a trial when they tried to restrain him; he perhaps felt a fellow feeling for the abbots of San Gregorio and Saint Paul's, but not for all autocrats. He also is said to have said, at the same time, that things would go well in the world when there were no kings but only barons. This Boniface knew what power he wanted divided.[45] If Boniface joked, there is no reason to think that his jokes were not significant. If the witnesses created in his mouth what they or their tutors believed a really perverse mouth would say, there is no reason to ignore the revealing shape of the imagined perversion. And clearly Boniface was seen by some contemporaries to be perverse. Fra Jacopone da Todi who had reason to know as well as to hate Boniface wrote:

Como la salamandra se renova nel fuoco
Cusi par che gli scandali sian solazo e giuoco.[46]

The Boniface of the trial can certainly be seen as this salamander Boniface renewed in his scandals. He is either the real Boniface burlesqued and misunderstood or the real Boniface stripped of his, and the official sources', protective reticence—Boniface speaking, as a witness said, in the vernacular as well as in Latin.[47] The sharpness of the tongue (aggravated, it has been suggested, by the pain of the stone) and the wit are much the same as those of the Boniface of other sources.[48] So is the disregard for appearance, or the regard for only some sorts of appearance, the violent thrust of ambition, the contempt for those who received, or did not receive, provisions: "We have heard good of you, but we do not like your name." Boniface was unscrupulous—promoting the bishop of Conza as a bribe or reward for his help in the extension of Caetani holdings; distorting a pious form to permit the bishop and chapter of Anagni to alienate to a Caetani; toying with the senatorial offer to a Caetani who was being kept for better office; planning perhaps a central Italian state for his nephew Pietro.[49]

In all this Boniface is immensely sad. When he loses to death two favorite nephews, he says, "Now what reason have I to live?" [50] He is the great man at the head of the universal church, a man of superb wit and intelligence and power, an able ruler, a man who expresses the extreme theocratic claims of the church, but who does not (one can easily think) believe.[51] He is encased in the most extreme version of the trappings of a great religion, but it is not, it seems, his religion. And those trappings were in fact suffocating; who could care about an image or a sacred vessel or a relic when the papal treasury was crammed full and bursting with so many of them deadened by their shells of inert gold and muffled in acres of drapery? (And when Boniface fell, according to a Lincoln proctor writing home, no one cared any more about him than they would for the most ordinary, even defective, denizens of Lincoln.) This is tragedy. It is high tragedy because of Boniface's dignity, his presence (caught beautifully in his seated figure at the Duomo in Florence). The great, variously remembered scene at Anagni has been told too well and too often to be repeated, but it is a great scene in every version: the old man in his palace, sitting upon his throne, dressed in his pontificals, wearing his tiara, holding his gold cross, waiting for death (or lying, infirm, clutching his cross, a recumbent Thomas), crying out, as the Lincoln proctor wrote, "Ec le col! Ec le cape!" [52] Less dignified, but in its lack of dignity and belief, equally poignant and very pointed, is the scene of a November day in Rieti: there was an earthquake, and Boniface fled from his Mass in the Duomo, ran outside and down the hill—the pope in danger ran from Mass and the church.[53]

The Anagni scene, like Boniface's policies (his family, his theocracy), recalls Innocent III. Innocent's calm face against the howling Poli and Boniface's against Sciarra Colonna encircle the century, the young man and the old man from Anagni. But the differences are staggering: Innocent's celebration of the Eucharist and Transubstantiation against Boniface's cynical jokes about the sacrament (or, if the old priest and the butcher and their cohorts were lying, against his running from the celebration of mass); [54] Innocent's definition of orthodoxy at the Fourth Lateran against Boniface's making the Colonna his heretics. One has come in this century of popes from a man who could not know disillusionment to a man whose essence is disillusionment, from shrewd practicality to baroquely flamboyant perversion.

The Popes

There is not a straight line of development from Innocent to Boniface, nor does Boniface proclaim a continuing new type. Gregory X (1271–1276), with his sanctity, his care for elections, his interest in the unity of a holy, spiritual church, his dreams of a reunion of East and West, could hardly be fitted into a simple, essentially deteriorating pattern—although he does represent a related evolving sophistication. Nor, to turn another way, could Clement IV (1265–1268), with his French patriotism, be fitted into an exactly straight line, although his high and horrid Holy Thursday and Ascension Thursday denunciations of his political enemies coincide with some part of the Innocent-Boniface line—nor could Hadrian V (1276), a great diplomat (although Boniface had been a diplomat), nor Celestine V (1294), a queer saint, nor the relatively worthless Martin IV (1281–1285).[55] Alexander IV (1254–1261), in his being "a man placid, sanguine, stocky, humble, jolly, laughing, affable, benign" seems apart. But he was a providing Conti nephew, and he could oppose enemies.[56] Nicholas III's (1277–1280) devotion to the city of Rome and to the Vatican does not particularly predict Boniface; but his care for the Orsini does follow the great dominant line of the century, the subordination of too much, if not all, to enhancing the territorial and castral domination of papal family and faction (or of Angevin domination). The theme of the century's popes is expressed by the cartoon of Nicholas IV (1288–1292) between two columns, his Colonna.[57]

But, although the contrasting personalities of Innocent and Boniface do not describe a constant, general development of papal personality, the going inside from Innocent's Subiaco to Boniface's Lateran does symbolize an important set of developments. Innocent III seems to rule in person as visible pope. He holds his consistories and talks informally to Marlborough and Clipston. By at least the time of Honorius IV (1285–1287) the pope seems much more screened, much more hidden by cardinals and officials. The nature of his real presence has become less secure, less visible. He has, within his growing bureaucracy, been depersonalized. He is in a cloud of unknowing of what he and his successors will do and have done, a cloud of *non obstante* and *non obstantibus* clauses.[58] Something of Boniface VIII's extravagance, and of the extravagance with which his personal behavior was imagined and depicted, must be a frantic, hopeless reaction to bureaucratization and depersonalization—an expression in some ways parallel to the contemporary development of the northern sentimental domestic family (like Henry III's) and of the loud personal religion of the friars.

There is a long change in the position of the personal leader in government from Gregory VII to Boniface VIII or Clement V—or in another government, from William the Conqueror to Edward I or Edward II. It is a change that must have been extremely unnerving to its royal and papal participants who in general can have been no more than subsconsciously aware of it. The heroic age, the period from William and Gregory to Becket, and perhaps even to Innocent and Francis, had clearly passed by Boniface's time; a new sort of community as well as a new sort of government had developed. Sometimes one can watch the painful elements of change, as, for example, when one sees the inadequate personal family of Henry II in England fumblingly doing the jobs appropriate to impersonal institutions. But the changes which were destructive to human leaders were not limited to those on the governmental crust of society. In important ways the thirteenth was a century in which warriors were no longer warriors, monks were no longer monks, and perhaps men were no longer men. Boniface VIII can be seen as a man desperately afraid of losing his manhood. This context lends a bitter resonance to those scandalizing icons of Boniface which he and those who wished to please or thank him set up during his lifetime in Rome, Anagni, Orvieto, Florence, and Bologna.

The pope, the really important pope, was in very many ways not a man at all. In some ways the personal scent may merely make, and have made, a quite impersonal office more palatable, more bearable, more comprehensible to the human mind. In this sense Innocent's brightness and Boniface's extravagance are distractions. If Boniface's perceived behavior is a rebellion against this condition, it is a rebellion against a continuing and old (necessary) malady as well as against a depressing (and necessary) change.

The fraility of the man Boniface is more devastatingly exposed if he is looked at in another guise, as Benedetto Caetani, holder of some part of the Caetani tenements. This Boniface or Benedetto is clearly exposed in a long list of his acquired holdings in Selvamolle which was compiled in the year 1294.[59] Benedetto becomes a figure through which a large number of tenements and rents, woods, vines, and fields of grain are integrated; he becomes a socially binding structure. All landlords are a convenience to, a function of, their holdings. Their personal and familial greed and ambitions allow the natural economy to express itself. So Benedetto Caetani, who thinks he is advancing his family as a free agent, can be seen as a tool, a machine, which allows diverse sorts of holdings to be worked cooperatively, or, at a slightly more human level, which expresses needed social unity for a group of diverse tenants. In this, as cardinal and then as pope, Boniface was like

The Popes

other landlords. But in this as in all else Boniface was extreme; he is thought as pope to have spent half a million florins buying land for his family.

Boniface, like other popes, moved around. Honorius IV, for example, although his place was Santa Sabina, spent the summers of both 1285 and 1286 in Tivoli; Clement IV spent about one year of his papacy in Perugia and two and one-half years in Tivoli.[60] Even as they moved, the popes were surrounded by a huge court, a huge body of officials. This bigness bursts from the papal accounts of money paid for curial supplies. In 1299 and 1300 great sums were given to Giovanni Zotti (or Zotto or Zatto) and other suppliers for sea-fish, for meat, for wood, for portage, for spices, for sugar, for saffron, for sweet-water, for candles, for coasters to put under candelabra, for wine, for people who went to Anticoli for water, for lots of paper, for lead and silk for bulls, for wine (forty-six *lire*, eleven *soldi*, ten *denari* in January 1300 for the buttery for wine, Latin and Greek, for the lord chamberlain and his clerks), for sugar and spices for the claret (the mulled wine) of the lord chamberlain, for barley, for medicines. There was a great miscellany of expense: for servants who took care of curial horses; for army wine; for the house of a master of theology; for repairing the house of the auditor of criminal appeals; for a horse to give to the lord chamberlain when he went to Zagarolo; for houses in the Lateran and for Greek clerics in Grottaferrata. Boniface bought curtains for his house in Anagni (forty-five *bracci* of linen for seven *lire*, five *soldi*). In 1302 he bought much stuff, cotton and Irish serge, silk, cloth of gold. He also bought a quart of pearls.[61]

In some of this, perhaps in spending so much for the curtains at Anagni, Boniface may at first seem personally extravagant. But the office of pope was interpreted as a ceremonial office. One of the pope's quite impersonal duties was to be splendid. He had to enhance with splendor the central office of the church. He had to carry on a sacred and beautifully embellished, and so encouraging, masque at the very center of the world's attentions. If the function of Rome can be seen as being a setting for processions, the function of the pope can certainly be seen as being the center of processions. This is apparent in the "barbarously involved" descriptions of Cardinal Jacopo Stefaneschi.[62] It is apparent in the expenditures of a discreet pope like Honorius IV as well as in those of Boniface VIII.

In 1285 Honorius needed to, felt the need to, improve and finish the façade of Santa Maria in Turribus, a Vatican church or chapel important in the coronation ceremony. His accounts survive in detail, the detail of thirty *denari* for the eggs for tempering the colors for painting the panels of the "frontispiece." Honorius paid for nails and planks and joists and builders and carpenters. He also paid men for working on a fountain, and he bought a key. He bought a horn lantern for lighting up Saint Peter's at night, and he paid Master Alberto for grafting trees in the Vatican garden. He paid twenty *soldi* to the house barber for his year's salary. He paid the expenses required for the twelve canons of Saint Peter's annual carrying of the Veronica (the *sudarium*) to Santo Spirito. He paid money to each canon (fifty *soldi* all together) and, besides, five *lire* (five times the barber's annual salary) for the canons' lunch, and also seventeen *lire* for the preceptor and brothers of Santo Spirito for feeding the poor according to Santo Spirito's privilege.[63] It may seem, the division of the money spent, a queer distribution of the available money; the succoring of Christ's poor seems muffled in the canons' lunch. But it is revealing of a rather general interpretation of curial and ecclesiastical responsibilities. Under Honorius, under Boniface, the central offices of the church had to be big and grandly kept.

The curia in Rome and in Perugia and the other papal towns was surrounded by shoemakers, wine-merchants, doctors, lawyers, notaries, scribes, cooks, taverners, water-carriers, candlemakers, a huge, bulky, engulfing establishment—the people who crowded the lodgings in the streets around Saint Peter's. Boniface VIII, however he felt about it, short of a real reformation of the idea of the papacy, had to move surrounded by this swarm of people. At the center of them, and here Boniface did show his resentment articulately, he was surrounded by his cardinals, his sacred chapter, without whose advice and counsel he should have done nothing important. And, although the specific political moments of curial history which are best known may show the thirteenth-century pope either acting independently of his cardinals or controlling them, casual curial references—the story in the chronicle, talking of something else, like what a saint did in Rome—show how constantly and normally the thirteenth-century pope was surrounded by his cardinals, eating with them, joking with them, absorbing their ideas, and, in this way at least, being controlled by them. Besides, some of them had made him pope.

Between the far periphery of shoemakers and the eventually purple college at the center of things stood the huge bureaucracy, offices and courts full of administrators and clerks, with common forms, fixed schedules, and set fees. Again, all these things and people hid the personal pope. His personality could only with the greatest effort and concentration of violence affect them. The pope was a lead seal, a tiara

The Popes

stand, not a man. This, again, should help explain the behavior of Boniface VIII.

It would, however, be unfair and misleading to conclude a discussion of the thirteenth-century popes with a view of their position so unflattering to it and so unlike their own. It would be better to look again at two of them, the omnipresent Nicholas III as he looked to men who saw him, and perhaps the greatest, Innocent IV, as he looks to us. We ought to try to see what a pope was like.

Nicholas III (1277–1280), Giangaetano Orsini, was a man and a pope of whom his contemporaries were very much aware. He was before people's eyes, talked about, and remembered. One of the continuators of Martin of Poland's Chronicle, a man writing, perhaps in the fourteenth century, short, paragraph-length lives of thirteenth-century popes, summarizes Nicholas thus:

Nicholas, by nation a Roman and an Orsini, was elected in the palazzo in Viterbo, and sat for two years, eight months and twenty-eight days, and he lingered (*cessavit*) for six months. He made one creation of cardinals in which he made eight new cardinals of whom many were his own relatives. In secular matters he was extremely prudent. He was a lover of the religious orders and particularly of the friars minor, of whom he was earlier the lord. In a little time he did a lot. He provided many bishops to various places. He enlarged the papal palace at Saint Peter's and he made there a very big garden.

A second continuator begins similarly (except in the count of days, the lingering, and the reference to nation), but after the talk of friars minor he continues:

He very much loved his relatives. Through Fra Latino, his nephew, the (cardinal) bishop of Ostia, he made peace in Florence between the Guelfs and the Ghibellines. In his time an earthquake toppled many castles in Tuscany and the Romagna with great injury to men. He died at Soriano not far from Viterbo and was buried at Rome in the church of Saint Peter's, in the chapel of Saint Nicholas which he himself had constructed. They say he would not have had his equal in the world, if he had been without relatives whom he too much favored.

This is what a condensed Nicholas III looked like.[64]

After the interval of Celestine IV's brief reign, Innocent IV (1243–1254) succeeded the second Conti pope, Gregory IX, the savage old warrior who had fought the Emperor Frederick II and who had, in his youth, felt himself, perhaps condescendingly, peculiarly capable of understanding and adjusting to their own interests the impulses of the enthusiastic Franciscans. Gregory IX, rather tragically, overestimated his own sensitivity, his own perceptiveness, and even his own intelligence. This was not the case with Innocent IV. He appreciated himself: "Well done, Papa," he said in his *Apparatus* of one of his own decretals (as Professor Gerard Caspary has pointed out). Innocent was sharp (even *furbo*) rather than *dolce,* religious rather than spiritual; and he overestimated nothing. Of thirteenth-century popes, who lived any length of time as pope, Innocent was clearly the most academically intelligent. He was a great writing lawyer, exact, subtle, intricate, brief. A modern historian who knows well Innocent's commentary on the Decretals has called it "terse, shrewd, and brilliant." [65]

Much, but far from all, of Innocent's energy was absorbed by the ugly Hohenstaufen war which he inherited from his predecessor. Innocent's pursuit of the war has been condemned. But without a violent revolution, a reformation, in the way it was assumed that men and popes behave, what could he have done? Frederick II, it should be clearly seen, was a deeply anti-Christian man, who intended through force to crush freedom of religion even in the limited institutional sense that then at best prevailed—and Frederick's flirtation with peripheral Moslems should not obscure this fact. The war with Frederick was not just a war of Italian politics (although, unfortunately, it was certainly also that); had Frederick not threatened, at all, papal temporal power, or had popes had no liking for that sort of power, they would still have been obliged to resist Frederick. The ferocity with which they resisted him, and the manner in which they did, with plowshares turned into swords, contributed to the barbarization, the secularization, the destruction of their office—and Innocent IV was deeply involved in this bloody degradation. But in it he could clearly see himself following Becket's path of duty rather than preparing for the (of course unknown to him) nasty, facile accommodation of men like Clement V. Innocent's mistake was not a petty or incidental one. It would have taken fantastic heroism of imagination to break the pattern which imprisoned him. Within the pattern he behaved very well.

Innocent IV's family, the noble Fieschi of Genoa, and their friends and cousins, have a bad reputation.[66] And Innocent clearly loved "his own." But his nephew Ottobuono, whom he made a cardinal, can have caused him no shame. Ottobuono can be seen as the man who as legate was most helpful in saving England from the wreck of its civil war. He promised, as Hadrian V, to be an impressive although not unpolitical pope. Men from Innocent's curia who were moved into provincial bishoprics were capable and could prove themselves real men of religion, as did, for example, his corrector Thomas, sent to Rieti as bishop in 1252, or as did his domestic chaplain Federigo Visconti, archbishop-elect of Pisa from 1254. Innocent was, moreover, sensitive to the intricate damaging difficulties of the provincial church, as his register and his recorded acts show. He was aware of sanctity, and its practical uses, as in his quick (and antiheretical) canonization of Peter Martyr or in his sending the saintly ascetic Franciscan, Rainaldo, to Rieti before Thomas, trying enthusiasm as a model of reform before he resorted to the more mundane expedient of an efficient and personally proved administrator.[67] Innocent's biography by Niccolò da Calvi is not a deeply revealing work, but its Innocent is not narrow, not merely an academic figure or a political one.[68] Every source shows Innocent to be violently, intellectually alive, active, and virtuous with a public and ruling virtue.

Innocent's limitations are thus the more striking. He was imprisoned within an office that could only be superficial in its concept of reformation and only occasionally more than conventional or less than formal in its interpretation of Christianity. Innocent himself was swathed in administrative bureaucracy. His chancery is particularly observable to us: it has been studied as other chanceries have not.[69] He gave it life, perhaps, as few men could have, but it too was a conventional machine molding his acts to a respectable pattern.

What did this brilliant imprisoned creature have to do with Rome? Not in the end very much. He did not have the roots and ties of Roman relatives. He did not rule Rome in any very serious way— when present in Rome, he can be seen as one of Brancaleone's attributes (and responsibilites), and a commercial one. He was not, moreover, much in Rome.

The pope as pope (and Innocent IV is a good example of a very papal pope) was not necessarily of very particular Roman significance, even in a century as relatively full of Roman popes as the thirteenth. He could use the papacy and implement its latent powers, as Innocent III and Nicholas III did, to rule the city, at least in part, if his local connections and the attendant circumstances were very good and his personality was very strong. As ruler of the papal states he was a temporary local magnate of peculiar importance, if not always of great strength.[70] He could, as Alexander IV reputedly did, demand the rights of a Roman citizen. Although without doubt (except perhaps in the presence of the emperor or the king of Naples) when he was in Rome he was Rome's greatest man, although he is the figure from thirteenth-century Rome who can be placed most sharply and personally before our eyes, and although he was Rome's great patron, the center of its greatest court, upon whom its wealth depended, he is also strangely insignificant and alien.

Seen from the other side, all looks different. Seen as the Roman of his year, when, as frequently, he was a Roman, who had attained the tiara, he was the year's great success. And his family will live in riches in the palazzi Orsini or Caetani forever.

CHAPTER

V

The Natural Family

I‌N A WONDERFULLY EFFECTIVE
passage, Gregorovius, approaching his description of Brancaleone's attempts to smash the power of the Roman nobles, talks of the nature and behavior of those nobles. "The curse of the city lay not in the turbulent spirit of the democracy but in the lawless nature of the feudal nobles . . . they sat entrenched within fortified monuments, as it were in quarters, warring daily with one another from motives of revenge or ambition, and mocking at the Capitol." [1] As often, Gregorovius gives a brilliant piece of the truth. The Roman nobles were warriors in their noble ruins; their nobility needed and, in fact, made ruins. It is fair to see these nobles, one should see them, in their towers "built and fashioned amid brawls and tumult" from which they could "hurl stones on one another with the savage rage of uncouth Lapithae." One should see them in their multitudinous towers, their forests of towers, before Brancaleone supposedly destroyed over 140 of them at mid-century, or again in the sort of fortifications and towers and ruins of towers that Cardinal Giacomo Savelli, thinking that he was perhaps about to die, but in fact about to become pope, parcelled out in his will.[2] These noble Romans were brutal, bloody, avaricious clusters of men and women, organized in "families," bent upon the destruction of their enemies, upon conquest and the acquisition of riches, and perhaps upon enjoying the pure delights of passionate disorder. They were, in all this, the enemies, at least superficially, of civil discipline, of peace, and of civilization.

But much of the interest of these people lies in the fact that some of them were so queerly and flamboyantly not only this—that this does not at all sum them up. They were also other very contradictory things. Even the Orsini can hardly be made to fit this picture. Their acquisition of property was too smooth, planned, intelligent. They seem more like the great financial families of the nineteenth century than like Gregorovius's nineteenth-century picture of romantic riot. The point of complexity, of contradiction within this pattern of nobility and

The Natural Family

family life, is made with harsh clarity if Gregorovius's generalized picture of sordid disorder is cast against a scene from the life of one of the members of probably the most notoriously disorderly and, at the same time, greatest of all thirteenth-century Roman families, the Colonna —the Colonna, who in one of their toughest representatives, Sciarra, supposedly found Nogarest queasy and weak when he balked at the killing of Boniface VIII.

The scene that makes the contrast is from one of the biographies of Margherita Colonna. Margherita was the saintly Colonna girl, the sister of Giovanni Colonna, repeatedly a Roman senator, and of Giacomo (or Jacopo) Colonna, a cardinal and one of the great enemies of Boniface VIII. Margherita herself, whose life and the nature of whose sanctity it is important to know in trying to grasp the nature of this family and type of family, was a "saint," was after her death considered one by the people in exactly the areas of Colonna power, in the stretch of territory from Subiaco to Palestrina and from Palestrina to Anagni.[3] She was very much a family saint; and her sanctity had flowered, in life, on Colonna hills.

The "life" from which the scene is taken is probably Colonna too, thought, on respectable grounds by the historian who knows it best, to have been written by Giovanni Colonna, the senator himself, by the father of wild Sciarra. It is thus odd to find in the "life" the scene that recalls, as no other thirteenth-century scene does, the dialogue between Augustine and Monica at Ostia. Augustine and Monica, it will be remembered, had gone to Ostia, the port of Rome, in 387 on what they thought was their way back to Africa, although Monica was in fact to die in Ostia. Mother and son, tired from traveling, and blockaded by civil war, had stopped and had taken a house in a quiet place in the port and on the Tiber, as they waited to take ship. One day mother and son leaned from a window of their house, from a window from which, as Augustine says, they could see the internal garden. They were alone and they talked together; and their colloquy lifted them through earthly things until together they reached the spirit. It is, in its beauty and seriousness, in context, a dazzling scene. In just such a way (but without Augustinian reference), as Giovanni Colonna tells us, his brother Giacomo and his sister Margherita excited each other to the love of Christ with the dialogue of their sacred conversations. They roused each other, first the brother the sister, then the sister the brother; they ascended toward Christ.[4] (The scene shares with Augustine's that daring composure in the face of "unnatural" eroticism that might frighten little, pious, hagiographic eyes which had never seen and understood, and temporarily at least rejected, the passion of ordinary life.) It is a glowing scene, and its existence ought certainly to make seem more complicated those men in towers who sallied out occasionally to kill.

There is, however, a crucial difference between Augustine and Giovanni. Augustine, who says that he and Monica spoke of the nature of the eternal life of the saints—a subject naturally to pull one heavenward—does not recall exactly much of the matter of discussion, so that each reader may create his own spirituality. Giovanni gives crisp example, and it is rather stunning. Giacomo and Margherita discussed whether or not Thomas, the doubting apostle, had actually put his hand and finger in Christ's wounds. It was a venerable question, and Augustine himself had had an opinion on it. Still, at first glance, it hardly seems the way to heaven. It is, however—the consideration of how exactly Thomas felt Christ—a very thirteenth-century way, the evocation, as if surprising Him, through Scripture, of the living, bleeding Christ, surprising Him in the tangle of disputation, and touching His body. And later a dream or vision brought to Margherita, still living in her holy colloquy, the sound of the apostle's own voice crying, "My Lord and my God," and the sight of the wounds of Christ.[5]

Margaret lived in a world of vision. Christ and His saints moved about the almond tree in her cloister garden in the hills. Her life was transformed by vision; it was a life lived, and increasingly, just at the border of the supernatural. But it was a life which could be, and she as a person could be, charmingly workaday. Driven by the need to see again the face of Christ "which she sought," Margherita came down from Palestrina to Rome to see the Veronica, the great Vatican relic which, one will recall, holds the image of the real face on Veronica's towel. It was, again, a relic which moved this century strongly: Innocent III's processions to Santo Spirito; Matthew Paris's drawings of it in his history; the monk of York dying, crushed by the crowd surrounding it in the Jubilee year. It moved Margherita—as did the substance of Christ in the sacrament at Mass. When Margherita came down from the hills (with her brother's permission) to see this real face of Christ, she stayed at Rome in the house of a holy old woman and semirecluse called Altruda of the Poor, who had taken her odd cognomen from the poor she served and also that she might be one of Christ's poor. It was Altruda's custom to rise early and, with a companion, set off. Having left the house, Altruda went first to Mass at the friars' church, perhaps Aracoeli. She then made her daily holy tour to the doorsteps of the

The Natural Family

saints (*limina visitabat sanctorum*), until noon, or even later, depending upon the length and complexity (*prolixitas*) of the day's itinerary. It was Margherita's delight in staying with Altruda to be her servant, to stay home and take care of the house, to wash the dishes, to sweep with a broom, and to fix lunch so that the holy women could eat when they came home tired. Thus her brother Giovanni could marvel, in telling of it, at the wonder of a Colonna turning domestic—indulging at the same time his century's taste for the paradox of Franciscan denial and his own Colonna pride.[6]

The air of simple domesticity (and work) around Margherita is not limited to this Colonna-glorifying *exemplum*. It is again apparent in another "life" of Margherita, a life written by a follower at Castel San Pietro, later a nun in the convent of San Silvestro in Capite in Rome, perhaps its first abbess and a Colonna relative. About vesper time on Christmas Eve the maidservant Prenestina hurriedly prepared to wash the clothes. Under the supervision of the authoress, she too quickly prepared the water, which was not yet hot, for the *lasceva* (*lisscivium, lisciva*), a mixture of supposedly boiling water and fine ashes used in washing clothes in Italy. Margherita, in the distance, knew, and smilingly she told the authoress that she knew that the water was not yet even tepid and that Prenestina should stop pouring.[7]

Margherita's life was a short one. She was born in about 1255, and she died on December 30, 1280. Her father, Oddone Colonna, died when she was about two years old.[8] Her mother, who was also named Margherita and who was the sister of Matteo Rosso Orsini and a pious woman, died when Margherita was probably a little over ten years old. Margherita was left to the tender mercies of her older brothers—and their mercies turned out in fact to be tender. In Margherita's late teens there was talk of an appropriate marriage for her, but it was not forced.

This seems the pattern not just of the Colonna but of the higher Roman nobility in general in this era, that is to have restricted but not forced the marriages of their children. The pressure was negative. All those noble Roman cardinals and nuns must be the expression of, at the very least, a subconscious attempt to hoard the wealth (at a level sufficient for serious investment) and increase it through office rather than to let it be subdivided and spent among a great brood of descendants. There was danger because, in spite of the evil reputation of the Roman climate, Roman noblewomen seem to have been fecund and Roman babies to have lived (certainly Orsini ones did). Sometimes Roman families conserved property and gathered money; they did not always force themselves to marry and (once they had arrived) align themselves with other powerful families. This is another argument (in a constant disputation) that the *provisino* and not the *spada* ruled Rome—flowing wine, not flowing blood. It also argues that Margherita's brothers' gentle attitude about her marriage was not necessarily impractical.

Margherita had, according to her brother's "life," always lived in peculiarly sweet piety. She had long wanted to retire from the world. Her desire to become some sort of nun was permitted and confirmed, according to Giovanni, by a vision that appeared to her brother Giacomo. Giacomo's vision, like Margherita's, is fresh, concrete, and homely. The anthropomorphic quality of the visions described in the "life" has seemed to its editor evidence of its author's not being a cleric, but actually the way in which mysticism, the supernatural, and the exact physical places—almond trees, and laundry—meet in these visions defines nicely the tone of Margherita's sanctity and also the spiritual space in which the three Colonnas lived. This spirituality of dishwashing is thirteenth-century Rome.

Giacomo was at the university of Bologna studying law. It was the feast of Saint Margaret (a saint, as the "life" says, then already long written in the catalog of saints—but, alas, no longer), his sister's name day, July 20.[9] Giacomo had eaten with the friars that day. He had solemnized the feast of the blessed virgin Margaret and poured out his devotion to her. He had fed the vagrants and pilgrims in his own hospice. He had done all that should be done. It was midday. Everyone else had gone off to siesta. Giacomo sat alone in the garden. He read the *legenda* of the virgin whose feast he had been celebrating. He had come to that passage which says, "Come virgin Margaret to the peace of your Christ," when things about him seemed to change; it was as if Giacomo were at a play (*quasi ad spectaculum*). As Giacomo looked he saw his sister carried by two guardian angels, between them, in the air. Wondering, he watched the scene and examined it as long and carefully as he could. He was able to look until he could recognize his sister quite clearly. After some time, an hour, the vision moved itself so that Giacomo's view of it was destroyed by the house's being between it and him. So, following the vision, he ran right through the house and out into the piazza before the house. Again he saw the vision and recognized his sister, but then the vision fled away into the sky. This happened, it ought to be remembered, in the heat of a Bologna early afternoon in July.

At first Giacomo thought that the vision meant that Margherita had been carried from this earthly prison. Later he realized that it was

The Natural Family

a sign of her rightful passing out of the bondage of secular Egypt into the Israel of the religious life. So in spite of the marital coaxings, for others, by some Dominican friars, Margherita was allowed to retire from the world.[10] In 1273 she withdrew to Castel San Pietro above Palestrina. Around the beginning of 1274 she began to dress herself like a Claress. Later she got permission to go to Assisi from the Franciscan minister general, Jerome of Ascoli (who was to become in turn cardinal bishop of Palestrina and Pope Nicholas IV—the Colonna's pope). Margherita tried setting up a convent away from home, but then returned to Colonna-Palestrina Castel San Pietro. Thence on occasion Margherita descended to Rome or to charitable work among the sick at Zagarolo or the leprous at Poli.

In 1278 and 1280 Margherita had two destructive, but elevating visions, of the wounded Christ and of Christ-pilgrim, after the first of which she became ill, and after the second moved toward death. They were visions which changed to greater mystical spirituality the tenor of her parting life. (In the year of her earlier vision, in a parallel movement, her Orsini cousin, Pope Nicholas III, made her brother Giacomo a cardinal—supposedly to stave off the pretensions of the powerful Annibaldi, the Orsini's prime enemies of the moment.) On December 20, 1280, Margherita fell into a fever. Ten days later she died.[11] Her death was like Francis's. She wanted to be on the bare ground under the free sky. She wanted the open air, or at least the opened window. The life this death ended, like the lives of all saints, is twisted by the pattern of interested, and at the same time conventional, hagiography. Admitting possible distortion, it brings an odd picture to mind of one of those noble Roman families living not only in country seats like Palestrina, but in Rome in those notorious towers, living like "uncouth Lapithae." This hagiographic Giacomo is just recognizable as the same man as that despicable young Giacomo, unworthily raised to the cardinalate, described by Boniface VIII.[12]

Margherita was, nevertheless, a Colonna saint and thought to be one. All accounts of her stress her noble birth. After her death she attracted Colonna gifts and Colonna nuns. Her Colonna-ness is emphasized rather humorously in the story of one of her miracles.[13] A man, sick and disgruntled, with an extremely painful ulcer, sent a messenger for his doctor. He discovered that the doctor was off climbing the hill above Palestrina to Margherita's tomb with Giovanni Colonna. The neglected patient waited all day and grew furious with frustration, and he said to his wife that it was an absurd disgrace. "Look here," he said, "it's embarrassing and degrading the way the lords of Colonna overdo it—absurd, disgraceful—trying to violently shoot their sister into the choir of saints." His wife was scandalized by his speech. He himself changed his attitude—and was cured. (Saints could do more than doctors could.) A Colonna propagandist was able to tell the story.

It would be absurdly wrong to think that Margherita was only something made up because the family needed her—so that a great family who had had no popes (unless the ancient, disreputable, and distantly connected Tusculan popes are counted) should have instead a saint to decorate it, to give it spiritual respectability, a sort of Christian or rational unity, a banner. It would certainly be wrong to think that Margherita was only that, crudely and consciously conceived, but it would be unwise not to notice that she was also that.

The complex of holdings and cousins, lands, rents, marrriages, and arms, of country towns and fortifications and parts of Rome ("astride the routes leading out of Rome to the east and southeast"),[14] which Margherita embellished, was a very grand one, and very potent. In some ways at least the Colonna can be considered the first family of thirteenth-century Rome.[15]

The family, reasonably securely connected with the old family of Tusculum, appears in modern form at the beginning of the twelfth century. By mid-twelfth century it was rich and powerful. It maintained its power and riches through the thirteenth century without the aid of an actual family pope. Its great early-century cardinal was a mixed blessing because of his "Ghibelline" connections, his imprisonment, his incurring the enmity of the "Guelf" tyrant Matteo Rosso Orsini. This Colonna air of continuing old wealth and old strength (heavily city as well as country) is unlike, or at least more striking than, that of any other major Roman thirteenth-century family: the Savelli, whom a pope raised from obscurity; or the Orsini, aggressively *arriviste,* not ignoble in background (carrying old Boveschi properties) but forced forward by a recent nepotistic pope and by a continuingly increasing and aggressive brood; or the Conti, rustic nobility given power in Rome by local marriages and a family pope and his successors; or the Annibaldi, of distinguished antecedents, perhaps, but brought to a position of real power by a Conti marriage; or the Caetani who arrived late and were obviously connected with one pope; or the slowly falling families, Pierleone and Frangipane; or the increasingly "feudal" and distant Vico or, their other name, "de Prefectis."

A division of Colonna property was arranged between Oddone di Giordano, the senator, and Pietro di Oddone in February 1252.[16] Oddone's share, the share of the city Colonna, Margherita's immediate

The Natural Family

family, included the country towns and tenements of Palestrina, Zagarolo, Capranica, and Colonna (from which it is thought that the family had acquired its name). In the city Oddone took a stretch of land from the Mausoleum of Augustus near the Tiber, through Montecitorio (now the governmental center, Piazza Colonna, part of Rome), to the Colonna center by the church of the Twelve Apostles. It was a broad but not solid band of holdings, cutting through the center of the walled city and encompassing the northeastern part of heavily inhabited thirteenth-century Rome, expressed in another way, from the Quirinal to the river through San Marcello and the area of San Silvestro.

The early thirteenth-century Colonna cardinal was Giovanni. Innocent III made him cardinal priest of Santa Prassede in 1212 (and his difficult tenure ended in 1245). In the early 1220s this Giovanni returned from the East and is supposed to have brought with him the greatest of Colonna relics (at least before Margherita), the column which was believed to be that at which Christ had been scourged. (The column was, and is, a beautiful little object, although it is very difficult to understand how a man could have been scourged at it.) It was placed in Santa Prassede. It was, and is, much venerated; and it did, and does, bring the Colonna honor. Giovanni Colonna brought his family more honor if, in fact, as it is believed, it was he who influenced Innocent III to accept the Franciscans; and he also brought to his family an extremely early connection with the order with which they would become increasingly involved, in its primitive period, the period to which later Colonna protégés believed they were trying to return. Cardinal Giovanni was powerful and long lived. He would have been an unqualifiedly good investment for his house except for his support of Frederick II.

Giovanni's importance in the church was reflected by the importance of Oddone Colonna as senator in the city. Oddone was an enemy of Brancaleone's, and he married an Orsini. This Oddone was Margherita's father, and Giovanni's and Giacomo's; and his temporary patching over of the Orsini enmity bore fruit when Giacomo was made cardinal deacon of Santa Maria in Via Lata by Nicholas III. Giacomo's cardinalate, buttressed by the existence of a Colonna senator, his brother Giovanni, and soon by a second Colonna cardinal, Pietro (Giacomo's nephew, Giovanni's son), pressed the Colonna to new eminence, until the time of their disastrous wars with Boniface VIII. These wars were, however, only in part and temporarily disastrous. By the pontificate of John XXII the Colonna were in a very strong position again. Those Colonna cardinal's hats and senator's costumes which are visible in their mosaics had returned. When Cardinal Pietro and Giacomo Sciarra died in 1326 and 1327, they were the powerful leaders of a toughly reestablished and heavily possessed family (which had rewarded in new prosperity its adherents in old adversity), prepared for the romantic nobility that the mid-fourteenth-century family, the enemies of Cola, would find in the eyes of Petrarch and, after him, of Gregorovius.[17]

The Colonna war with Boniface VIII brought the family a unique dramatic intensity; the two cardinals falling from grace, being taken as prisoners to Tivoli, escaping; the rough vigor of Sciarra at Anagni. Boniface was extraordinarily violent and vile in his hate-filled effort to break them completely, so that he broke the monument and tampered with the cult of their saintly virgin Margherita, broke even Cardinal Giacomo's little shrine in Colonna Rome. Boniface is said, too, to have sown salt in Palestrina. He not only broke the Colonna hold on Colonna properties, but he actually tore Colonna buildings down. Again, they were his heretics.

Cardinal Pietro Colonna looked back in complaint, after Boniface's whirlwind was spent, and wrote a sort of elegy, a financially interested elegy, on Colonna losses: [18]

The city of Palestrina itself was turned into a ruin, with its noble and ancient palaces and its great and solemn temple dedicated to the Blessed Virgin. They were buildings built by Julius Caesar, the emperor, whose city Palestrina was in antiquity. There was a grand staircase made of the noblest marble, splendid and large, so that one could ride up on one's horse to palace and temple. And the steps of the staircase were more than one hundred. The palace built by Caesar had the shape of the letter C after the first letter of his name. The temple attached to the palace was sumptuously and nobly built in the manner of Santa Maria Rotunda [the Pantheon] in the City. [Boniface and his people destroyed] these and all the other palaces and buildings of Palestrina, and the very ancient walls of Saracenic work, made of great squared stones. No skill or money can rebuild them ever; they were too old and too noble.

Pietro also wrote of the *castro* of Monte Prenestino (Castel San Pietro) "where there had been the noblest *rocca,* and the prettiest palace, and walls of the most ancient Saracenic work" and "the noble church of San Pietro [Margherita's place] which was once a monastery," all destroyed, with the other palaces and houses of the place "of which there were about two hundred." He wrote of Colonna with its beautiful *rocca,* its towers and palaces, its pretty buildings and tenants' houses. It was such a rich and imposing complex, Cardinal Pietro

The Natural Family

wrote, that Cardinal Giovanni had spent on its fabric, not counting the tenants' houses, something between twenty and thirty thousand *lire* of the senate in the currency current in his day.

Pietro wrote also of Rome. In Rome Boniface had destroyed, at Montecitorio, at the "Fornitariis," so that, "alas, in the City of Rome, the Colonna live in strange houses" (*immo habitant Columpnenses in domibus alienis in Urbe*).

Pietro's is a lament with an odd taste. Perhaps it is a Colonna taste. Perhaps it is the peculiar taste of Pietro himself. He was a patron of hospitals (San Giacomo in Augusta, San Salvatore at the Lateran) and the owner of a great library with a Chrysostom, an Avicenna, a Maimonides, a Seneca, a pseudo-Turpin, a huge and various collection. Pietro's lament is queerly romantic and nostalgic, all the lovely stonework fallen into ruin, the toppled towers, the pretty fortresses, the great squared stones, Caesar's C—in it the sense of the Anglo-Saxon poet of "The Ruin" and that of the writer of the *Mirabilia* seem to join and distort or enhance both the Colonna hatred for the heretic pope and also Colonna family greed. The most violent clash between an old and a new family in this part of Roman history produced this odd epitaph. From the sad ruin that it describes not only the Colonna but the Caetani recovered.

In looking at this first Roman family, the Colonna, one should be convinced of its reality, of its quick peculiarity, that it lived. One should not, however, assume that it was typical; nor should one assume that it was a clearly and neatly defined unit. One, of course, speaks reasonably of a group of people related through blood and sharing a common name as the family Colonna. One can, with no difficulty, go further. In the later thirteenth century Cardinal Giacomo and Cardinal Pietro and their siblings, two generations, with their attendant households and followers, formed a relatively tight, bound unit, the center of a recognizable and active faction. But there were also various sorts of divisions within the Colonna family. Not all Colonna were Roman Colonna. There were the Colonna of Gallicano, of Genazzano, of Riofreddo. There were Colonna, important, central ones (although not normally in line for major inheritance), who supported Boniface VIII, were "loyal to him," and received confiscated possessions of the "treacherous" Colonna. One of Boniface's charges against the deprived cardinals was that they defrauded other Colonna of their goods. A preserved letter from Boniface to Landulfo Colonna in 1297 tries to rouse him against the Colonna faction; and Landulfo's will of 1300 speaks of "the very holy father, my lord benefactor, Boniface." [19]

The neatness of the family Colonna is sharply questioned in another direction. In blood, the Colonna of the great generation (Giovanni, Giacomo, Margherita) were as much Orsini as Colonna. Their Orsini grandfather was as much their grandfather as their Colonna grandfather was; and Giacomo's very name, like Margherita's, was probably an Orsini legacy. It has been suggested that Margherita's Franciscan piety was stimulated by the Orsini Franciscan connection —her grandfather, for example, left his best bed and two feather coverlets to Franciscans; she is thought particularly to have been influenced by Matteo's house, where Francis had been when little Giangaetano (Nicholas III), Matteo's son, played the games of a child.[20] In other circumstances, with the same blood relationships, Colonna-Orsini could have seemed one house. In fact the contemporary Franciscan chronicler Salimbene complained of Nicholas III's nepotism in appointing his own relatives cardinals partly because of the appointment of Giacomo Colonna.[21] At the outset one should realize that the great Roman family, although a—perhaps the—dominant feature of the Roman landscape, was not a clearly defined thing. Interest, economics, politics, even sentiment could weld various parts of it together, or split its parts, in genetically quite unexpected ways. The family name represents a specific, but extended, nexus within the broad network of common family relationships.

Some of the lessons and problems of the Colonna are pointed up or clarified by returning for a moment to the Savelli, whom we have seen before as the family of popes Honorius III and Honorius IV. Giacomo Savelli's will of 1279, composed while he was cardinal deacon of Santa Maria in Cosmedin, is a particularly helpful document.[22] First it shows the massive extent of the Roman properties of one of the central Savelli inheritances: houses, towers, and ruins of towers which stretched out in two directions from the church of Santa Maria de Gradellis (Santa Maria Egiziaca, the temple wrongly called *Fortuna Virile*), on one side to the Theater of Marcellus and on the other to the Rocca Savelli on the Aventine—a strong (but, again, not solid) stretch of territory along the Tiber from the Ponte Fabricio to the Ponte Aventino or Sublicio, from a point across from the island to a point across from the Porta Portese, along the bank of the Aventine and the Marmorata. (The areas of family dominance within the city, areas in which family properties were concentrated but which were not held at all exclusively by the family, recall in their structure the composition of family faction in which actual members of the blood-family dominated a total group which included members not of the kin-

The Natural Family

183

group.) This neighborhood did not include all the urban areas that were, or were to be, connected with the Savelli; the Vicolo Savelli suggests a specific outrider—the family, after all, remained important for a long time, from its at least partial replacement of the Pierleoni in the thirteenth century, through its own weakening in the fifteenth century, until the death of the last of the Savelli of the direct line in 1721, when the papal marshal's conclave key, Savelli for four and one-half centuries, passed to Chigi strangers.[23]

Giacomo Savelli made his brother Pandulfo and his nephew Luca, son of Giovanni, his heirs. He tried to exclude from inheritance female collateral descendants and their lines. He wanted the inheritance to remain relatively compact and, in name, Savelli. But should male collateral descendants fail, he contemplated descent through the female line. Giacomo's will gives a helpfully neat, although still flexible, definition of family; it emphasizes name and the male line. It thus helps explain by analogy the division of Colonna and Orsini. This definition and explanation must, however, be treated cautiously. One of the most noticeable of the "Savelli" in the thirteenth century is the prelate Giovanni Boccamazzi, a nephew and cousin not called Savelli. The Savelli were obviously the patron family closest to Boccamazzi or the close family best able and most willing to help him—in them, for him (and he was one of many, it will be remembered), power and proximity met most profitably.

The point about the Colonna that the Savelli most clearly reiterate is in quite another direction. The Savelli say, as the Colonna do, that if you take off the cover of a great familial collection of urban towers, houses, ruins, and fortifications, and look at the actual people inside, what you find is not necessarily morally squalid. It is not necessarily depressing—it is not in the case of this family whom Stefaneschi called in his list of processing nobles "the mild (*mitis*) Savelli," whose pious devotion to city and church is physically recalled in the great bell (now in the Vatican museum) which Pandulfo, for the redemption of his soul, had cast for Saint Peter's in 1289 by Guidotto Pisano and his son Andrea.[24]

It has seemed wise to introduce direct consideration of the greater Roman families with the eccentric Colonna and the mild Savelli so that the families, as a group, should not be morally or intellectually underestimated or treated too mechanically. But for a more central view, to see a family behaving more characteristically (or in a way, at least, that would generally seem more characteristic), one ought to turn to the Orsini.

27. Deposition, from Tivoli, see p. 265.

28.   Cloister of Saint Paul's outside the Walls, see p. 216.

29. Bronze Saint Peter
within Saint Peter's
in the Vatican,
see p. 67.

30. Fresco from Moses
cycle at Grottaferrata,
see p. 278.

31. The Cross of San Tommaso dei Cenci, from the
Aracoeli, now in the Palazzo Venezia, see p. 285.

32.   Head of the Crucified Christ from the
Cross of San Tommaso dei Cenci, see pp. 86 and 285.

33. Grieving Virgin from the Cross
of San Tommaso dei Cenci, see p. 285.

34. Detail of the Cross
of San Tommaso dei Cenci,
cloth, see p. 285.

35. Feet of Saint John,
from the Cross of
San Tommaso dei Cenci,
see p. 285.

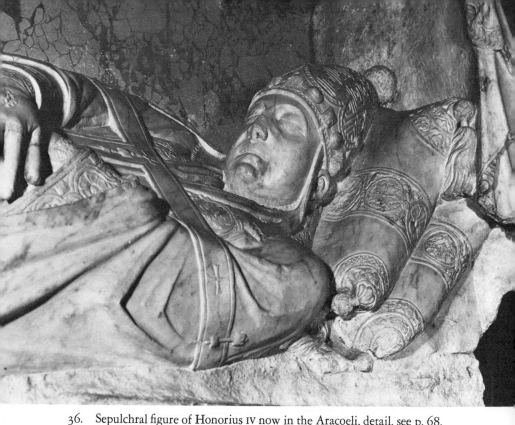

36. Sepulchral figure of Honorius IV now in the Aracoeli, detail, see p. 68.

37. Boniface VIII, recumbent, in the Grotte of Saint Peter's in the Vatican, see p. 269.

BONIFACIVS
PAPA · VIII
✝

38.   Boniface VIII
at the Lateran, see p. 162.

39.   Boniface VIII in the
Musei Civico at Bologna,
see p. 162.

40. Boniface VIII in the
Grotte of Saint Peter's
in the Vatican, see p. 162.

41. Boniface VIII in the
Museo del Duomo at
Florence, see p. 162.

42. Ciborium of Saint Paul's outside the Walls,
Gothic cupola with rose and Adam and Eve,
see p. 216.

43. Campanile of San Silvestro
in Capite, see p. 231.

44. Glorious Virgin
of Sant'Angelo in Pescheria,
restored, detail, see p. 44.

45. Tre Fontane, series of frescos,
second storey, facing fields,
see p. 67.

46. San Georgio in Velabio,
see pp. 195-196.

47. Torre delle Milizie during restoration, see p. 190.

48. Medieval House near Santa Cecilia in Trastevere, see p. 38.

Angels from the Last Judgment by Cavallini, fresco in Santa Cecilia in Trastevere, see p. 67.

50. Sant'Angelo in Pescheria during restoration, see pp. 39-46.

On October 21, 1286, the noble lord Matteo Rosso Orsini (the son of the tyrant Matteo Rosso Orsini and himself senatorial vicar of Rome under Nicholas III) sat on a hill in Rome now called Monte Giordano (the Orsini hill, named for a Giordano Orsini, although in 1286 still called by its old name, Monte di Giovanni di Roncione, where by 1262 there was already an Orsini loggia as well as more formidable buildings) and sold tenements to his nephews.[25] They were tenements, some of them, bordering on those of Pandulfo Savelli and his nephew Luca. There were four Orsini nephews, Napoleone, Matteo, Orso, and Giovanni; they were sons of Rainaldo, Matteo Orso's brother and the founder of the Monterotondo branch of the family. The nephews paid 12,500 florins for the collection of holdings which they bought. The sale took place in the house where lodged the cardinal deacon of Sant'Eustachio, Giordano Orsini (in a personal family place some distance from Giordano's titular church). Giordano was, as the document says, Matteo Rosso's brother and the paternal uncle of the four nephews. The proceedings were watched by a little assembly of witnesses: the cardinal himself; the archdeacon of Messina; Master Giovanni of Zagarolo, the cardinal's chamberlain; Matteo of Naples, a papal auditor and chaplain; Master Angelo of Trastevere and two other members of the cardinal's household; a Fra Pantaleone of the city; a clerk, another witness, and one of those medical doctors (in this case, Giovanni di Pietro di Messer Luca) who seem to appear everywhere in large Roman households in this era of fear and fever.

In the scene on Monte Giordano a great Roman family assembles to do business. The business has to do with extensive tenements and lots of money, both of which are being kept within the larger family. The scene is both an example of family solidarity and of family fragmentation: an uncle as an individual sells to his nephews as a group. A cardinal uncle (as familiar a figure as the attendant physician, less widespread but infinitely more powerful and desirable) looks on as a presumably impartial but interested observer-arbiter. ("My uncle is a cardinal" is the device of strength in thirteenth-century Rome; "My uncle is the pope" is the device of victory.)

In 1300, in a scene strikingly similar to that on Monte Giordano, in the hospice of Cardinal Francesco Orsini, the cardinal presides over and witnesses the division of properties between his nephews. In document after document, groups of Orsini are seen dividing, acquiring, reassigning property, concerned in family property action.

The ordinariness of members of a family's acting together cannot be overstressed, but neither can the ordinariness of two or more com-

The Natural Family

posite groups within a family's dealing with each other as they might with a group from another family. The point is made perfectly clear by an act of 1293 in which the sisters in a sibling group (Orsini neighbors rather than Orsini) sell a share of property next to a garden of San Salvatore in Campo, in the heart of Rome to their brothers. (And this act recalls the constantly observable ambiguity in the importance of women in family and city; they were exposed, consulted, valuable, active—but they were not really like men.) The complexity of intrafamily action, and something of its possible, occasional motivation, is suggested in 1262 when Orsini nephews alienate property to a cardinal uncle, "for benefits received or hoped to be received," and in 1267 when the nobleman, lord Napoleone di Matteo Rosso, citizen of Rome, with the consent of two brothers and two nephews, gives to his brother the cardinal, and future pope, Giangaetano, a fourth of Rocca Galeria, freely, not because of fear or threat, "not in consideration of his church or cardinalate, but because of the love and brotherhood he felt for him." [26]

The Orsini themselves are the success story of thirteenth-century Rome. They are, in a way, its composite Horatio Alger. (The qualification is due to the fact that they did not start from nothing—the Boboneschi Boboni were a Roman noble family—and to the fact that there is no implication of homely virtue in their story.) From the point of view of a Colonna sympathizer, the Orsini seem to have been single-minded. Their aim was the Lateran (or in their case the Vatican, the papal residence and church in their part of Rome) and the various profits it represented and brought. Occasionally, it should be noted, even by the least sympathetic observer, Orsini wavered from the straight and narrow path and cast covert glances at religious enthusiasm. Latino Malabranca, who generally seems, as a cousin and nephew, to have elevated the Orsini connection, is, for example, the cardinal whose inspired speech, according to Stefaneschi, suggested Peter Celestine as a candidate for the papacy as well as the legate, who, according to Salimbene, so disturbed the ladies of Bologna by restraining their fashion.[27] Like the families of tyrants elsewhere, the Orsini were involved with the Franciscans, and they were great patrons of the arts. Nicholas III, although very Orsini, was an essentially capable, beneficent, and serious pope. In general, however, the Orsini were distinguished by their straightforward aggressiveness, by their success, and by their clannishness.

The rise of the Orsini was due to the fact that Celestine III, the

Bobone uncle of the first actual generation of prominent Orsini, advanced them to power. He himself had been advanced by the great Alexander III. Celestine was, by the time of his papacy, old in importance (and perhaps virtue), and he had come from a family known in Rome. The family was forcefully pressed forward by Matteo Rosso, the "Guelf" dictator (the beast of the Septizonium). It was made both more powerful and more respectable by Matteo Rosso's nephew, Cardinal Matteo Rosso. This clerical Matteo Rosso (in what is thought to have been the fashion for pretentious young Roman nobles in mid-century) had gone to Paris for theology, and perhaps (in, again, what is thought to have been the fashion) to Bologna for law. Himself a theologian, brother of a theologian, connected with Aquinas, the scion of a house that favored Francis, chaplain to Innocent IV, cardinal, he can be made to sound very good; and he has been (as the hero of a short essay). Besides giving his family another vote in the sacred college, Cardinal Matteo Rosso gave it better tone—not perhaps so high as his brother's, but more directly pertinent and noticeably helpful.

But the triumph of the Orsini was sealed by the election in 1277 of Giangaetano Orsini, son of the tyrant Matteo Rosso, as Pope Nicholas III. Giangaetano's election crowned a long, ambitious, and respectable career—he had been a cardinal since 1244.[28] Nicholas III's favoring his family is, as is by now surely clear, one of the most familiar of late thirteenth-century Roman themes. The Franciscan chronicler Salimbene drew attention to it in talking of the cluster of four cardinals who were related to Nicholas. Salimbene did not really speak badly of three of them: Matteo Rosso, Giacomo Colonna, or the Dominican Latino Malabranca. Salimbene achieved his effect with the fourth, Giordano Orsini, the man on the hill:

The fourth cardinal among the pope's relatives was lord Giordano, the pope's brother, a man of little learning, rather like a layman. But because flesh and blood told, he made him cardinal. So he made those four cardinals from his own family (*parentela*). So he built, as other Roman popes had, Sion on his own blood, of which Miceas III says, "who builds Sion on blood and Jerusalem on iniquity." I believe certainly on my conscience, I am persuaded, that there are a thousand friars minor of the Order of Saint Francis (of which order I am a minor and inferior friar) who are better suited to be raised to the cardinalate both by reason of their learning and their saintly lives, than many of those who by reason of their being related to the Roman pontiff are elevated.

The Natural Family

And Dante has Nicholas III say in hell:

> *e veramente fui figliuol dell'orsa*
> *cupido sì, per avanzar gli orsatti . . .*
> [And I was truly a son of the bear
> So anxious was I to advance the little bears . . .]

There is even a story, not very well authenticated, that Nicholas planned an Orsini kingdom in the north. He certainly used his relatives in governing the temporalities.[29]

The family brought to consolidated and obvious power during Nicholas's papacy was a family based in, and centered around, obvious geographical places. These places were not all in Rome. There were Orsini places in the *campagna* and in the Regno—and the Orsini grew into the great Vico estate around Lake Bracciano. But the Orsini places in Rome were very noticeable: the Theater of Pompey, the Campo dei Fiori, and the Piazza dei Satiri, Monte Giordano, Castel Sant'Angelo, the area around Saint Peter's—a strong (but not solid) band through the center of Rome. In talking of the Orsini's holding Boniface VIII their prisoner after Anagni, the chronicler Ferreto speaks of the complete way in which the Orsini controlled the area around Saint Peter's "with their buildings all around and close to the church in a place where they were really powerfully dominant." The church was in their district; it was their temple. Nicholas III glorified it, and he was carried from Soriano (which he had taken from the "heretic" Guastapane) to be buried in it. The Orsini consolidated their power over Saint Peter's in the early fourteenth century when the chapter was packed with members of the family and Cardinal Napoleone Orsini was its archpriest.[30]

In the Orsini, the machine of the great Roman family is exposed. In the thirteenth century it is the most efficient, and perhaps least charming, of the group of major families, but in many ways all of them are alike and have recognizable common characteristics and ways of behavior. The scene on Monte Giordano is a scene that, with a change of names, would have seemed natural in any of their similar seats. In his poetic celebration of the coronation of Boniface VIII at the end of the century, Stefaneschi identified a cluster of really great Roman families —those of the Roman nobles who press forward and lead the brilliant band of ritual merrymakers: the Orsini, the Colonna, the Savelli, the Stefaneschi, the Conti, the Annibaldi, and the Vico prefect of the city. The Stefaneschi, a Trastevere family like Innocent II's declined Papareschi, although noble and well connected, may have been mentioned rather than other families of similar pretensions—Normanni, Sant'Eustachio, Suburra, Papazurri, others—because the author was himself a Stefaneschi. But surely more than any of the other comparable families the Stefaneschi dominated a distinct area, that around Santa Maria in Trastevere; and as their patronage shows they were very rich. No other family's reputation is so concentratedly one of *mecenatismo*, of the lavish and informed patronage of the arts. The importance of Bertoldo Stefaneschi as a patron of real taste is apparent not only in the Cavallini apse mosaics in Santa Maria in Trastevere but also in Bertoldo's exquisite floor-tomb within the same church. Cardinal Jacopo Stefaneschi, the poet, is, however, the central figure in Stefaneschi patronage. His largesse was spread broadly but with discrimination, as one can see in his selection and description of Giotto, the artist who created the Stefaneschi altarpiece and "Navicella" for the Vatican. Jacopo was also a patron of illuminated manuscripts, including three manuscripts of his own works, of which one volume of his *Opus Metricum* contains a very elegant painting of the beginning of the procession from Saint Peter's at Boniface's coronation, the procession which allowed him to include the name Stefaneschi among those of the leading Roman nobles. One seems to be watching the very process of the Stefaneschi (already intermarried with the Orsini) buying and giving, writing and patronizing themselves into the highest charmed circle.

Over and over again, however, one must force oneself to realize that the circle was not a closed one; it was not isolated. The families in Stefaneschi's list were closely connected with lots of other families in lots of ways of which marriage is only the most obvious. Families in and out of the list cluster together in institutions like the Orsini's Saint Peter's where one finds Frangipane, Ponte, Annibaldi, Cenci, and a Suburra chaplain.[31] In any court, moreover, two families of great thirteenth-century importance must, moreover, be added to the Stefaneschi list: the popular, or relatively popular, Capocci, and, after the coronation, Boniface VIII's own Caetani—and, in addition, the fading Pierleoni, and the Frangipane, apparent in fascinating and, at least in its widow Giacomina (or Jacoba or Giacoma), sometimes splendid decay. Of all these families, two others ought to be looked at a little more closely, two nonpapal families whose pattern is centered boldly around the figure of a single man, the Annibaldi and the Capocci.

The Annibaldi had a more pretentious background than some of their thirteenth-century peers, but not one so pretentious as they in the Roman manner have claimed. Descended presumably not from the pretended great Carthaginian Hannibal but rather from a German

The Natural Family

189

count named Annaldus or Anwaldus, they increased their importance with the marriage of Annibale Annibaldi to a daughter of a count of Tusculum in the later eleventh century. Their actual political importance in Rome rose, in large part it seems, from the marriage of Annibaldo senior to a sister of Innocent III (although the marriage must also have seemed advantageous to the Conti as they moved into Rome). The Annibaldi rose on the back of Conti greatness by strongly supporting, with Pandulfo of the Suburra, Riccardo Conti and the pope in their violent Roman disputes at the beginning of the thirteenth century.

As the Savelli physically replaced the declining Pierleoni, the Annibaldi replaced the Frangipane. By the early 1240s the Annibaldi had succeeded in getting Enrico and Giacomo Frangipane to grant them half of the Colosseum—a transaction connected with Frederick II's involvement with Roman tenures, monuments, and families. Innocent IV later quashed the grant, claiming the Colosseum as a papal fief. In the 1250s, nevertheless, the Annibaldi controlled the Colosseum; it was still their area of Rome in 1303 when Boniface VIII supposedly wanted to flee to them (or their neutrality) from his Orsini-Vatican captivity. Even now the tower of the Annibaldi points out that this was once their part of Rome. The Annibaldi also came to control the great Torre delle Milizie until at the turn of the century it was transferred to the Caetani (a family more successfully and permanently, although interruptedly, acquisitive). The Annibaldi towers controlled the roads that led toward their possessions in the country, in the *campagna*, and the Marittima. They were a town and country family with major "feudal" holdings. In this they were typical as they were in their rise from relative obscurity through a familial connection with the papacy.[32] And one ought not in thirteenth-century Rome to lose a sense of a surrounding, in many ways conventional, "feudal" world, in which Vico, Frangipane, Conti, Colonna, Orsini, Caetani play other, seignorial and feudal, games.

The Annibaldi were also typical in breaking, in the thirteenth century, into branches. The Annibaldi de Urbe became the Annibaldi de Coliseo and the Annibaldi de Militiis. They, again typically, were divided between opposing parties, "Guelfs" and "Ghibellines"— perhaps a kind of family insurance, perhaps in extension of family dispute. Typically, too, members of the family held important positions in both the Roman church and the Roman state. Annibaldo senior was senator in 1221 and again in 1231 (and it is he who is thought, with Gregory IX, to have been responsible for a sordid secular inquisition against heretics). This Annibaldo was probably the father of Cardinal Riccardo Annibaldi, the repeated Annibaldi candidate for the papacy and a great building member of his family.[33]

Riccardo Annibaldi was by blood as much Conti as Annibaldi; Innocent III was his uncle. In 1238 Riccardo was created cardinal deacon of Sant'Angelo in Pescheria by his realtive Gregory IX to replace a cardinal who had been promoted to the see of Porto, and when Riccardo died, toward the end of 1276, he was still cardinal deacon of Sant'Angelo. From 1240 to 1248 Riccardo was papal rector of Campagna and Marittima, an important shrieval job. He was in general charge of a large area of papal territory with fiscal, military, and judicial responsibilities. He adjudicated problems, as in the case of one town's (Piperno's) trying to build an aqueduct in another's (Terracina's) territory. He mobilized one town (Terracina) to capture back a castle from another (Setino). He settled a dispute between the nobles and commons of another town (Anagni), and demanded that a further town (Alatri) keep peace with another (Ferentino). These jobs were important, difficult (particularly in Gregory IX's reign), and, potentially at least, very profitable. (The sort of profit that might be reaped is suggested by, for example, the determination of the council of Perugia in 1276 to offer to its hoped "defenders" at the papal court, six cardinals, eight hundred florins apiece to help Perugia in a dispute with Gubbio; of the cardinals suggested, two were Savelli, one Orsini, one Fieschi, and one was Riccardo.)

When Gregory IX's successor, Innocent IV, went north to Lyons in 1244, he left a triumvirate of ruling cardinals in Italy; one was Riccardo. After 1249, when the dynamic young cardinal Pietro Capocci took over real control of papal Italian government, Riccardo was made vicar of the city of Rome; and the vicariate, with its considerable local powers, particularly in arranging the settlements of disputes, could be a profitable office for any holder who wanted to make it so. Riccardo was also then much involved with the "English party" (perhaps because of earlier contacts with the English court) in the advancement of Richard of Cornwall to the throne of Sicily through the senatorship of Rome; this proved profitable to his family in English livings and probably to the religious order he protected in houses.[34]

Under Alexander IV, Innocent's successor, and, like Innocent III and Gregory IX, a Conti relative, Riccardo was at full strength politically (although he may have been malarial), and he remained politically strong under Alexander's French successor.[35] From Alexander, Riccardo received the office of archpriest of Saint Peter's Basilica, a

The Natural Family

church which had a profitable trail of dependents: forty-four churches, fourteen hospitals, two monasteries, and twelve castles. Over all this the archpriest ruled with a relatively free hand. He dispensed the livings (although propriety as least would demand consultation with the canons). Riccardo seems to have been insufficiently discreet in appropriating the funds of the basilica. Successive popes tried to preserve appropriate amounts of the income of the church for the upkeep of the basilica itself: Alexander IV asked that a fourth be so used; under Urban IV, in 1263, the church's accounts were destroyed or lost, and in 1267 Clement IV was complaining about the use of the basilica's income.

In the early 1260s the French pope Urban IV decided to enlarge the college of cardinals. He appointed fourteen new cardinals, balancing two new Orsini against two new Annibaldi. In these appointments he recognized the central familial dispute in Rome in his period. He also created a strong literally Annibaldi node of cardinals in the curia —and the nephews accepted the position of being their uncle's followers. Meanwhile, having given up the hopeless English cause, Riccardo had become a central figure in the fight for Angevin supremacy, a policy of initial success. Riccardo had become a very political cardinal.

It is possible to find in Riccardo, as his biographer has, consistent principles—specifically, support of the freedom of Rome and of the Holy See. But these are principles that evidently allow much freedom of action—sufficient certainly to anger the actual holder of the Holy See, as they did Clement IV. They are also principles under which one can support one's family. Riccardo's strong opposition to Brancaleone might seem more disinterested if Brancaleone had not been replaced by Emanuele da Brescia, a man who was too much an Annibaldi tool to survive in the senatorship. The Annibaldi faction had about it a peculiar "aristocratic" and "clerical" quality that is hard to define and is best seen perhaps in its opposition to Brancaleone, for which the support of certain concepts of "freedom," particularly class and ecclesiastical freedom, seem a natural and pleasing face, or explanation. Riccardo's principles, in other words, do not seem too secure. (His switch to the Angevin party is notorious.) His family's principles were, of course, less secure than his own: Riccardo received as a specially courteous present from Charles of Anjou the life of a nephew who had followed Corradino to his defeat—that he might not die as had two other nephews, who had been hanged by Brancaleone upon his recovery of power.

The central theme of Annibaldi policy was its war with the Orsini. In conclave after conclave Giangaetano Orsini and Riccardo Annibaldi are thought to have kept each other from becoming pope. Repeatedly they were chosen electors to select the new pope. In the end Giangaetano won because Riccardo died first. Under Giangaetano, as Nicholas III, the Orsini got what the Annibaldi had tried to get; but the Annibaldi were not completely crushed. After Nicholas's death the Annibaldi instigated a rebellion against the Orsini in Viterbo and the patrimony, and Orsini holdings were raided, and Orsini cardinals were captured. When at the Sicilian Vespers the wheel of fortune turned for families all over Italy, and the French eventually died on the Roman Capitol (and four years after the Vespers Cardinal Ancher, Urban IV's favored and hated nephew—son some said—seems to have been murdered in his church of Santa Prassede), the Orsini rose again. Eventually, within the year 1284, the Orsini made peace with the then "penitent" Annibaldi (but not with the town Viterbo, which had implemented Annibaldi policy). By then the Orsini were changing their antipathies to other familial centers of rising strength. In fact one of Riccardo's nieces married an Orsini, although the two families were certainly not considered allies by the time of Boniface VIII's death.[36]

In the end Riccardo Annibaldi had failed to win the papacy and thus to carry his family to the pinnacle of Roman power. He had, however, over many years increased its prominence, its riches, and its power. He is an example of how a cardinal could help his kin. He made them the acquaintances not only of popes, but of kings and saints. In the period of Angevin-Annibaldi friendship there were banquets and receptions, royal entertainments at an Annibaldi country house—the same house at which Riccardo entertained Thomas Aquinas; Riccardo could play the intellectual host with Thomas and the gallant host with Charles. But particularly, in the great years, there was land and money.[37] In explaining Riccardo's influence, his biographer has said that it was due to his potent personality, to his constancy and his experience over a long period of time, but also to the "great personal riches that he had accumulated."[38] The riches were enhanced by a number of podestàships: Riccardo was podestà of Velletri in 1270; his family ruled in Terracina, Alatri, Todi, Viterbo. The riches were connected, perhaps primarily, with investment in land and associated rents, tenements, fortifications, and men. In 1271, for example, Riccardo, in what comes to seem a rather characteristic thirteenth-century sort of deal, with the Conti-Poli affair in mind, got Rocca di

The Natural Family

Papa from the poor Frangipane. Frangipane heirs had pledged Rocca di Papa as a surety for a loan of 6,500 *lire*. The Frangipane lost their surety and the Annibaldi fortified the Rocca.

More than twenty years after Riccardo's death, when the money (perhaps 200,000 florins) was collected for the great Caetani purchase of Annibaldi properties, including Sermoneta and country around Ninfa, eighty *salme* of gold and silver and brass, it was said, were carried for it through the countryside on a train of mules and horses—mules and horses not safe, however, from armed (Colonna) attack.[39] The eighty *salme* of precious metals should be seen as a century's accomplishment, what a shrewd, hard-working cardinal deacon could help his family achieve. The gold is also a token of the family's defeat, after death had defeated their cardinal leader. It stood for the triumph of the Caetani over the Annibaldi, the reduction of the Annibaldi, as in the case of the Torre delle Milizie, to a more purely financial power, less an armed, territorial power, at a time when arms could still be important against gold, as, ironically, the Colonna attack clearly showed. But the Annibaldi achievement, the gold which was carried on the mules and the horses, was not presented to themselves or their contemporaries completely unaccompanied by some connection with what is more conventionally considered religion. The Annibaldi did not produce, as the Colonna did, a saint, but they did have a religious order (and a scholar).

Riccardo Annibaldi, more than anyone else, formed the Augustinian Hermits into the order they became. It was, in a way, his order until his death. Although he was very helpful to it, he was sometimes heavy-handed in dealing with it. In 1244 he became corrector of the Tuscan Hermits. In the years between 1244 and 1256 he collected his Hermits and similar fragmentary and half-formed orders into the order of Augustinian Hermits—brought their ecstatic indiscipline and enthusiasm into a conventional pattern—and got for them a Roman seat in the church of Santa Maria del Popolo. From 1257 until his death he was their Cardinal Protector. He was to the Augustinians something of what the Orsini were to the more conventional wing of the Franciscans. A difference lay in the fact that although his order was less brilliant and powerful than theirs, he controlled it more completely—he was Ugolino and Orsini combined. But although Riccardo's connection with his order was more personal than the Orsini connection, it does not seem to have been a bond of spiritual sympathy of the sort that tied the Colonna to the less conventional wing of the Franciscans. Among the chaplains within Riccardo's richly diverse household there seems to have been no Augustinian Hermit (although there was a Capocci).[40] If, however, every great Roman collection of wealth, land, and blood needed, in order for it to become a really significant Roman family, a talisman of religion, the talisman the Annibaldi held was the order of Hermits.

Pietro Capocci, cardinal deacon of San Giorgio in Velabro, brought his family (or was instrumental in bringing it) another sort of religious talisman. The Capocci talisman was a miracle. During the night of September 26, 1256, it was later said, the horses in the Capocci stables, behind what is now the gallery of the Piazza Colonna, were noisily restless. Their grooms, who went to see what was bothering them, found that the well from which the horses were watered was overflowing. On its surface floated, unnaturally, a tile or slate, and on the slate was a painting of the Madonna. The grooms were unable to snatch the painting from the water. They called the cardinal. The cardinal prayed and retrieved the painting. It was immediately much revered; but the incident was investigated thoroughly by Pope Alexander IV before he permitted the formal veneration of the image. Alexander approved the miracle, and Pietro Capocci built a chapel in which the painting might be kept and honored. All this is legend, but legend that winds its way around historical fact. The legend's truth seems undoubted by the pious faithful who still crowd around to drink little glasses of water from the sacred well in the miraculous Madonna's chapel in the church of Santa Maria in Via.

The miracle was not Pietro's only method (if in fact it was his) of making more ecclesiastically respectable and attractive, at its end, an aggressively secular cardinal's career, and one that moved out of a recently antipapal and "popular" family. Pietro is remembered in his own church of San Giorgio as well as in Santa Maria in Via. There Pietro was one of a group of thirteenth-century patrons. On the architrave of the porch the rebuilding of "Stephanus ex Stella" is still celebrated in monumental thirteenth-century letters. Inside the church the munificence of Jacopo Stefaneschi is recalled in the damaged but still existing Cavallini frescos of the apse. To the altar's right, on an inscribed stone dated 1259, after the Roman manner of putting the details of gifts and holdings in stone (as on the wall of San Nicola in Carcere), is the record of Pietro's inalienable legacy to the church of land next to its tower (called "Advallaran" or in a place called "Vallaran," that is, the Vallaranum) and of the chapter's oath on May 20 always to celebrate Pietro's anniversary. Pietro's memorial concludes, before its final date, with the injunction: "Whoever reads this pray for

The Natural Family

him." Beneath it another stone tells that in 1621 Vincenzo Capocci of the cardinal's same *gens* had the stone placed in a safer and more honorable place to protect it from the ravages of time. Pietro was a patron elsewhere. He founded a hospital, Sant'Andrea (Sant'Antonio Abbate), near Santa Maria Maggiore, where the contemporary inscription of his cardinal executors remains above the door of what is now the church of the Russian college; and he was a patron of Santa Prassede and Santi Silvestro and Martino (San Martino ai Monti)—churches in the Capocci part of Rome.[41]

The Capocci, like the Annibaldi, were deeply involved in the Conti disputes at the beginning of the century. Cardinal Pietro's father, Giovanni, was a leader in the violent attacks upon Riccardo Conti and Innocent III's other adherents, perhaps because some of them (Riccardo, Pandulfo of the Suburra, even the Annibaldi) were contenders for power in parts of Rome close to the Capocci's own. Although the Capocci had outlying possessions, as in the Trevi and Colonna near Santa Maria in Via, the center of their holdings was an area stretching from the Suburra to the Biberatica, between Santa Maria Maggiore and the Forum, the area of the towers that still bear their name, and of Santa Prassede and San Martino with the towers at the Santa Maria Maggiore edge of their area. Capocci patronage spread to chapels in Santa Maria Maggiore (where Pietro was buried in the Capocci chapel of Santa Barbara) and Santi Apostoli. A Fra Pietro Capocci is remembered in the book of anniversaries of Santo Spirito in Sassia.[42] By the end of the thirteenth century Capocci holdings spread out along the Salaria, the Nomentana, and the Tiburtina. Giovanni Capocci himself was twice podestà of Perugia—Capocci fame spread even farther than their Roman holdings. The Capocci were allied through marriage with the Savelli and Colonna and, by the end of the century, with the Orsini. Pietro's brother Ottone was, like Pietro, favored by Innocent IV Fieschi.

In 1267, as captain of the people (and the tale recalls the recurring popular tone of the Capocci, their not being a "noble" family), Angelo Capocci was able to lead the commune against the "Guelf" forces, to recall the policies of Giovanni Capocci less violently but more effectively. The strength of his family's position had been increased by the family investment in clerical careers, those of Ottone and especially Pietro. Pietro's relatively early death, in 1259 when he himself was about fifty-nine, kept him from being, as he might have been, the sort of constant contender for the papacy that Riccardo Annibaldi and Giangaetano Orsini were. As in the case of the Annibaldi, but not the Orsini, the Capocci gains from having in their house a powerful cardinal were not extended and solidified by their also having a pope.

The actual career of Pietro Capocci, favored by his Savelli relative Honorius III and advanced by his friend Innocent IV, was, in spite of its clerical name, remarkably secular. Pietro was the bright young man who was promoted to the sacred college in 1244 by Innocent IV and in 1249 replaced the triumvirate of cardinals (including Riccardo Annibaldi) then ruling papal Italy. As early as 1231 Pietro had acted as a papal military commander; and in his own period of secular ascendancy, from 1249 to 1251, his main job (the focus of his general rule) was to organize the papal effort at military invasion and conquest of the Regno. Apart from his use in military affairs, Pietro was used particularly as a legate—an office that demanded a very general sort of competence. Competent, efficient if not successful, a dispassionate official, he seems to have been; there is nothing to indicate that there was anything particularly religious or ecclesiastical about him beyond the area of his service, the office of his superior. His position, though, in Savelli and Fieschi confidences undoubtedly allowed him to retrieve the losses (at least in reputation) of his father's opposition to the Conti. He, like the Orsini cardinals and Riccardo Annibaldi, was one of those important counters or chessmen who moved their families' positions forward in the thirteenth century—without dissipating their augmented properties among (at least known and legitimate) progeny. Pietro did, however, disperse some reasonably large part of his property in ecclesiastical bequests, a sort of conspicuous consumption (or lack of consumption), which added to the status and thus, in various ways (like marriageability, potential leadership), to the power of the family.

Abundant physical evidence of family self-consciousness remains from the later thirteenth century. Family arms (*stemme*) decorate and identify walls and tombs and, in records, cloth. Stefaneschi *stemme* proclaim family importance in Santa Maria in Trastevere, Caetani in Santa Maria in Cosmedin, Savelli in Santa Maria in Aracoeli; and in the warehouse of the Vatican treasury lay two pieces of red Lucchese cloth decorated with griffins and the Savelli arms.[43] But, in spite of this clearly expressed self-consciousness, it must repeatedly be emphasized that it is artificial to separate sharply any small band of Roman families (like, again, that of Stefaneschi's list, slightly augmented, or of Saba Malaspina's list of leading "Guelf" and "Ghibelline" nobles) from their neighbors who were only slightly less rich or prominent or powerful. If there was, in the thirteenth century, any clear institutional

The Natural Family

division beyond that of offices actually held (and perhaps set positions at formal occasions, in processions), between one group of families and the rest, it has been well hidden (except for the occasional use of honorific titles: proconsul, *magnificus, baro, miles*) from our eyes. What evidence remains is ambiguous, like Stefaneschi himself, with his great nobles pressing forward from the throng of the procession which passed other noble houses, or like the antimagnate legislation of 1305, which specifies the opposed families presumably because a caste name was insufficient. The mid-fourteenth century statutes, on the other hand, talk of recognizable classes of commoners, knights, and barons or magnates, recognizable enough for graduated fines; and they suggest particularly clearly an expected horsed, game-playing, office-holding caste. Still the central point is clear enough: the class structure of Rome in the thirteenth century, particularly in its upper reaches, was still in a dynamic stage and was flexible and responsive. Any family which got for itself a cardinal, or of course a pope, could move with its new wealth and power to a position as elevated as its local and provincial properties could sustain.

The closeness of other families to those which have been treated here as of the first magnitude (with whom they married) is quickly illustrated by the family of Pandulfo of the Suburra. The importance and wealth of Pandulfo at the beginning of the century is made perfectly clear by his position and that of his properties in the Conti disputes. Pandulfo's senatorial eminence and his role of connecting the Conti with the city were the official and active expressions of his towered wealth so important both to attack and to defend. At the end of the century Pandulfo's descendant Filippo di Matteo di Pandulfo de Suburra was still very rich. In his will of 1301 he left five hundred florins (from the two thousand florins to be gotten from houses for specific legacies) to the paupers of the hospital of San Matteo on the Merulana.[44] In June 1270, when, as a plaque in the Pantheon portico proclaims, a new bell tower was erected there, Filippo's relative, another Pandulfo de Suburra, was archpriest of Santa Maria Rotunda, the Pantheon—and in the same decade a daughter of Pandulfo de Suburra was abbess of San Cosimato in Trastevere.[45] If the Pantheon's Pandulfo had been moved from archpriest to cardinal priest of a Roman titulus, the name of the Suburra would almost surely have joined those in Stefaneschi's list. Of course cardinalates were not accidental gifts; the Suburra position in the 1270s may well not have been strong enough to demand elevation. If it was strong enough to suggest elevation, if some surface accident of appointment had brought elevation, with reasonable luck and ability Pandulfo might have done again for his family what Riccardo Annibaldi and Pietro Capocci did for theirs.

Among those magnates proscribed in 1305 were the family of Oddone of Sant'Eustachio, a family prominent in the twelfth century and still, in the thirteenth century, clustered around the Roman church which presumably gave it its name (although a family hard to distinguish definitively from other people who are spoken of as coming from its area of Rome, of Sant'Eustachio). The Sant'Eustachio can be seen in 1279 involved in arranging, selling, and thus dividing a major inheritance. Oddone di Angelo de Sant'Eustachio is seen selling for ten thousand *lire* to his brother Paolo his share of a number of holdings including those he held together with his brothers, Paolo and Matteo. (This Matteo is just possibly the Franciscan priest Matteo de Sant'Eustachio who appears in friar's habit on an incised floor tomb now on the wall of the north aisle of the Aracoeli.)

The inheritance was various. There were holdings out beyond the Milvian bridge, next to a place called "Three Columns," and a number of houses in the center of town in the *rioni* Sant'Eustachio and Santa Maria in Aquiro. These latter included: half the house in which Matteo lived, with half its cupola and arch, adjoining land of Sant'Eustachio, and with public roads on two of its sides; half the house in which Oddone lived (like Matteo in the *rione* San'Eustachio) again with roads on two sides; a great house with tower and arch in the *rione* Santa Maria in Aquiro bounded by a road, the wall of the Pantheon, and the property of Compagio di Giovanni Lucidi; another house inhabited by the shoemaker Thomaraccio di Giacomo (or possibly a man named Calzolario) with the public road on one side, the portico of the Pantheon on another, and the Via di Portico (the Via di Rotondo). The Sant'Eustachio had half a palazzo and a number of other half houses twisting around the property of their neighbors, like Oddone Bruno the notary, and their tenants, like Angelo di Pietro di Stefano, and around the Pantheon and Sant'Eustachio, in the center of Rome.[46] They were an old family, rich, and in the early fourteenth century they could still seem dangerous. The Sant'Eustachio have been paired with the Normanni in twelfth-century position, and the Normanni appear with them in the list of 1305.

The family di Giudice, "de Iudice," was an imposing and senatorial one, important around Sant'Angelo in Pescheria and elsewhere, which, related to the Orsini, attached itself through marriage to the Conti and eventually to the Frangipane. Another family, the Parenzi, has a peculiar distinction. The Pietro di Parenzio who was senator in 1245 was the great-grandson of Saint Pietro di Parenzio, the podestà of Orvieto murdered there by heretics on May 21, 1199 (an event which he and Innocent III had half-predicted amid the tears of Pietro's wife and mother)—a nice orthodox saintly-governmental line.[47]

An extensive Roman nobility proclaimed the pope as he rode under their arches from the Vatican to the Lateran on his coronation day, as he passed their palaces, their complexes of houses and towers. Propertied families sat entrenched, as presumably did the Massimi in their immemorial place, although this was one of their dormant centuries. The Margana seem already to be beginning to gather around the area of their tower under the Capitol near the market, the Cenci were in the Giudia. Around them move the almost ghostly but still elegant declining Frangipane, with their exquisite Greek connection: the marriage to an imperial Comnena at Veroli in 1170; the name of the Magnifico Saba Frangipane in the Rocca di Papa sale of 1272; the bequests of Giovanni Frangipane to the monks of San Saba and Grottaferrata before 1266; it is even conceivable that Saint Francis seemed "Greek" to the eyes of this family of the Septizonium—perhaps Francis's friend Giacomina, widow of a Frangipane of the Septizonium, recognized in Francis a return to Anthony Abbot—perhaps Frederick II looked "Greek" to them, both as a rejection of currently dominant Roman power and wealth and as hope for change.[48]

The thirteenth-century Frangipane, although they survived and would continue to survive indefinitely, always bring to mind the mutability of Roman family fortunes, their rise and fall. A number of specific tenements follow the rise and fall of individual families. San Felice Circeo (where Circe sang) on the coast south of Rome, on the point between Terracina and what is now Sabaudia, was in the hands of the Frangipane in the twelfth century. At the beginning of the thirteenth century Innocent III gave it to his Annibaldi connections. After a mixed century, during which it was sometimes in Templar hands, it was sold in November 1301. Riccardo de Militiis degli Annibaldi sold San Felice and its appurtenances to Pietro II Caetani (the pope's nephew, the really aggressive figure in building the Caetani empire) for twenty thousand florins.[49] To that date, 1301, the names Frangipane, Conti-Annibaldi, Caetani represent two centuries' turn of the wheel of fortune.

The name of this Riccardo's branch of the Annibaldi, "of the Torre delle Milizie," recalls a similar pattern. Built on antique foundations in its present form under Gregory IX (a Conti), the *torre* seems to be on or near the site of fortifications of Pandulfo of the Suburra, a Conti supporter, and to have been later in the hands of partisans of Pandulfo. By 1296 it was held by Riccardo Annibaldi; within a few years it was bought by Pietro Caetani. Pietro's purchase of some of the area surrounding it is interestingly recorded. In April 1301 Simone "de Tarquinio," the inquisitor at the Franciscan house at the Aracoeli on the Capitol, in the presence of his associate Fra Gionata "de Tarquinio," sold to Pietro's proctor, for one thousand florins of gold, the possessions of Federicozzo da Roma, in the *rione* Biberatica, in the *contrada* Milizie, next to the houses of the Milizie itself which had formerly been possessions of Riccardo de Militiis (that is, Annibaldi), by the Via "Silicis," by the road from San Basilio (in the house of the Knights of Rhodes) to the Milizie in front of San Salvatore de Divitiis. Federicozzo's property was in Simone's hands because it had been confiscated by the inquisitor Alamanno da Bagnoregio because of Federicozzo's aid to "the schismatic Colonna." Simone's proctor, Crescenzo di Don Angelo Silvestri de Archionibus, actually took Pietro's proctor, Guido di Don Pietro da Orvieto, a knight, down to the property for Crescenzo to induct Guido. Crescenzo seized Guido's right hand and they perambulated the place. Guido went into the houses, buildings, and things sold. He walked through them. He picked up pieces of sod. He touched the stones that were there. And as he walked and picked and touched, Guido said that he wanted to take possession of these things for Pietro—the formalities of taking seisin occurred, the formulaic act was enacted, as these two men and their witnesses (Franciscans, a knight from Orvieto, a banker from Viterbo, the great notary Nicola Novello da Vico) wandered through the ruins of, or by, the Market of Trajan.[50] As the Torre delle Milizie moved so did the Theater of Marcellus, in this case from Pierleoni to Savelli, and so too did the Colosseum, from Frangipane to Annibaldi. These places witness the change of family names, names standing for different eras, but seeking their stability, seeking to anchor their prominence (and to protect the roads which led to their profit-making possessions) in the same sorts of property as their predecessors, in the same stones at the same places (and once, it will be remembered, in the affair Poli-Conti, the same name).

The wheel of fortune was often very quick in its turning. In 1301 Boniface VIII and Pietro Caetani seemed to be carrying all before them. In February 1304, understandably, Pietro was retrenching. He was then selling half a castle south of Rome for thirty thousand florins; and he was selling it to Pietro Frangipane.[51] The replacement of one fam-

The Natural Family

ily by another, however, does not seem necessarily to have implied any lasting enmity between them. In April 1310, in a Savelli fortress in the country, when that *Magnificus vir dominus Johannes de Sabello Romanus proconsul* sold properties to Donna Agnese, who was abbess of the Augustinian nunnery of Santa Maria della Rotonda in Albano, his first witness, a man pretty clearly present to visit him, was that noble Roman Uguicione di Don Sisto Pierleoni.[52] The Pierleoni and Savelli could obviously sit down together—as the Orsini and Colonna could rule together.

Something of the structure of the great Roman family has been revealed, one hopes, through the description of a number of individual families. First of all, the great families of thirteenth-century Rome were all rich. Their riches came from urban and rural land (and other real property) and office and the ability to supply armed men. Each source supplemented the others, helped the base of the other sources of wealth to grow. These families were all either Roman families who moved out to gain possessions in the country surrounding Rome or provincial seignorial and "feudal" families who moved into the city and added to their holdings a Roman base, although the movement is seldom clearly and visibly in a single direction.

Each family had a base in, and dominated, a section or sections of Rome: as the Savelli did the stretch of land from the Theater of Marcellus to Santa Sabina; the Orsini that stretch from Saint Peter's to the Campo dei Fiori, the Theater of Pompey, and the Giudia; the Colonna from the Mausoleum of Augustus to the Dodici Apostoli end of the Quirinal. Each family had consolidated holdings, and scattered ones, in the surrounding country: so the Orsini swamped the area around Viterbo and Bracciano; the Colonna around Palestrina; the Annibaldi around Rocca di Papa, Molara, and Campagnano; the Savelli in the regions around Palombara, Albano, and Ariccia.

The family's city fortifications not only provided it with a place for actual confrontation with its enemies, and security from them, but also guarded the roads that led to its country properties. The country properties could provide, and their money buy, troops; the men in the country could be attached to their lords in a much more conventional "feudal" way than the minor urban members of their faction were. Money, not arms, seems to rule thirteenth-century Rome; but the residual importance of arms is clearly visible at the end of the century, and the Annibaldi's exchange of land and fortresses for money is again, I think, a clear sign of their decline. The sort of income that Roman country properties actually provided in the thirteenth century can be studied and understood in detail in the miscellany of various families' documents preserved in the Caetani archives (and extensively edited). One can see exactly what it meant, for example, to be lord of San Vito in the year 1292, to know what the lord was owed by each of his local tenants, and by them all together.[53] This knowledge defines the term "lord of San Vito"—and on a larger scale the definition of any family's holdings defines, in one very major way, itself.

The great families were families of office. Office made them great, and office—potent like the senatorship, the cardinalate, the papacy, or honorific like the proconsulship, the prefecture, the "chancellorship"—was the natural formal expression of their greatness. Some of their sons and daughters were cut off from marriage, the boys sent like nineteenth-century New Englanders off to the China trade (at least like Robert Acton in *The Europeans*) in their time of mating, sent to the church to gather its treasure and build their houses and fill their churches with the equivalent of chinoiserie and with money. The girls were often sent into convents, sometimes neighborhood convents or faction ones, respectable havens from expensive marriages (when their marriages were not needed) and wastefully disparate descendants (where descendants were not needed). One sees, for example, at the Dominican house of San Sisto in 1331, a collection of Roman noblewomen, remnants of great thirteenth-century houses: a Colonna, two Pierleoni, a de Giudice and her daughter, and three Annibaldi (in an Annibaldi neighborhood and not in the Annibaldi's Augustinian order—but the Annibaldi had had Dominican connections at least since the time of their Dominican scholar Annibaldo in the mid-thirteenth century).[54] The Roman career in ecclesiastical office which brought the family profit is fully illustrated by Riccardo Annibaldi's career. The full, great dignity of office—the expression of the Roman family's doing what is was in a way designed to do, remembering a great Roman past—is finely illustrated by the Savelli pope and the Savelli senator ruling together. Their position could only have been enhanced by a great abbess at their side.

Refraining from marriage could help the family, but so could marriage. Marriage was one of the businesses of the Roman family; and every Roman family showed that it knew the value of marriage. Wealth and prominence were presumably always desirable in a spouse—they are what Margherita Colonna rejected in rejecting marriage. All the great Roman families married other great Roman families. The Caetani, rising to power, married both Conti (once from their own town, Anagni) and Orsini. The Savelli married the Orsini;

The Natural Family

the Annibaldi the Conti; the Capocci the Savelli. The great families were related to one another in a great skein of intermarriage. Sometimes prominent Romans saved the name of mother or grandmother in their own names: so Stefania Rubea, or Rossa, is echoed in many Matteo Rossos in the Orsini family. Romans, however, did not marry only their own, as the heated quest for Margherita Aldobrandesca's inheritance makes clear.[55] The names of many husbands and wives are unknown; theoretically each may have been anyone (and one must always remember that there was no caste rule about marriage). But there is actually nothing to suggest that Romans married at all erratically. Marriage, like office and land, was a road to increased wealth, prominence, and respectability; and it was taken seriously.

The seriousness of marriage and its close connection with property is stressed, in a 1274 document recalling acts of 1248, by the involvement of notable witnesses (Normanni, Frangipane, Surdi, Annibaldi, probably Suburra) and distinguished senators (Pietro Annibaldi and Angelo Malabranca) in the arrangements and effects of a dower settlement made between Giacomo di Napoleone Orsini and Matteo Annibaldi for the marriage of Giacomo's son Fortebraccio and Matteo's daughter Golitia—an important attempt at alliance. And it is stressed in a more specific way by Martin IV (or his chancery or its petitioners) in a papal dispensation for a marriage between an Annibaldi and a Conti (Giovanni and Perna) who were too closely related to marry freely. Martin granted the dispensation, he said, because he believed that if these houses (*domus*) were joined by a blending of blood it would be advantageous in many ways (*multipliciter*) to both houses and to the factions surrounding them (*ac partibus circumvicinis*) —an interesting eruption of party into the papal register.[56]

Increased respectability (and possibility of office) was also achieved through education, perhaps particularly at a certain stage in each family's rise, perhaps particularly at a certain stage in the rise of the university. Matteo Rosso Orsini, Giacomo Colonna, Latino Malabranca, Annibaldo Annibaldi went off to universities. Like education, religious connections enhanced respectability and perhaps gave the family, at least at times, a sort of banner of unity: the saint, the order, the miracle, the connection with Francis.

The great families divide the city of Rome, make it into a composite of coherent communities, geographically touching or intertwined, connected with one another, tied together through marriage and also through their own similarity to one another. The city's government, its politics, its movement are all written in the terms of these families. Family names are the names that must be used in the history of Rome. But what, in fact, were these great families? They seem to have been groups, factions, dominated at their center by a nucleus of closely related kin, with extensions of the kin laced through the peripheral parts, but including only some members of almost any kin pattern, and often including other genetically surprising relations. They seem to have included in their more general factions dependents and adherents of very various sorts, like Cardinal Giangaetano Orsini's household (and part of Cardinal Matteo Rosso Orsini's) clustered around Giangaetano in 1269 in Viterbo, in the chamber in the episcopal palace where he was staying waiting "for the making" of a pope and making grants to relatives. In some ways these families seem more like the Mafia "family" than the family of the novelist or the anthropologist. But the distinguishing thing about them as social groups and presumably about their counterparts in other cities is that the basic relationship around which the whole structure extends (the relationship as central as that between lord and man in conventional feudalism) is the relationship within the family, the small, nuclear, domestic family, between brothers growing up in the same houses (or towers) and holding together as a group the same properties. Common blood, a common household of origin, and, perhaps particularly, commonly held property lay at the center of the group. The actual structure of the central family recalled in part Roman law, and it certainly, sometimes at least (as in the case of Giacomo Savelli's will), in spite of people like Giovanni Boccamazzi and Latino Malabranca, showed a clear preference for the male line and the family name (itself a peculiarity). The fourteenth-century Roman statutes protected the father's disciplinary rights over his family, and they recalled the ancient cliché that shedding one's own family's blood is "abominable to God and man." [57]

Seen as a unit in a total social structure the family looks a machine, and a not inefficient one, a machine whose coordinated group self-interest, following recognizable patterns, shaped, with its fellows, the larger society. It obviously did not look this way from the inside. The Colonna who show us the machine at work also repeatedly make us see this different, inside view: Margherita, Sciarra, Giovanni, Giacomo, Pietro on the ruins of Palestrina. These people loved and hated, were pious and impious, and acted from a great variety of what must have seemed very natural motives. When Boniface VIII wept at the death of his nephews, he was at some level of mind and behavior

The Natural Family

weeping at a failure of the Caetani faction, at a social failure, but it would be insane not to see that he also wept because he was sad, because death comes, and because it had to come to those he loved.

Again, one must not only remember the internal sensitivity of this seeming social machine but also its complexity. One will recall that members of the family Orsini were leaders in both Saba Malaspina's "Guelf" and his "Ghibelline" faction—a matter that is confusing enough in itself and not unlike, on the surface, the divided family of Colonna in the time of Boniface VIII. But the division of the Colonna is, in the material around Boniface, relatively easily understandable. Boniface obviously wooed some Colonna to help him against others; he wanted some Colonna to look dispossessed (whether or not they were), so that he might seem a Colonna protector in destroying other Colonna. But the Orsini of the opposing factions under Arrigo of Castille were closely related men, with closely intertwined tenements in Rome; both sides were part of what one thinks of as the Orsini party, the Orsini property interest in mid-thirteenth-century Rome. There is no evidence that they were competing for inheritance. Why then were they opposed to each other? Did they believe different things, have differently oriented personalities? Did they personally annoy one another, or had they had different experiences with Arrigo or with Pietro Vico, or did one and not the others find Charles unbearable? Were they, however consciously, insuring Orsini success or survival regardless of the outcome of Charles's wars? The answers are not forthcoming. But the warning against assuming that the family machine was too simple an affair is clear.

As one descends the social and financial ladder from the great family to those below it, one of course observes major changes. One moves from the world of leaders and protectors to that of dependents and clients, a change in direction which must have altered the kind of shape and coherence the family had. One also moves from relatively decent documentation to almost no written documentation at all. (And particularly with nameless people one fails to see complexity because one fails to see connection.) Still, as far down as one can go, short of the actual familially displaced servants in other peoples' houses—people who have left *parenti* for a *familia* (quickly to change tongues)—the family seems the dominant social unit in Roman society, a unit in the lower strata as in the higher held together by commonly held property and common familial households.

Again and again in thirteenth-century Rome, family property—

the property of minor families without a traceable lineage or a recognizable cognomen, who appear in written history only for the length of one transaction, perhaps as neighbors in the description of a boundary—is described as being held in common. It belongs to the widow and her son, the two sisters, the three brothers, or the two brothers and their three nephews of whom one of the brothers is the guardian, or, for example, the widow and children, some of them minors, of Leone di Francolino who had to sell their commonly held land outside the Porta San Pancrazio on the Gianicolo in order to pay debts and buy food.[58] Familial group holdings were very common. They point to the reality of the family (including its women) as a unit. Without an important sort of reality could it have been the model for that which held land in a community in which land was such an important source of wealth?

One ought to recall the descendants of Locrerengo selling their *pezza* of land in the Bravi outside the Porta San Pancrazio; over thirty people from four generations were involved, some dead, most only consenting. It is a family act as compelling, and as property bound, as the Orsini sale on Monte Giordano. One ought also to remember that one of the five shares of the Bravi land was held by the church of Santa Maria in Trastevere—here, too, as in the neighborhood, household, and faction of the great family, the outsider is involved. One ought also to recall those households revealed in San Silvestro leases and particularly in Nicholas III's Vatican purchases. It is clear that the families that lived in their houses were often extended ones (but only slightly extended ones), of several generations, of more than one collateral branch. It is also clear that these families were often, collectively, the owners of more than one property—they chose to live together in one central household. These revealed households of people actually living together are a degree smaller than the families caught in the sale of Orsini and Locrerengo properties. The Sant'Eustachio brothers' disentangling themselves offers an obscure hint about how the moving out, the moving apart, took place. This moving apart is more sharply exposed in a 1335 act in which Napoleone di Orso "Milex" Orsini makes his sons (Orso, Giovanni, Francesco, Matteo, Bertullo) his "universal" heirs and divides his properties, for the future, into two distinct and inventoried parts. (And all these operations clearly suggest how wars within families could begin in the division of property.) [59]

It is a queer and fascinating stage in familial development, a stage that has not been completely abandoned in Italian, or, certainly,

Roman life; the mother-in-law of the thirteenth century is still present, as are the large household and the complication of common property. (*Divorce Italian Style* revealed the delights of this sort of living in Sicily.) It is a stage of development made more piquant in the thirteenth century by the contemporary development of the sentimental domestic family, developed perhaps partly as a disguise, a sublimation, as old potent units of society changed and declined. The Colonna were clearly attached to each other sentimentally. Yet like all their lowlier brethren they were caught together on their common property like flies on a piece of fly paper.

In any event, the thirteenth-century Roman did not naturally face life and society as an individual alone. A great many mawkish things have been written about the associative tendencies of medieval men; but it is certainly true that thirteenth-century Romans moved in groups and that those groups were essentially based on, and shaped by, family. It is a superficial paradox that this clustering corporate behavior, this hesitance about living and acting singly, should have been the dominant mode of social behavior at exactly the time when great actors like Boniface VIII strode on stage to play their single roles. Actually, however, clustering and role-playing seem naturally to complement each other. The individual Roman stepping forth from his group to act alone did it not only perhaps with compensatory brio, but also with the aid of a great type role to support him in his unwonted loneliness. More conventionally, the individual Roman, often in family interest but not only in family interest, could, drawn by a powerful continuing surge of religious feeling, take himself from his natural family into the family of a convent or religious order, or, as everywhere, from the service of his lord to the service of the Lord. Margherita Colonna, always a Colonna (as Boniface was always a Caetani), still stepped forth alone, supported by her role as a recognized sort of "saint," but also supported by her realization of the presence of the living Christ.

Rome belonged to her great families. When, for final exact illustration, the Colonna fell, they lost their leases to the Palestrina property of the church of Sant'Eustachio (near the Pantheon). Francesco Orsini, cardinal deacon of Santa Lucia in Selcis, the then administrator of Sant'Eustachio, transferred the leases to Orsini hands (for three generations with eighteen-year renewals).[60] As a distinguished historian has pointed out, even the "communal revolution offered the Roman nobility a new field for their political wars." [61] The great families were, as both the Caetani's and the Colonna's surviving Boniface VIII's policies illustrates, more lastingly powerful than any pope.

> Orsin, Colonna, Cenci e Frangipani
> Riscuoton oggi e pagano domani.
> Più assai che peste, papa ed Imperiali,
> Più a Roma sono assai crudi e fatali,
> Più assai che fame, Galli e Aragonesi,
> Savelli, Orsini, Cenci e Colonnesi.[62]

So go the later verses when Rome had tasted more fully the Aragonese and the Cenci. But with hardly a change of name they make their point for the thirteenth century:

> [Orsini, Colonna, Cenci, Frangipani
> They collect today but pay *domani*.
> More than plague, pope, or those of the Empire,
> More, at Rome, crudely raw and more deadly,
> More even than hunger, or the French or the Spanish,
> Savelli, Orsini, Cenci, Colonna.]

—these constituents of social order and enemies of discipline, the same. They married, they fought, and they lent themselves at profit to foreign contenders. But their control is not quite so simple. They too were dependent. Almost all of them rose to power through a pope, a family pope, or a pope related by marriage, or a pope who made family cardinals—the business of the circle. Because they profited from, they also depended upon, the Empire and the French. Their importance, their prominence in the world, the value of their city properties and their suburban vineyards, depended upon Rome's being the pilgrim's and curial center of Saint Peter and the pope.

The Natural Family

CHAPTER

VI

The Spiritual Family

THE THIRTEENTH-CENTURY CITY
was lashed together, made a unit, and at the same time divided, broken
into faction by the powerful families who lived in and around it. The
fact of the Colonna, the Orsini, the Conti, the Savelli, the Annibaldi,
the Capocci glares from any history of thirteenth-century Rome. The
physical quality of their tough familial reality still, through their relics,
shapes neighborhoods in Christian-Democratic Rome: the Orsini in
their queerly rustic garden hill of Monte Giordano; the Savelli on the
Aventine and in the Theater of Marcellus; the Colonna in the area of
their palazzo and their churches and their Mausoleum of Augustus; the
Conti, the Capocci, and the Annibaldi in those towers around which
the modern city moves. Family names and family signs are on alleys
and in chapels. No one with an eye or ear can forget their past importance.

But there are families, spiritual ones, against which these carnal
families, families of blood and marriage, can be made to seem transitory,
almost momentary, aberrations, shuffled with later families, the more
recent arrivals, the Borghese, the Odeschalchi, and the rest, and lost or
shaded into relative obscurity. These tougher or longer lasting spiritual
families (although subject, too, to fluctuation and failure even before
the nineteenth-century suppressions) were, like the carnal ones, encrusted with property, subsisting on it and at the same time holding it
together and giving it reason—like insects on wood, like relatively
invisible insects on very visible wood. But in spiritual families the
property was held together at its center, not by the relation between
parents and children, uncles and nephews, husbands and wives, broth-
ers-in-law and sisters-in-law, but rather by that between abbot and
monk, abbess and nun, monk and monk, clerk and clerk, nun and nun,
and by their relations with aspirants and devotees. The saint's cult and
religious sentiment replaced the marriage bed and controlled lineage.
Some of these religious families came very close to being immortal.

It is not simple modern affectation to talk of monasteries as fam-

The Spiritual Family

213

ilies. Benedictine monasticism was an heir (as was the thirteenth-century Roman carnal family) of the old Roman family. The abbot was a father. Postulants rejected their natural families for spiritual families in one of the oldest (and sometimes most beautiful and sometimes most tiresome) pieces of repeated medieval rhetoric.

The rejection, in fact, was often incomplete. It will by now have become clear that the Colonna in a house of religion was still a Colonna. The noble families of Rome had a great deal to do with shaping the rich religious life of the thirteenth-century city, including the life of its religious communities. Individual families could dominate religious houses; although few dominated and infiltrated any house so completely as the Colonna came to dominate the house of San Silvestro in Capite.

The number of monasteries and collegiate religious houses in most medieval cities, and certainly in Rome, is startling to modern eyes. There is an early fourteenth-century (almost surely from between 1313 and 1339) list of Roman churches and their attendant religious and clerics which is called the Catalog of Turin (again, because it is deposited in the National Library in Turin).[1] The list seems to have been composed by or for the "Roman Fraternity," the association of resident Roman clergy, a presumably knowledgeable group, so that, although there is no necessity to believe that it is perfectly accurate, there is no reason to believe that it is seriously inaccurate. The list is not quite of the period of this book; it was put together after the pope's withdrawal to Avignon. There are in it some additions to, as well as some lamented decay from (the reason, perhaps, for the list), the list that would have been composed in 1300. It seems safe, however, to assume that the Catalog is a fair outline of the religious settlement of Rome around the year 1300—again after a few decades of the lamented decay, due probably in large part to papal absence. In conclusion the Catalog sums up its findings: there were 414 churches in Rome; of these twenty-eight had houses of religious men or monks, and eighteen had houses of nuns. Twenty-five had hospitals. There were the churches of cardinals and papal chapels; and there were twenty-one collegiate chapels of three, four, five, or six canons. In all the churches there were 785 secular clerks, 317 men who were members of religious orders, eight abbots, 126 monks, and 470 nuns. With miscellaneous additions, 1,803 men and women of religion (excluding recluses) were to be found in Rome, probably five percent of the total population and a considerably higher percent of the adult population.

The entire city of Rome was divided by the fraternity into three general divisions (Holy Apostles, Saints Cosmas and Damian, Saint Thomas), and these were divided, perhaps less formally, into regions of parishes and neighborhoods. The church which was the seat of one of the general divisions, San Tommaso dei Cenci, was called San Tommaso *de fraternitate*. San Tommaso is now never open, but a modern photograph shows that it contains a roughly incised stone which records its consecration by a cardinal with attendant bishops and its being indulgenced in 1243; the church was at least modestly active and appreciated in the thirteenth century. The fraternity itself was ruled over by rectors. A canon of Santa Maria in Trastevere, for example, was one at about the time of the composition of the Catalog, and a list of ten rectors survives from 1212. As a body, the clergy was sufficiently coherent to engage in legal dispute, as it did under Alexander IV with the basilica of Saint Peter's. Less formally, one would talk about "the Roman clergy," or "the clergy of Rome" assembled, for example, on the steps of the Capitol, as a 1266 document did and went on to enumerate: the prior and three Dominicans, the lector and many Augustinian hermits, and brothers of San Giacomo in Septimiano, many canons of Santa Maria Nova and the Lateran, many others, including named friars—a concept "the clergy of Rome" and very "virtual" representation.[2]

In any neighborhood the cluster of resident religious was noticeable, sometimes in relative isolation, surrounded by vineyards and towers and other religious houses as on the Aventine, sometimes in the thick stew of the city, as in the Campo Marzio. The neighborhood of Santa Maria in Trastevere is a good example. It held a number of collegiate and religious churches. Santa Maria itself, besides its cardinal priest, had twelve canons (although one need not assume that all resided nearby). San Crisogono, a little to the east and closer to the river, had eight clerks besides its cardinal priest. San Callisto, just to the southeast of Santa Maria, had four clerks. A little farther to the east, San Cosimato had thirty-six Claresses and two friars minor; and up the hill (by the time of the Catalog) San Pietro in Montorio had eight brothers of the order of Peter of Morrone (Celestine V). Outside the gate, San Pancrazio had thirty-five Cistercian nuns. San Bartolomeo on the island had five clerks. In southeastern Trastevere, in the Ripa, San Francesco had fifteen friars, and Santa Cecilia, besides its cardinal priest, had ten canons. This was a neighborhood thick with clusters of friars, nuns, and secular clergy, but not of monks. This last absence is in part a Trastevere and a fourteenth-century accident; on the Aventine and Celian things still looked different, with monks at San Saba, Santa

The Spiritual Family

Prisca, San Gregorio, and Premonstratensian canons at Sant'Alessio —monks, perhaps, required space. But also, on the whole, by the beginning of the fourteenth century, the less formal and restricting, and expensive, colleges of clerks, the more enthusiastic houses of friars, and the women's houses were more widespread, scattered more thickly, in Rome, than were the great old houses of monks. Still, the monks and their houses set the formal patterns for the others. They were at the heart of the old residual religious expression.

What was a Roman monastery? Saint Paul's outside the Walls (and once across a creek and by the river) remains to answer the questioner. It remains, admittedly, in strange condition. The great old basilica, rich with its ancient treasures, in size the second church in Rome, was burned and almost totally destroyed on a July night (between the fifteenth and sixteenth) in 1823 (as, in the familiar, dramatic story, Pius VII, who loved the church, lay dying and untold in the Quirinal).[3] The basilica was rebuilt and reconsecrated: Gregory XVI consecrated the transept in 1840, and Pius IX the basilica in 1854; the atrium was added later. Of the huge church little, beyond approximate dimension, that remains is actually medieval. But the modern church, except in its sense of columned vastness, is so meticulously fatuous that it does little to divert the eye from that which is in fact medieval. The paschal candle, the remnant mosaic, the ciborium, and the cloister are vibrant cynosures within a vague linedrawing ecclesiastical sketch which itself gives, again, only a sense of line. At the head of the nave just to the east of the *confessio* where lies (it is piously believed) the body of the great Saint Paul (who was martyred nearby), before Honorius III's mosaics (for which the pope summoned mosaicists from Venice), stands the thirteenth-century ciborium, or altar canopy. It is an Arnolfo di Cambio ciborium, done it says with the help of Pietro his assistant (the names Pietro Oderisi and Pietro Cavallini have been suggested without particular conviction), a collaborative work, completed in 1285, and suggestively, as in the case of Santa Cecilia in Trastevere, closely connected with important Cavallini frescos (in this case lost).[4]

The ciborium seen casually, at first, from the distance, can seem peekily frail and fussily insignificant. Watched, even from the distance, it grows impressive and aggressive and powerful. Seen close, with some patience and time, the ciborium becomes a fantastic work, capable of holding forever some part of the viewer's imagination. It encases its altar as finely as (but at greater distance than) a shell an egg. It is, though, a shell that opens in roses and bursts into pinnacle. At its inner corners, where vault springs from column, are attendant angels.

From the two western corners the angels hang, head down toward the altar, censing. The surprise and agitation of their movement, these visually, humanly articulate angels, is echoed in the twisted leaves, perhaps windblown, of one of the eastern ciborium capitals—leaves, some of them, very naturalistic after the late thirteenth-century style —and also in the twisted bodies of other angels who hold on the four sides the ciborium roses. In the four corners, on the outside of the ciborium, with their backs to the interior angels, stand men identified as Peter, Paul, Luke, and Benedict; and in bas-relief across the ogives from each other are, north and south, Adam and Eve in shame and Cain and Abel sacrificing. To the east are two figures called Constantine and Theodosius (a Biblical, imperial, monastic medley) and to the west, the abbot Bartolomeo offers a tabernacle to Saint Paul, for it was Bartolomeo who had the ciborium built here, fifteen years before the Jubilee when the famous coins would be raked at almost exactly this spot.

South from the thirteenth-century ciborium and mosaic, and outside, are the thirteenth-century cloisters. The cloisters (a gift, it would seem, from Cardinal Pietro Capuano of Amalfi) were completed under Abbot Giovanni before the year 1214. They were in part the work of Vassalletto who was also responsible for the similar and equally, or almost equally, beautiful cloisters at Saint John Lateran. The two sets of Vassalletto cloisters, like the two Arnolfo ciboria, help define (and in part, with them, chronologically, the beginning and the end) thirteenth-century Roman style. The Saint Paul's cloisters make a richly encased small formal garden (at least now), surrounded by its rows of double columns twisted after the manner of the literarily remembered temple at Jerusalem and encrusted with jewellike tesserae. (Early thirteenth-century surface seems hardly to have been able to resist tesserae.)

It is still possible to visualize the monks gathered there on a spring morning, perhaps on the feast of the Annunciation, March 25—the sun catching them in the cloister's northwest corner, the huge church behind them, they themselves safe, warm, and refreshed by the enclosed beauty, in their precious little coffer of a cloister, twisted and decorated, but clearly defined. On the feast of the Annunciation the countryside bursts with the season: the ground is white with daisies; the trees are turning green; the sun is strong but fitful; and the wind is strong and gusty. From the strong winds, in the strong sun, the monks could shelter in their cloister. There were, in the early fourteenth century, according to the Catalog of Turin, forty of them, if one

The Spiritual Family

counted also those monks stationed in outlying manors and holdings. With this crowded claustral prettiness clearly in mind one should complicate Saint Paul's tone by remembering the evidence which Nicola of the City, a monk of Saint Paul's, gave at Boniface VIII's posthumous trial. Four monks had come to Boniface to denounce their abbot, Gauberto, as a heretic, accused even by the inquisitor Simone da Tarquinia; and Boniface had sent them away saying, "You are idiots; you know nothing." The testimony hardly admits the possibility of the constant harmony at Saint Paul's which a spring day urges one to re-create. The repeatedly reported, heavy Saint Paul's debts—involved, no doubt with its building—although not unusual for a thirteenth-century monastery, must sometimes, too, have darkened the sun.[5]

Fortunately, two major documents survive from thirteenth-century Saint Paul's. They form a pleasing complement to the church's two major thirteenth-century architectural remains. One is approximately contemporary with the cloisters and one with the ciborium. With the earlier document, from May 1218, Honorius III took the abbey under his protection and confirmed the privileges which Innocent III had granted the monastery.[6] These privileges included the abbot's right to his miter and his sandals, his right to bless, to choose ordaining bishops, and to be elected—his rights to a peacock's tail of showy and sometimes profitable honors; but privilege in this sense meant particularly the right to one's sources of income (although these sources were often, even usually, entangled with more spiritual or at least jurisdictional rights).

Honorius III's privilege starts with a central point: the monastic order, after the rule of Saint Benedict, should always be observed in the place of Saint Paul's. The privilege moves, eventually, to its enumeration of Saint Paul's holdings, going out from "the place itself, in which is located the most sacred monastery in which the holy body of the celebrated saint rests, and its *borgo*" and its mills and appurtenances, to holdings in Rome—the churches of Santi Sergio e Bacco in Suburra and San Nicola de Formis, rights to fish in the river at the Marmorata, other sources of income—to Ostia—the church of Santa Maria, saltpans, one-half of one pond and one-third of another—out to the Alban hills (to Albano and Ariccia), out the Nomentana, north and east to Terni, Todi, Amiterno, Civita Castellana, south to Velletri and Anagni. Saint Paul's was, and importantly so, a great network of properties, houses and towers, offerings to altars, villages, vineyards, and olive groves—a complex of receivable rents. Carlyle was absurdly mistaken in despising talk of carucates and charters in the description of a monastery—they were monastic reality. What was a Roman monastery? It was a holder of properties. It is not the only answer, but it is an essential one.

A Roman monastery was always an owner of Roman property from which it got rents in both money and kind. But a Roman monastery was also, probably without any exceptions, an owner of property in and around provincial towns, particularly, but not only, in Lazio. As the owner of a hospital in Amelia, a house in Anagni, or mills in Tarquinia, the monastery of Saint Paul's played in specific instances a universal role; and it was a role integral to the very being of a thirteenth-century Roman monastery. The Honorius III privilege thus exposes Saint Paul's the great landlord; it also shows the vicar of Christ protecting Saint Paul's in this role.

The 1218 document reveals the great web of property stretching out from the monastery, attached to it, and supporting it, but in many ways external to it. The document from the other end of the century, from December 1287, two years after the completion of the ciborium, reveals the monastery's interior.[7] Honorius IV had committed to Peregrino, bishop of Oviedo, and Paparono, bishop of Spoleto, the job of visiting and correcting the monastery of Saint Paul's. The two bishops, having examined the place, produced for it a set of constitutions. The constitutions do not of course describe the whole internal workings of the monastery; they speak only of those things about which the bishops found it advisable to say something constructive, about areas of laxity or areas of importance. The initial ordinance of the constitutions is in some ways its most central clause, because it deals with the old essence of monasteries—their round of liturgical prayer. The sacristan is ordered to provide sufficient artificial light for reading in the choir whenever it should prove necessary. This provision is central, but, except for its concern with expenditure, not at all representative of the whole body of the constitutions.

What is characteristic of the statutes is the emphasis on the well-ordered communal life, in which an interest in propriety and modesty is combined with an interest in the welfare of the individual, fairness, economy, and, to a certain extent, charity. The concern with the monks' habits is a fair example of the tone of the statutes. Eleven *lire* (of the senate) were to be spent on the habits of each monk each year. On a certain day the cloth for all the habits was to be bought, all of the same value and color; the cloth bought should not exceed in value ten *soldi* the *braccio*. The habits should be conventional and conservative. Monks should not wander out of the monastery to get their own habits,

The Spiritual Family

but rather the dean, joined by two or three others, should go out and bring back the cloth. The habits of obedientiaries, the monastery's officers, should not be of different color or better quality than those of the cloister monks. Old habits were to be given to the poor, unless there happened to be some among the monks who really needed them. If, by chance, the old habits were sold, the dean was bound to notify the convent of the price they had brought, and he was bound to use the money for the monastery's needs.

The constitutions are concerned with the state of the infirmary, an interesting office in the thirteenth century. This particular infirmary was on the advancing border of the malarial countryside south of Rome, so that it must have been important as a house for the sick. The infirmary was, quite beyond this, the office in many thirteenth-century monasteries where monastic restrictions on the eating of meat were evaded. Although there was a clear prohibition in the rule against the eating of meat, except for monks who were sick, many thirteenth-century monks, well and of full appetite, wanted their meat. A way of getting it to them was being worked out between the promulgation of Innocent III's renewed prohibition (*Quum ad monasterium*) of 1202 and Benedict XII's considerable, in this matter, relaxation of the rule (*Summi Magistri*) of 1336. Benedict XII permitted and regulated one of the ingenious evasions that carnivorous monks had evolved.[8] (There were several which allowed them their desire, at least occasionally, without actually staining the monastic refectory with blood-gravy: a second dining hall, or *misericord,* for merciful eating; eating with the abbot; eating with the sick.) Benedict permitted the healthy, in relays, to join the sick to eat meat in the infirmary, and as a corollary permitted half the monks of a monastery to be absent from refectory meals at a given time. It is thus not surprising to find the infirmary prominent in monastic ordinances drawn up during the course of this development.

At Saint Paul's it was demanded that the infirmarian should be a fit and proper person from the monastery, appointed by the abbot with the consent of the convent. If his carefully audited income (controlled in a semiannual public accounting) should prove inadequate, other income was to be diverted to the infirmary. (A special statute deals with the problem of inflation and deflation so that the sums provided would be adjusted to the real not the nominal value of 1287.) If money was left over from the budget for medicines it was to be distributed for petty necessities. Dying monks were to leave their habits to the infirmary.

The cellarer and his Saturday accounts for the buying of the week's wood, garbanzo beans (chick peas), salt, pepper, saffron and other necessities were regulated. The diet, which the cellarer's purchases imply, sounds more palatable, even luxurious, when to it is added the diet implied by the statement that, although during abbatial vacancies business and change were to be restricted and waste avoided, the monks could eat things that were then given to them: chickens, hares, kids, wild boar, goats, and other meat. This diet was not to be consumed without thought for the poor. Two containers were to be placed in the refectory in which leftovers of food and wine from the monks' meals were to be put and saved for the poor; and the containers were never to be used for any other business. Presumably this arrangement was supervised by the lay brothers, or *conversi,* who alone were to serve in the kitchen.

The statutes also provided in a conventional way against the corruption which might come from monks' staying too long and improperly accompanied outside the monastery's walls and its communal atmosphere. All monks and obedientiaries and guardians of monastic properties were, if they possibly could, to live within the cloister, to return there after their day's work. If they could not return nightly, they were to live by twos or more. They were not to remain in any external administration more than a year at a time; other monks were to take their places so that they could return to the monastery. When they slept outside the monastery, two monks were to sleep in the same chamber, with no gaping space between their beds. (No woman was to enter any house or office of monks or abbot.)

Considerable space is devoted to punishment in the statutes. Lawless obedientiaries were to fast or to be deprived of office, some never to reassume it. Any monk or lay brother who conspired against the abbot or another monk was to spend two years in the monastery prison, the first year on bread and water, the second on bread and wine and beans or olives (unless the abbot granted a dispensation on grounds of health). At the end of the two years, if it seemed wise to the monks, the prisoner was to be expelled to another house of the order, if not, if he was to stay at Saint Paul's, he was to fast for two more years on bread and water on Saturdays and to be disciplined according to the magnitude of his crime. He was to have no voice in chapter or administration, and to take the last seat in choir and refectory for the whole of his life (or presumably until his sins were surpassed). Still, the accused monk or administrator was to be protected. No monk, on pain of excommunication to those who imprisoned him, was to be put in jail

The Spiritual Family

or gravely punished unless his crimes were notorious and manifest or he had legitimately confessed or been convicted. (There is a little, a very little, air of Magna Carta in the close, conventional community.) Monks who ought to be jailed, moreover, were to be jailed in the monastery's own jail not farmed out. The jailbreaker was excommunicated *ipso facto*.

At the other end of the scale there was concern for learning, in a very mild sense of the word. The abbot was to provide a learned master, either secular or religious, to teach the monks some grammar. Illiterates and idiots (*pace* Boniface VIII) were not to be admitted to the community.

The abbot was bound to listen patiently to anything any monk felt it was important for him to say. The abbot was not himself to hold the abbey's seal. It was to be committed to keepers who would use it only with the convent's consent. The abbot was not, even with the consent of the abbey, free to grant away pensions. At an abbot's death an immediate inventory of his goods was to be taken so that his whole property could be presented to the new abbot, except his clothes, which could be given to the poor, and his bed, which could remain in the infirmary. His seal was to be broken.

The forty monks of Saint Paul's with their privileges and restrictions, their huge church, their pretty ciborium and cloister, made an abbey. It was a complex piece of property and of ordered relationships, a thing of vestments, of jail and infirmary, of pots of second-hand food for the poor. It aspired to (or it was hoped by others that it would aspire to) a conventional, habitual, prayerful, communal life of moderation in which the poor were remembered, literacy was maintained, and sex (and widely separated beds) avoided. The abbey was meant to maintain and order properly its income and expenditure. This large and slightly rustic example of a Roman Benedictine house maintained its cult—the important cult of Paul—around its relics. It was the sort of house to which smaller urban Benedictine houses like San Cosimato or San Silvestro in Capite might, in the early thirteenth century, have looked with envy—feeling that with sufficient income and numbers of postulants, and perhaps more imposing relics, they might approach its stature. Except for the abbot's supposed heresy and his enemies (and an effort at externally imposed reform thought to be overly harsh even by Innocent III), there is no evidence to suggest (unless one finds fresh conviction in its sculpture) that there was any sort of spiritual enthusiasm at Saint Paul's in the thirteenth century.

Saint Paul's shows no signs of enthusiasm. In this it is like most other old, Italian, and, certainly, Roman monasteries. Their aspirations, when they are at all noticeable, were to decency and order. But what was true of the older monasteries was not at all true of the church, in its broader sense, as a whole. The thirteenth-century Italian church at its most extravagant peaks of shrill spirituality is the church of Margaret of Cortona and even of Angela of Foligno. Angela was born just before the middle of the century (slightly after Margaret).[9] She came from one of those prosperous urban families which seem particularly characteristic of the little cities of mid-century Italy, and which seem to have been prolific, and naturally so, in producing saints who, joyously and with extreme expression, renounced their riches. Before Angela died in 1309 she had been wife and orphaned, childless widow. But happily so, although with some grief too, because, that she might follow the path of the spirit, she said, she had prayed God, Who did, to take from her the distractions of mother, husband, and children. (She stood naked of close relatives on the bare ground.) She traveled to Rome and prayed to Peter for true poverty. She walked with the Holy Ghost through the beautiful Umbrian countryside between Spello and Assisi; and He told her that He loved her better than anyone else in the Vale of Spoleto. Angela "burned" with the love of God. She was an extreme case, but the country which sustained her produced many gentler variants of her crazed enthusiasm—they swarmed like swallows in a Roman spring. Some of them were eventually considered heretics and some saints. Some can seem horrible as she herself can; some were as glorious and beautiful as (as Dante said) the sun. If their enthusiasm had a geographical center it was Italian Umbria, but it touched all of western Europe, and even, and variously (as Margherita Colonna and Celestine v testify), the jaded heart of western Europe, Rome.

To understand thirteenth-century enthusiasm, to transcend its natural limitations and see what in the imagination, in perfection, it could seem, one must look at Francis. To look at him, the most Christlike of Western saints, always brings pleasure, the experience of still palpable joy. It is also an invigorating historical experience. With him one has in hand an intricately informative but explosive key to the understanding of the substance of the thirteenth century. And one does not know which of the fragments that form our memory of him are factually exact about him. The sensation of Francis was so violent, perhaps, during and just after his life that his more hesitant followers who became his apologists and biographers seem to have been confused, seem to have been forced in their fallibility to screen, if unintentionally, from

The Spiritual Family

human eyes the real Francis, to hide him behind their variously conventionalizing biographies, and also behind their own more, but variously, conventional, following lives, and sometimes to stylize him into an ideologue. Much of the narrative material from which a biography of Francis must be built is, in fact, so inextricably shuffled and confused that it can probably never be properly disentangled.[10] Still a real man, although an extraordinary one, and perhaps a slightly different one for each reader, is irrepressibly present. One can say something more perhaps from the remembered stories of his close companions, written down two decades after his death, now carefully disentangled from surrounding material. One can see Francis, if in fragment, fairly clearly: Francis at the Portiuncula, at the last cell behind the house, next to the garden hedge, where the gardener had lived when he was alive, going by the fig tree near the cell, reaching to the cicada on the branch, saying "Come to me, my sister cicada," and "Sing, my sister cicada," an hour of singing then the placing of the little sister back on the fig tree, eight days, then freedom for the cicada and freedom from vainglory for Francis—ideas, faith, charity, religion, emotion, all intellectually muted, acted out, and twisted in sound and touch around the natural reality of the fig's branch.[11]

For Francis knowledge was action. To know him one must see him kissing the leper, talking to the birds, driving the friars from their house in Bologna, surprising at their Christmas feasting the friars of Rieti, begging, eating grapes so that a sick friar would eat them, sick and wanting parsley in the evening, asking for Roman marzipan, making his Testament, greeting with joy his Sister Death. He must be seen keeping the morrow's vegetables from being soaked in hot water that no thought be taken for the morrow. He was, and it is crucial to see this in order to understand him at all, a man who believed literally in the message of Christ's life, believed in acting it out again physically, without too much worry about the study of texts, without the hindering "bag" of learning. Francis built at Greccio a real Christmas *presepio* with living animals. "He taught by example and by acted parables. . . ."[12] He was a tactile and visible saint who loved God in the sun and the larks that flew, and who himself left stunned for their lives those men and women who had seen him, heard him preach, touched his clothes in the crowded square at Bologna. (Federigo Visconti, a distinguished archbishop of Pisa, member of a powerful family, friend to the world's great, could remember all his life the time in his youth when he and Francis were in the same Bologna piazza.) Francis's charisma was tactile and visual and full of wild joy.

ROME BEFORE AVIGNON

Francis was a romantic. Quite literally, as has been brilliantly demonstrated, his life after Christ acted out the principles enshrined in medieval romance, the popular literature of his youth—"knights of the round table," the Emperor Charles, Roland and Oliver, and all the paladins. The personifications with which he surrounded himself—Sister Death—are very close to the allegorical figures of the enchanted forests. Most familiarly his Lady, the Lady Poverty, is the inverted figure of the Fair Lady of romance. His heroic acts—kissing the leper—are the inverted acts of heroism of the romantic hero (or, if the hero is Lancelot and the lady is Guinevere and the poet Chrétien, the inversion precedes the translation, and Francis is the act to the word in literally spiritual terms).[13] To the more usual romance Francis is a deflection, but, in the retained idiom, about as sharp as "Auccasin and Nicolette."

Francis was born in about 1182 if not to riches at least to the relatively heavy and obvious wealth of an Assisi merchant's family. After his young conversion he found bourgeois wealth and bourgeois family both cloying and filled with guilt; but, in rejecting both, he retained, as in the case of the romance, their idiom. In the newly obtrusive jingling of money in his era's Italy, Francis worried with property and espoused poverty with constant violence.[14] He said, "I do not want this skin [rug] over me anymore since through my avarice I did not want brother fire to eat it." Francis practiced at begging with the beggars at Saint Peter's in Rome; and he became a beggar. He would not be a thief, he said once on the way from Siena, and keep for himself clothes that another needed more; they were only a loan. (There is in this a tantalizing suggestion of Aquinas's related, if more arid, view of property—as well as of that of ancient Saint Martin.) Francis, and after him Clare, fought desperately against the forces of gift-bearing, conventional, embarrassed society, fought to keep the poverty of their orders absolute and tangible (but the lilies of the field are an affront to men in houses).[15]

Preoccupation with poverty was a personal characteristic of Francis's, and in him it found a sort of Christian perfection. But it was also (and naturally in the obtrusive jingling) a preoccupation of his age and country. There was a current taste not only for extreme wealth and extreme poverty but for sensing and viewing the contrast between the two—between the begging Franciscans and the cardinals at their rich dinners in close juxtaposition to each other. "Lazarus and Dives" was a contemporary theme in painting; and the contrast seems to have been savored by real-life Dives rather in the way that in the fifteenth-

The Spiritual Family

century North the living savored viewing the sculpted dead.[16] Beyond being a stimulating and also reassuring sensation it gave some sense of sanctity to the sensation seeker.

A case in point is that of the Blessed Rainaldo, bishop of Nocera Umbra, who died in 1222, four years before Francis's own death.[17] Rainaldo, in what he considered his compassion, took a little orphan pauper to raise in the episcopal palace. The boy was taught everyday before dinner to come to the bishop and then to each of his guests to beg from them and say, "For the love of God and the Blessed Virgin Mary, give alms to me a little pauper"; and thus, his holy legend goes, the saintly man kept alive the memory of Christ in his poor. This praised action of Rainaldo is surely related to that in rich and powerful men like Cardinal Ugolino (who became Gregory IX) which caused them to smile on Francis and his followers and to be pleased with their existence. It is a sentiment both conflicting with, and in a complicated way, complementary to, that which caused them, the rich and the powerful to quench the blazing poverty which sometimes seemed to threaten them (to threaten particularly their view of themselves), not merely to make it rich with gifts but also to discourage its existence.

Francis's poverty both met a taste and threatened the security of his own time; in both these ways it heightened and extended and symbolized existing or already potential tastes and fears and tensions. In this and other ways Francis was a phenomenon connected specifically with his own time. His literalness, externalness, both in his taste for action and in his appreciation of the visible, the outsides of things, was certainly related to a turning of sensitive and intelligent men's attentions to the actual nature of observable external forms. There is a clear bond, although certainly not a direct causal one, between Francis's viewing the outside world, as he moves through his various reported "lives," and Innocent III's proposed remedies for an always ailing church in his great council of 1215. For Innocent III thought was action too.

For both Innocent and Francis thought was also moral action. In this (in one case consciously and in one case not) they echoed the ideas taught by great teachers at Paris, the center of the intellectual world, at the turn of the century. Their Italian morality was clearly and closely connected with the practical morality, the pastoral theology, of Peter the Chanter and his circle.[18]

Francis is thus in part explicable as a product of his time; he answered its needs and threatened its fears. He was the essence of its most revisionist manner of perceiving. But his genius (and it was an intellectual, although not academic, genius with its superb understanding of the human psyche) formed the answer to temporal needs into something quite extraordinary, too attractive and too threatening to be borne. The whole succeeding century can be seen as an answer to him, an effort to emulate and at the same time to repress. Sometimes the reactions are joined—as in the building of the heavily materialistic basilica at Assisi, the weight and cost of which make sure (in irresistible image) that Francis is dead beneath it, or as in the career of Saint Bonaventura, in its spiritual and intellectual abstraction (not without beautiful paradox), its movement to contemplation.[19] In the course of the century Francis became pious allegory, acceptable, fanciful, and relatively safe; his memory could be carried, at times, like a sweetly scented or lightly colored piece of cloth to distinguish the men of his order from those of other orders—although it never lost its potential revolutionary danger, its power to incite extremists or to elevate the lives of quite ordinary men. Submission to clerical power had been a central theme in Francis's teaching, and reverence for the Eucharist, the function of priestly office, but so had the consideration of office as a sort of property. There must always have been in some papal circles a nervous, haunting sense that Francis's possibly Christlike rejection of office was a threat; there must too have been queer, subconscious reactions to Francis's having bridged so beautifully what almost seems the natural chasm between morality and religion.

The essential connection between Francis and thirteenth-century Rome is that the ideas and people connected with him dominated movement in the religion both of the city and of the world. Franciscan persuasions presented themselves both to popes and to people (and held up the Lateran); they moved into monastic cloisters and even threatened religion's sleep in monasteries like Saint Paul's—although perhaps never in Saint Paul's itself. But Francis and Rome are also more personally, if less importantly, connected through the saint's having come to the city and spent some time there.

He came at the end of the first decade of the century for Innocent III's approval of the friars' way of life. It is said that he stayed in the hospice of San Biagio in Trastevere, near the Porta Portese by what was by then the old Jewish cemetery. His cell exists still (it is piously believed) in what became San Francesco a Ripa, and also the stone on which he lay his head. The room is now dominated (insofar as it is not dominated by a baroque machine in which relics revolve) by a thirteenth-century portrait of Francis, attributed to Margaritone d'Arezzo,

The Spiritual Family

227

and according to tradition commissioned by Giacomina de' Settesoli. Giacomina, it will be remembered, was Francis's Roman patrician disciple who, according to the "Mirror of Perfection" (after the writings of the companions), made him marzipan, after the Roman recipe and then called *mostaccioli* or *mortariolum*, for which he had asked as he lay dying ("She made it for me many times when I was at Rome")— marzipan made perhaps (one may at least think romantically) from a recipe from the kitchen of the family ruin (acquired in 1145) where old Cencio Frangipane had kept a leopard as a symbol of a quite recent but very different sort of Frangipane strength. Marzipan ties Francis to Rome as an orange tree at Santa Sabina does Dominic, and more credibly.[20]

One of the ways in which the idealized Saint Francis was brought to Rome, and specifically to Roman monasteries, was simple and direct—much simpler, in fact, and more direct than really effective. Thirteenth-century popes, as guardians of the entire church, felt strongly the responsibility for maintaining in good order houses of religion, places especially devoted to the service of God. Innocent III, Honorius III, and Gregory IX were all distressed by the presence within the community of monastic houses, particularly old Benedictine houses, in various states of physical and spiritual decay. Gregory IX stirred up Roman nunneries; Innocent III, and Honorius III after him, hoped that the Benedictines could be reorganized rather after the fashion of the Cistercians, so that with connections between individual houses, with chapters and visitations, a real, constitutional order might develop. They hoped that this in turn might brace the slipping minor houses and keep them up to the standards of the better houses of the order. In Italy these popes and their successors went further. They not infrequently replaced the Benedictines in decayed monasteries with religious from orders which they considered reforming or reformed, who would, at least incidentally, again attract gifts to their churches and again "spiritualize" the monastery's existence as a nexus of property. The popes particularly favored the Cistercians and the Claresses or Minoresses, two female branches of the Franciscan order. To Rome for which (as the repeated harangues of papal letters tell us) they had a particular responsibility, the popes brought Franciscan women.[21]

Saint Clare, from whom the Franciscan order or orders for women sprang, was in her more secluded way almost as extraordinary a figure as Francis. She was probably born at Assisi in about 1194. When she was seventeen or eighteen, in the lent of 1212, she heard Francis, "the herald of the Great King," preaching in Assisi. She met Francis several times. On Palm Sunday, in full spring finery, she went to Mass at San Rufino, the Duomo (in part, and beautiful part, now as it was then); that night she fled into religion. After a series of flights and alarms in which she made a clear point of the rejection of earthly family (as Francis had and as he demanded of his own followers), but in which she was joined by her younger sister, and during which she was sheltered by Benedictines, Clare was settled by Francis at the chapel of San Damiano. Like Francis she drew followers, hers to San Damiano, and "she taught her sisters to forget their families, their home, and their country, that they might please Christ." [22]

From the beginning, Clare, like Francis, was devoted to the most absolute conceivable poverty which limited possession to that which was needed for immediate and minimal subsistence. But although Clare encouraged Francis to active rather than secluded contemplation, her own house became one of strict enclosure (because women were not men). At first the nuns seem to have tried to live by their own labor, then by begging, but, in their essential enclosure, begging aided by attendant friars. Clare, like Francis, was full of Christian joy and courage, of optimism. She believed that God would help her, insofar as she needed help and insofar as it was to her spiritual advantage. She was not preoccupied with rules about food and dress (which in real poverty were, after all, absurd), nor was she morbidly afraid of the presence of men (friars) at the house, if they were a help, particularly a spiritual help. Her one great fear seems to have been that her nuns would be forced to give up poverty and accept property; and this fear shows her intelligence and her sense of the way the real world really works. In opposing property she could, if the stories are true, face cardinals as bravely as she could in other circumstances face Saracens. When Cardinal Ugolino said that if her vow stood in her way, that is if it were keeping her from accepting property, he could absolve her, she said, "Holy Father, absolve me from my sins if you will, but I wish not to be absolved from following Jesus Christ." [23]

Clare's fight for poverty is reflected in the tangled history of the rules of the Claresses and the Minoresses. After an initial "formula" by Saint Francis, Ugolino produced a rule in 1218–1219 which took away any peculiar restrictions to poverty; this rule was followed by Innocent IV's rule of 1247. But in 1253, just before her death, Clare got Innocent IV's approval of her own rule, a rule of real poverty, at least for her own house. This rule, it is said, she kissed many times before she died. Victory seemed hers, and in a way Innocent's. But the French pope Urban IV, a man who has been admired for his shrewd and tough

The Spiritual Family

administrative abilities, while referring to the acts of his Conti predecessors, Ugolino and Alexander IV, defeated Clare after her death.

In 1263 Urban promulgated two separate rules for followers of Clare, for the Damianites as they were generally called, and from his rules the nuns of the two Damianite persuasions took their different names—Claresses and Minoresses. Of the two rules, the one for Claresses was the general rule (called the Urbanist rule); that for Minoresses (called the Isabella rule) was designed for the French royal house of Longchamp, founded by Louis IX in 1255 and presided over by his sister (with the future French pope, Martin IV, as guardian). The Isabella rule was later extended to some other houses, the houses in England, the Colonna house in Rome. Besides a difference in tone between the two rules—the Isabella rule seems to be phrased more politely and has perhaps a little of that dreadful sycophantic air with which medieval clerks sometimes disgraced their communications to royal persons not currently excommunicate—there are some differences in the calendar of fasting and communicating, in the wording of the vows: the Claresses, for example, promised that they would observe the rule specifically to Clare as well as to God, the Virgin, Francis, and all the saints; the Minoresses did not. In general, however, the rules (elaborate for dress and fasting) were very similar for the two sets of plainly clad, black-veiled, cord-cinctured, generally enclosed nuns—and most noticeably similar in omitting talk of communal property. Looking back at this development of rule, one of the historians of the Minoresses has said, "Except then for those few houses which followed the practice of St. Clare, the Damianites . . . or Claresses were no more poor than any ordinary Benedictine nuns." [24] They had been domesticated. They no longer threatened common pious assumptions. They still managed to look, however, in their relative newness, relatively pure, so that they still excited secular enthusiasm, and they attracted the donations that they might now accept. Thus they came to, or stayed in, Rome.

Two of the Roman houses to which Franciscan women came were great keepers of records. They are thus unusually visible both for the time before and after their change of order. They tell what monasteries were, in Rome, and how they were changed by the pressure of Franciscan enthusiasm. Both monasteries, or their places, still exist, although altered in form and without their Franciscans. Both are within the city walls and in very Roman-seeming—though distantly different— parts of Rome. One of the monasteries, San Silvestro in Capite, was at the northern edge of the thickly populated part of the city just a little east of the Via Lata or the Via Flaminia (now the Corso) as it moved away from the Capitol toward the Porta Flaminia or the Porta del Popolo or, as the gate was called most frequently in San Silvestro documents, the Porta San Valentino. San Silvestro was in the *rione* Colonna (almost at the border of Campo Marzio) and near the Column of Marcus Aurelius, which it once claimed as its own. The second of these two monasteries, Santi Cosma e Damiano, or San Cosimato, as it is now known and was beginning then to be popularly called, was in Trastevere, south of Santa Maria, the great church of Trastevere, about halfway between it and the city wall, and so, of course, in the huge and unevenly populated *rione* Trastevere.

San Silvestro in Capite was one of those Roman monasteries which had been built by a papal family on one of its own tenements. In 757 Pope Paul I had succeeded his brother Stephen II to the papacy. It was in their house that the two popes, and particularly Paul, established the church and convent that would constantly honor and protect the relics of saints removed from the catacombs to the safety of the city—and particularly the bodies of the saintly popes Silvester and Stephen.[25]

The church also received in its dedicatory title the name of Saint Denis perhaps in memory of a Frankish gift and connection. Thus the most regular name of the convent at the beginning of the thirteenth century was probably the *monasterium Sanctorum Stephani, Dionisii adque Silvestri quod ponitur cata Pauli* (or *quod vocatur cata Pauli quondam pape*). But before the beginning of the century a relic more significant even than the two popes had come into the church's possession. The church held, it was, and is, piously believed, the head of one of the greatest of the saints, John the Baptist. In honor of this very special relic it is thought (although the sound *cata Pauli* ought to be kept in mind) the church's name changed and varied so that it was also called *ecclesia Santi Silvestri lo capo de Roma* or more regularly *Santi Silvestri de capo*, or *de capitis*, or, a little more grammatically, *de capite*.[26]

By the thirteenth century the convent can be seen essentially, again, as an event, a continual celebration of its own preservation of two great relics, the body of Saint Silvester and the head of Saint John, with these surrounded by an impressive hoard of other relics, most prominent among them the body of the other pope, Stephen. This continuing celebration is emphasized by a long series of the monastery's documents dealing with their properties. Rent after rent was to be paid on the feast of Saint Silvester (New Year's Eve) or its vigil, or, more

The Spiritual Family

often, on the feast of Saint John in the summertime (June 24). John's importance in the saintly galaxy is obvious. Silvester's importance, one may remember, was due to the significance and success which medieval ecclesiastics were able to see in his dealings with Constantine, as in the narrative frescos of the oratory of Saint Silvester at the church of the Quattro Coronati—repeatedly, in the thirteenth century, a governmental place.

Much of the thirteenth-century church and convent of San Silvestro has disappeared, as has certainly the semirustic air of a place now caught between the central post office and public transport *capolinea*. There are fragments of the old church outside, and the many-storeyed campanile surmounted by its brazen cock. It at least still pins to earth the old place. The church was drastically modernized by its cardinal (Johann Dietrichstein) at the end of the sixteenth century. Fortunately for antiquarians, as Dietrichstein was modernizing, Cardinal Cesare Baronio, the historian, was antiquing his titular church of Santi Nereo e Achilleo. Baronio got the old pulpit of San Silvestro. One can find it now either there or in neighboring San Cesare, it is not quite certain which. So for a sense of the old interior of San Silvestro one can go across Rome to the pleasant, ancient-looking little church near the Baths of Caracalla, surrounded by privet and laurel, by pine, locust, and oleander, and by the roar of traffic. Inside (besides graceful sixteenth-century paintings, Leonine mosaics, talk of Gregory I, and evidence of sixteenth-century archaeology) one finds cosmatesque work that would recall the thirteenth century were it still in San Silvestro.[27]

In the thirteenth century San Silvestro was, until 1285, a house of Benedictine monks. For it, as for any thirteenth-century monastery, it is seldom possible to establish the exact number of resident monks, although one can establish a minimum number from the names of monks who witnessed and consented to monastic grants; and sometimes the list of consenters does seem to imply that it includes the names of all the monks. In an 1198 list (which does not make this implication) one finds at San Silvestro the abbot, two priests, three deacons, one subdeacon, and three men whose orders are not specified and who therefore were probably not in major orders. There were then at least ten members of this little hierarchical group. In 1247 there were nine, in 1236, 1246, and 1249 eight, and in 1214 and 1269 seven. Often, even from very early in the century, there is evidence of only four or five monks, and occasionally of only three. It is thus not necessary to assume on the basis of the lists of only four monks in the 1280s, alone, that the monastery was then unusually small and decrepit. It was always, in the later twelfth and thirteenth centuries, small.[28]

Except for two peculiar periods, the thirteenth-century Benedictine house was ruled over by an abbot. In the years from 1207 to 1210, Sergio, who had been called abbot in 1205, administered the abbey as "rector" and "administrator"; between 1269 and 1277 (after the translation of Abbot Gregorio to San Gregorio on the Celian and before the preferment of Abbot Matteo, who had been prior of the Sacro Speco in Subiaco) Giovanni di Monticelli presided over the abbey as prior. There were at least ten abbots in the thirteenth century, but of these the last three presided during the abbey's unstable years, 1277–1285. The length of earlier abbacies had varied considerably from that of Giorgio, which, at the end of the 1250s, cannot have lasted as long as four years, to that of Silvestro, who was abbot at least from 1220 to 1244, that is, at least twenty-four years.

This abbot Silvestro, who ruled the abbey for a quarter of a century catches one's attention. Unfortunately he cannot be recovered in biographical detail. He had a nephew who was also named Silvestro and who sometimes accompanied him; and the coincidence of names perhaps suggests a family particularly dedicated to the saint's own church and possibly a family from the immediate neighborhood. Abbot Silvestro himself was a priest of the house by 1218 and the house's chamberlain by 1219; he had been a monk since 1207. Silvestro's long abbacy was supported by the stewardship of Giovanni Muti, a monk by 1217, steward (*yconomus*) by 1218 and again from 1226 or 1227 to 1236; by 1243 he had been replaced by Ylperino. At the end of its period of Silverstrine stability, the abbey was relatively full of monks, although shortly afterward, in 1249, it was fighting to recover from serious financial difficulties.[29]

Silvestro and Giovanni Muti were not the only monks who were long connected with the abbey. Benedetto, for instance, who was prior from at least April 1249 to January 1258, was probably a monk in the abbey by 1228 and surely was one in the 1230s. San Silvestro was an abbey to which one could devote one's life; and the abbots from before the late thirteenth-century period of provisions (from 1277) seem regularly to have been monks of the house. It was not a large but it was a constant community.[30]

The monks at the abbey's center were surrounded by other Romans attached to them by blood, devotion, occupation, tenure, or proximity. Besides the abbot Silvestro's nephew Silvestro, there was Gottifredo di Monticelli, who appeared as a witness under Prior Giovanni di

The Spiritual Family

Monticelli, perhaps a relative, perhaps a neighbor from the Monticelli in the Regula across Rome by the river. There are oblates like Fra Pietro Ciarfo and Fra Matteo, in 1275, who had given themselves to the monastery. A servant of the monastery, abbatial cooks, officers in charge of houses, and pages or squires of the abbot witness documents along with various other artisans and officials not specifically (or specifically described as) abbatial, porters, taverners, gardeners, a marble worker, warehousemen, and the everpresent shoemakers and scribes.[31]

The abbey's scribes, the notaries it used, are of special interest. They are visually present in their documents; the dog's head of Graziano's document of 1244; the crossed swords of Nicola Andree's document of 1236; the splendid *stemma* of Giovanni Berardi; the slovenly "S" of Sanguento; the elaborate monogram of Stefano di Lorenzo; and the curious buglike creatures of a series of notaries, of Angelo, Giovanni Rainaldi, and Giovanni. There is little to indicate that San Silvestro was the sole employer of its own notary, but the convent did use some of its notaries repeatedly so that they must have reckoned the convent among their regular customers. One of them, Castorio, who worked in the 1250s, was on occasion a business agent for the monastery. San Silvestro notaries seem to have been capable but not peculiarly so. They seem not to have kept a cartulary for the monastery, but they sometimes made multiple copies for greater security. San Silvestro's notaries, and its tenants and neighbors who were notaries, are, one will recall, representatives of a major Roman industry.[32]

San Silvestro was, then, a group of people as well as a celebration of relics; it was also and probably most importantly a collection of property and of rights that were hardly, if at all, distinguishable from property. Its possessions reached far outside the city to Gallese, Sabina, Sutri, Orte, Bassanello, Vitorchiano, Vallerano, Aliano, Palombara— the sort of towns in which a number of Roman religious houses had acquired interests. These holdings were not infrequently densely concentrated: in 1230 the monastery granted away property in Gallese "bounded on all sides by monastery property except that a road passed its head and a ditch its foot." [33] Closer to Rome, San Silvestro had property outside the Porta Pinciana in the Orte Pisce, or Pisscina, outside the Porta San Pietro near the Vatican at Oliveto, out the Via Tiburtina by the Ponte Mammole, on the Via Salaria at Silvaproba by the river, and, the property with which it dealt most often, its most important cluster of suburban holdings outside the Porta Flaminia, around its church of San Valentino, off the sharp western edge of Monte Parioli, between Rome and the Ponte Milvio. The monastery's suburban property was outside the city's northern gates—the gates of the monastery's own side of the city. Finally, the monastery held houses and gardens within the city itself; but, as in the country, its holdings were concentrated in specific neighborhoods, in the *rione* San Lorenzo in Lucina-Campo Marzio, the *rione* Trevi, and its own *rione* Colonna. So from itself and the garden against its very wall, the monastery's property spread out to the neighboring *rioni* in its part of Rome, to areas outside the northern gates, to country towns. The administrative center of this system of properties was the convent cloister and the prior's chamber, its recording heart, where obedientiaries met and transacted business—and heart and extremities depended upon each other.

In what sort of business and property was the monastery involved? A miscellany of example answers. In December 1231, in the piazza at Bassanello, the monastery as one party and two men from Bassanello as the other accepted as their arbiter the podestà of Bassanello to define the division of income and the responsibility for repair (in case of destruction by army or flood) of a mill; and in the same month the podestà rendered a decision.[34] In January 1261, before the Fonte Sepalis of Viterbo (the great Fonte Grande conventionally dated as having been begun in 1206 and finished in 1279), the steward of the monastery, with the notary Castorio present, granted to Rainucio, the son of Rainucio Carboni of Aliano, for three generations, for one hundred *soldi* and sixteen *denari* of Lucca to be paid each year on the feast of Saint Silvester, all the possessions which his late father had held in Aliano.[35] In January 1258 a man named Pietro di Giovanni granted to a man named Pietro Grasso ("Fat Peter") a part of a salt sluice in the Campo Maggiore in Serpentarola in Annito Maggiore; and the notary Castorio redacted an instrument in which Pietro Grasso promised to give, for each year that he worked the salt, a tub of salt as rent to the monastery.[36]

Early in 1279 the monastery's church of the Virgin at Cerreto, in the tenement of Castro Flaiano, had no archpriest because of the death of the last incumbent. According to what was then called the custom, the parishioners, as patrons of the church, were summoned at the sound of its bell for making a new archpriest, and, again according to the document, they unanimously conceded their power of election to the great lord Gentile, son of Bertoldo Orsini (himself within the year to become a senator of Rome). Gentile Orsini chose the subdeacon Enrico di Federico Sarraceni as archpriest; proctors were sent to the monks of San Silvestro for their confirmation of the election. The monks con-

The Spiritual Family

firmed and ordered the archpriest's induction; and he swore to give them each year on the feast of Saint Silvester two good hares and two *rubbi* of grain, and on the feast of Saint Denis ten pair of good fat doves.[37] In 1191 the monks granted away an olive grove, in the Sabina on the river Cutri and next to land that the abbey of Farfa held, for thirty *denari* Pavia and one-half barrel of olive oil each year on the feast of Saint Silvester (or by its octave), measuring the barrel at three palms around according to the palm measure on the stone at the gate of the monastery of San Silvestro itself.[38] The little Roman monastery of San Silvestro helped to manage the estate of central Italy, and for its efforts, and in defense of its rights, it received its pennies, its oil, its hares, its doves, and its salt.

Occasionally a document ties together the monastery's rents and its internal life. In January 1214 the prior Sergio and his monks granted to a man with the Roman name Enea, to work it for nine years, a tenement of arable land outside the Porta San Pietro in the Oliveto next to holdings of San Biagio and Santa Maria in Via Lata. The tenement had previously been granted, in 1198, to Leonardo di Stefano di Paolo di Leone to work, for a promised annual rent of twenty *rubbi* of grain. In 1214 the monastery extracted from Enea a promise of not only the yearly amount of twenty *rubbi* (measured according to the *rubbio* of the monastery) to be brought to the monastery on the feast of the Assumption at Enea's expense, but also each year on the vigil of Saint Silvester's day (the day before the New Year's Eve) a good pig (*unum bonum bustum porci—un buon' busto di porco*). The rents, the witnessing abbey cooks, the great feast of the sainted pope, the Christmas festival, the monastery beneath the campanile, all swim together in a savory image.[39]

The monastery was, in the middle half of the thirteenth century, constantly occupied with its holdings outside the Porta Flaminia, around its church of San Valentino. These lands were the vineyards or wasteland appended to vineyards against the sharp edge of Monte di San Valentino, as that edge of Parioli was then called. These suburban holdings were cultivated by an identifiable community of men and women—sometimes involved with one another as is clear, for example, when in 1278 the tenant Pietro Buccafusca acts as proctor in the sale of his immediate neighbor's land or in 1266 when Gregorio di Cesare does. The constant rent demanded from these holdings, repeated as each tenant alienates his land to another, is one-fourth of the new wine and a canister of grapes at the vintage, as well often as five *soldi,* initially, for the convent's consent. The tenant was generally protected by a clause which allowed him a minimal residue of wine, and he was sometimes made to promise that he would give food and drink, as well as a *vascatico,* to the representative of the monastery who came for the wine. He regularly promised to return, in the conventional phrase, one-half of the value of any precious metal found if it was worth more than twelve *denari* of the senate. After three years of neglect, his tenement would revert to the monastery.[40]

The monastery repeatedly tried to return deserted vineyards to production, as when in 1262 it allowed Giacomo Crescentii to alienate to Andrea di Giovanni di Andrea a waste vineyard next to his own vineyard with no rent to go to the abbey for the first four years of Andrea's tenancy so that he might restore the vines (after that Andrea was to return the conventional fourth part). In 1271, in a similar case, the monastery stipulated that if Nicola Musei should sow the deserted vineyard next to his own holding, a fifth part of the produce should come to the monastery. Certainly the monks were not feckless, or not completely feckless, landlords in dealing with their suburban vineyards. They used, moreover, their vineyards as a source of rent in kind, in wine, a constantly necessary commodity, which could help compensate should there be unfavorable fluctuations in the value of the currency of their fixed urban rents.[41]

The monastery's properties within the city walls were, for the most part, houses or half-houses or huts or cellars, with gardens, or little gardens, behind them and often beside them—and in one case with a piazza in front of it. They were, as we have noticed, at the edges of Rome's heavily populated areas, for example around the churches of San Giovanni Ficocia and Santo Stefano de Arcionibus and close to property, with a monument, of Santa Maria in Campo Marzio (near the fountains Trevi and Tritone). There was also property in the walled country: the monastery was proprietor of what seems to have been a rush house with a garden behind it, on one side of which was the road that went to the "chancellor's garden" and on another a garden.[42]

The monastery's houses were held by tenants—men, widows, families of heirs—for periods of three generations, nineteen or twenty-nine years, sometimes specifically in perpetuity. With the monastery's consent, for which the tenants paid, the tenants could alienate; and the new tenant accepted the old tenant's obligations to the monastery—normally the payment of a yearly rent on a monastery feast, particularly of Saint John, a yearly rent of something like four *denari* or pennies of the senate or two *denari* of Pavia (the currency

The Spiritual Family

237

maintained in recurring rents even though the sale prices changed to the more regular *provisini* of the senate as the distance between the value of the two sorts of pennies seems to have increased, to the monastery's advantage, during the thirteenth century). It makes a regular and comprehensible picture.

The monastery's relations with its tenants were not always this simple. Sometimes the monastery, like other landlords, became involved in its tenants' debts or their arrangements of dowry. Sometimes the tenements themselves were of surprising complexity, like a house, with a small garden behind it, surrounded on all sides by abbey property, which the abbey held with Giovanni Mardonis and the heirs of Andrea Mardonis. In 1268 the abbey permitted a tenant to make division of half of the house with a man named Giovanni di Nicola. The abbey got twelve *provisini* for its consent, and also a promise of four *provisini* each year on the feast of Saint John. In the next year the abbey can be seen consenting, for two *soldi* Pavia and a Saint John's rent of two *denari* Pavia, to the division made by Giovanni di Nicola with Giacomo di Giovanni Tasconis and Mingarda his wife, while Giovanni Mardonis and the heirs of Andrea Mardonis still held the other half undivided. The proprietary structure of San Silvestro's part of the Colonna was, one can see, intricate.[43]

But the regular action of San Silvestro was simple alienation of property, or the approval of simple alienation by its tenants; and in these simple alienations the abbey's notaries produced a series of sharp, simple snapshots of itself and its city. In September 1199, for example, the abbot Stefano, with a consenting monk, leased to Giovanni Petrioli, for three generations, a house with a small garden in the *rione* Trevi next to the church of San Giovanni della Ficocia (or Ficoccia or Ficozza) on the present Via de' Maroniti, between the Largo del Tritone and the Via della Panetteria. The house and garden were bounded by the holdings of Romana di Girardo, who held of the church of Sant'Agnese, by the church of San Giovanni, by a public road, and, at the back, by the tenement of the notarial scribe Giovanni Rainaldi. The witnesses include two brothers and the priest of San Giovanni and a Rainucio identified as a porter. The new tenant paid a sum of three *soldi* Pavia and promised a rent of a penny a year on the feast of Saint John. The document was written by a scribe repeatedly employed by San Silvestro, Stefano di Lorenzo (a quintessentially Roman, diaconal name). In 1246 Scotta, the widow of Pietro di Romano, sold to Matteo Vecclazolo, a mason, or *muratore,* a *pezza* of vineyard with half interest in a set of vats outside the Porta Flaminia at the Monte San Valentino, against the side of the Monte, for seven and one-half *lire* of the senate and a promise to give the monastery one-fourth of each year's new wine and a canister of grapes each year at the vintage. Abbot Gentile consented to the sale and received five *soldi* for his consent.[44]

In March 1250 Abbot Gentile and the convent renewed for nineteen years the lease to Angelo di Gregorio di Pietro Tosi, for himself and as a proctor for his brothers Paolo and Francesco, of a house-lot, with a shed on it and garden behind it and at its side, in the *rione* San Lorenzo next to a road, with the brothers themselves holding on one side and Giovanni di Giovanni Periculi on two other sides. For the renewal the abbey received two *soldi* of the senate and a promise of two *denari* of the senate yearly on the feast of Saint John. The witnesses included Biagio, the abbot's squire or page, his *scutifer*, Ognissanto, another *scutifer*, and Giacomo the usher. In October 1270, in the cloister of the monastery, Prior Giovanni and the convent renewed the lease to Angelo Paolo and Francesco, sons of Gregorio di Pietro Tosi, of a tenement this time described as a house with a garden beside and behind it, in San Lorenzo, by a street, and next to tenements of Giovanni di Giovanni Periculi on two sides, and a tenement of their own on another, again for two *soldi* of the senate (of "good" *provisini*) and again two *denari* on the feast of Saint John. The witnesses to this agreement, written by the notary Carlo, included Agostino, a man described as an agent and procurer. The real names of real people (and their fathers and grandfathers) recur, in real neighborhoods, and in this case in one of remarkable stability.[45]

These regular activities and gentle economic pleasures of San Silvestro, gentle perhaps to the point of insufficiency, were interrupted in 1285 when Pope Honorius IV cleaned out the monastery and scattered the surviving monks to other houses.[46] About to be replaced by Minoresses and Colonnas, the days of the Benedictine Benedicts, Silvesters, Stephens, and Johns were over—although, and perhaps significantly, the incumbents of the last few years had the relatively alien names of Gerardo, Sighilnulfo, Angelo, Rainaldo, Leonardo, Pietro, Tommaso, and Giacomo, nothing to indicate a continuing attachment to the ancient saints. Nevertheless, in their last preserved act the Benedictines still formed the center of a little local community; in it, the renewal of the lease of a house in the *rione* Trevi, they were accompanied by Sisto, priest of Sant'Ippolito (a neighboring church to San Giovanni), by Petruccio, a servant of San Silvestro, and by Paolo, its cook.[47]

Some of the signs that indicate approaching mortality—or change of hands—are apparent at San Silvestro in the half century

The Spiritual Family

239

before its fall (although the most important sign, the apparent papal policy of turning small Benedictine houses over to other orders, has nothing specifically to do with San Silvestro). The translation of an abbot to another Roman house, the translation of Gregorio to San Gregorio by Clement IV is one sign. The long period (1269–1277) when the abbey had no abbot, but was ruled by its prior, is another sign. The preferment of a foreign abbot, Matteo from Subiaco, by John XXI in 1277, and Matteo's subsequent removal to Saint Paul's by Nicholas III in 1279 is a double sign. A further one is Martin IV's concern with the abbatial vacancy and election of 1283 and 1284. One sees the dangerous nervous agitation of short abbacies and papal involvement.[48]

There is an earlier and more basic sort of sign preserved in an act from the year 1249. In that year, on April 27, the abbey alienated an important tenement. The monks listed themselves: Gentile, the abbot, Benedetto, the prior, Stefano, the chamberlain, Gregorio and Stefano and Nicola—all three subdeacons—the acolyte Gregorio, two priests—Placito and Giovanni—and Castorio, their scribe and proctor. They leased away a whole tenement of lands cultivated and uncultivated with the "Cripta Maria," vineyards, and gardens on the Via Tiburtina, east of Rome but on the Roman side of the Ponte Mammolo, to the Count Giovanni Poli (the Conti "heir" to the Poli estates). The tenement was next to one which Giovanni already held, and next to another which belonged to the church of San Lorenzo fuori le mura. For the alienation, the monastery was to receive fifty *lire* of the senate and a yearly rent of twelve *denari* and two containers of wax a year, each with an estimated value of twelve *denari*. Of the fifty *lire,* four less five *soldi* were given to Pietro Malabranca for the price of a horse; four *lire* were used for various necessities of the house, and five *soldi* were for the *refutationibus* of the *denari*. But the bulk of the sale money, thirty *lire* and twelve *lire* (and one notes a superficial failure of arithmetic), was used to pay back two loans for which the monastery had pawned belongings. Thirty *lire* were to go to the archpriest of Santa Maria Rotunda (the Pantheon) to recover the things the monastery had pawned to him: a gilt silver cross, another cross of silver with precious stones, two dorsals, two chasubles, and one cope; twelve *lire* were for another lender, Antonio, for the recovery of a gospel book, a book of epistles with silver binding-boards, and a red cope. The lease and the arrangements for recovery were recorded in an act drawn up in the curia of Stefano (de Normandis), cardinal deacon of Santa Maria in

Trastevere (1228–1254) and then papal vicar in the city, and before the vicar's vicar. Present were a number of men: Master Nicola of Trastevere; Bartolomeo, abbot of San Teodoro de Trebiano; Master Luca de Babuco; and Master Bartolomeo, medical doctor and chaplain to the cardinal vicar. The formality of this medically attended reordering of San Silvestro's possessions is an indication that the shame of its debt and pawning were publicly known—another danger signal for a small thirteenth-century Benedictine house. The sale also, in a different medium and at a different level, recalls the Annibaldi sale of lands—both houses were faltering, and both sold.[49]

But in spite of these rather ominous indications of decay, Benedictine San Silvestro also showed signs of vigor even later in its life. In the year 1267, with great ceremony—and, no doubt, silver cross and red cope—the relics of the church (stone from the Holy Sepulchre, a relic of Calvary) were rearranged and made the center of public celebration, as many of them were translated from the chapel of Saint Denis to the new altar erected in honor of Saints Paul and Nicholas.[50] It is possible of course that this was a last sad effort, this advertisement of relics, to attract veneration and support. Presumably, however, a monastery still this active, inhabited and endowed with precious and venerated relics, might have survived had it not been needed for other purposes, needed to house the followers of a new "saint." Presumably it was in the interest of the "saint," her followers, and her family that Honorius IV, in a surviving letter dated from Santa Sabina on November 2, 1285 could find that San Silvestro, already Minoress, had been vacant through the translation of its abbot to the Benedictine monastery of San Lorenzo fuori le mura, and had not been able easily to reform itself through the efforts of Benedictines. On the dorse of the papal letter, in that space where the name of the person or institution by, or for, whom the letter was procured customarily appears, in contemporary chancery hand appears "Cardinal' de Columpna"—the Cardinal Colonna.[51]

Margherita Colonna had died in 1280, on, coincidentally, the Eve of Saint Silvester, December 30. Although Margherita had considered herself a follower of Saint Francis, and more particularly of Saint Clare, she had not been in any formal way a Claress or Minoress, nor was her community bound by one of the Franciscan rules. The community had lived together on Monte Prenestino in Colonna territory, women of enthusiastic virtue in the freshness of their vocations gathered around and bound together by the leadership of their saintly young noblewoman

The Spiritual Family

—a princess, if not in name, on a hill, and a hill, too, over which she might have ridden in stately grandeur had she chosen. Margherita's position gave the community the security it needed.

With Margherita dead, things looked less certain, perhaps particularly to her own brothers and the more extended Colonna family. It was very much in their interest to keep her cult alive and clear, and to see that her community sustained itself in order. The nuns themselves, deprived of their protector, asked (at least formally) the aid of the cardinal bishop of Palestrina, Girolamo Masci, the future Nicholas IV, the former minister-general of the Franciscan order, a Colonna man.[52] The cardinal talked to the pope, who accepted the nuns into the Isabella order of Minoresses, confirmed their elected abbess, Herminia, put them into the Franciscan province of Rome, asked six friars to serve them, and gave them the house of San Silvestro in Capite.

Thus prompted by two cardinals, a Colonna and a Colonna sympathizer, Honorius IV Savelli brought the little community from the Alban hills inside the order and inside the city. As the body of Margherita was brought down from the hill to her new church in the city, the bells in San Silvestro's campanile, then already a century old, rang out on their own accord to greet her arrival. The old monastery (with its tenements, rents, and privileges) and the new nuns had of necessity to adapt themselves to each other. They managed with disappointingly little strain.

The nuns were quickly busy about the monastery's business. In October 1287 the abbess and nuns consented to a tenant's sale of his tenement of his *pezza* of vineyard outside the Porta Salaria (at the Monte di Sacco Guiderulfi where the monastery had a cluster of vineyard tenements). The nuns were to get five *soldi* of the senate for their consent and one-fourth of the wine at the vintage. Earlier in the same year the nuns were renewing a grant for three generations in Bassanello, accepting lands in Gallese, and securing their hold over the church of San Pietro in Vitorchiano with its rights of baptism, burial, and tithe. Again in 1293—all the old business—Herminia and the convent approved the sale of his tenement by a priest, Pietro di Stefano di Romano di Giovanni; the tenement was a house by the road with a garden behind it in the *rione* Colonna, bordered on three sides by tenements which Pietro and his brothers held of the monastery and on another side by a tenement which the buyer, Egidio di Angelo di Malabranca, held of the monastery. For their permission the nuns received twelve *denari* and a promise of a continuing rent of two *denari* each year on the feast of Saint John in the summertime.[53]

By 1298 the nuns were standing behind their grill, where, as they said, it was customary for them to stand and do their monastery's business, and consenting to this business. (By the early fifteenth century they would be explicit about which grill: the upper one next to and near the great altar, or the lower one by the gate of the monastery.) The effect is very odd. These nuns, who in the saint's lifetime seem like wildflowers blossoming on the Alban hills, sustained by visions and the love of God, around the almond tree in their garden, are now pressed behind their grill within the city walls. The effect is also odd when it's seen another way; these women, so withdrawn from the world that they look at the great altar of their church through a grill and sing behind it, come to the grill to meet witnesses and transact business. Cut away from the world almost completely, they are not cut away from just those affairs of the world which Francis and Clare had found, only a little time before, most corroding. The scene is a distillation of a great Christian tragedy. One feels in hearing it, too, something of the dissatisfaction one feels in hearing Madame de Cintré behind the grill of the Carmelite chapel in the Avenue de Messine, and at least for some of the same reasons.[54]

It would have been consistent with apparent papal policy if the house of San Silvestro had been taken from the Benedictines and given to the Minoresses partly so that waning gifts and resources would again be stimulated by the presence in the old house of a seemingly brilliant new order, capable of attracting financial support. San Silvestro, like other houses in the order, was protected by Nicholas IV in his general privilege of immunity; even San Silvestro's enemy, Boniface VIII, assured the house, in 1298, of the Claress privileges. Again one feels a curious sensation, as one watches the repeated papal privileges for the followers of Francis, who had so seriously enjoined his own followers not to seek or accept papal privilege.[55]

In fact San Silvestro was greatly enriched by its change of order, and almost immediately; but the enrichment was due primarily to Colonna not Franciscan connections. In 1290 a papal chamberlain, Pietro Colonna, son of Pietro Colonna, prepared his will.[56] In it a fortunate legatee was the house of San Silvestro, *ubi pauperes quedam spiritu moniales existunt;* Pietro's grant of major holdings in the Colonna's part of Lazio, in the areas of Gallicano, Pantano, and Campo d'Orazio, made the poverty of these nuns even more spiritual. It was a grant which came with strings attached. One testamentary injunction, perfectly ordinary, although through the vicissitudes of time no longer observed, was that a perpetual chaplain should be attached to an altar

The Spiritual Family

which should be erected and where he should be bound to say a Mass for the dead (including Pietro himself and his own) on Mondays and a Mass of the Virgin on Saturdays. Pietro also asked the nuns to receive as nuns, should the girls wish, his nieces, Bartolomea, daughter of his brother Fortebraccio, and the same brother's natural daughter, Angelella; Pietro also asked the nuns to accept Andrea, the daughter of a poor woman named Gemma from the Colonna town of Gallicano—perhaps another Colonna bastard, perhaps a poor girl who had attracted Colonna sympathies (and Pietro's attachment to Sant'Andrea in Gallicano where he chose to be buried may seem a clue). Pietro named as his executors Giacomo Colonna, cardinal deacon of Santa Maria in Via Lata (1278–1297, 1306–1318), Pietro Colonna, cardinal deacon of Sant'Eustachio (1288–1297) and of Sant'Angelo (1306–1326), and the Roman senator Giovanni Colonna (Margherita's biographer).

The Colonna neighborhood—their holdings at the mausoleum of Augustus, and the site of the hospital a member of the family was about to found at San Giacomo in Augusta, and the two cardinals' churches at Sant'Eustachio and Santa Maria in Via Lata—seems to close in around San Silvestro. It seems to, even more, when one considers some of the people attached to San Silvestro in these years. The nuns' notary and proctor and witness Antonino was clerk and canon of Santa Maria in Via Lata, Giacomo Colonna's church. Another proctor, Andrea, like one of the early nuns, is identified as coming from Affile in the Colonna hinterland between Tivoli and Palestrina.[57]

Another of the nuns' proctors, the monk Angelo of the monastery of San Gregorio on the Celian, introduces a more provocative and elaborate connection.[58] He may have been the Angelo who was a Benedictine monk of San Silvestro just before the Benedictine house fell. The San Silvestro-San Gregorio connection is indicated by the earlier transfer of Gregorio, abbot of San Silvestro, to San Gregorio. The Colonna connection is suggested by a story in Giovanni Colonna's life of Margherita. It is the story of a miracle which happened in a convent of nuns of Saint Benedict in Roiate near Subiaco (and again in the general Tivoli-Palestrina area of Colonna dominance). Through the invocation of Margherita, as the nuns watched, the boy, debilitated since infancy, was cured as an adolescent of the stone ("which shot out like a bean from a pea-shooter"). This boy, at the time of the telling, was Matteo da Roiate, a monk in the monastery of San Gregorio on the Celian in Rome; it may also be recalled that one of the last abbots of San Silvestro was a Matteo from Subiaco who passed through San Silvestro on the way to Saint Paul's.[59]

The convent of San Silvestro directly and indirectly, in sure evidence and by inference, seems caught in a Colonna web, in neighborhood great and small, in family and *familia* close and extended. This is seen in the list of eight consenting nuns on a document of 1296; one, a leader in the house, was Giovanna, daughter of Giovanni Colonna, and another was Margherita, daughter of Oddone Colonna. There is a constant sense of surrounding family, and Colonna connections remained with the house a long time: in 1368 a Maria Colonna was abbess; in 1417 there were two Colonna nuns, Maria and Egidia—but there were also two Orsini, an Annibaldi, an Anguillara, a Caffarelli, two d'Antiochas, one of whom was abbess, and a woman called Nardola Colae Pauli Judei.[60]

Cardinal Giacomo Colonna, the "founder" of the house, as a transcript of the pertinent document calls him, drew up rules for the house. They concern themselves with proper attendance at the Divine Office, with silence and propriety in dormitory and refectory, with the removal of the sick to the infirmary, and the care with which the cellarer should guard the food. The rules speak of a special privilege granted to the house of "conversations" between Mass and Terce and between siesta and Vespers. In spite of this laxness in terms of isolation and silence, the rules are very particularly concerned with the nuns' enclosure and their contact with the outside world. The use of the opening in the grill, through which the nuns received Communion, and particularly the *rota*, through which communications of various sorts came from, and went to, the outside world, were carefully regulated. The keepers of the *rota* were normally to appear at it together, and other nuns were not to dillydally near it. Nuns who went to the well to draw water were to do it silently, not with loud conversations or voices raised high enough to be heard by the outside world. Badly behaved nuns were to be punished, and for several offenses, by sitting upon the ground in front of the convent and eating a harshly restricted diet—a strange sort of punishment for followers of Francis.[61]

San Silvestro, like much else, was involved in the war between Boniface VIII and the Colonna. Enraged by the Colonna and greedy for their goods, Boniface turned upon San Silvestro as a Colonna house, and in turn, no doubt, further enraged the Colonna. In December 1297 Boniface made the great Franciscan Matthew of Acquasparta protector of San Silvestro after he had deprived Cardinal Giacomo Colonna of the office that had been granted to him by Nicholas IV. Boniface also deprived Giovanna, the niece of Giacomo and Margherita, of the office of abbess. Boniface further deprived the monastery of its aristocratic Isabella rule and gave it the plain Urbanist Claress rule—a change in some minor observances and an insult. The house was even further humiliated by being placed under the protection of the Orsini cardinal, Matteo Rosso, the protector general of the Franciscans. The convent resisted Boniface's injunctions and was excommunicated and placed under interdict by the pope.[62]

In September 1303 Giovanna and her nuns and convent were freed from the disabilities which they had incurred for disobeying Boniface, and Boniface's injunctions were revoked. In December 1303 Benedict XI further ordered the restoration of confiscated monastery goods. In August 1318 John XXII appointed Cardinal Pietro Colonna to succeed Giacomo, then deceased, as the nuns' protector. And in 1322 the house was described by the papal chancery (echoing a petition) as one in which "very many of the family Colonna and their relatives are known to exist." To its church in 1347 the grieving Colonna widows stole with their three slain Colonna husbands, victims of Cola di Rienzo and objects of his wrath.[63]

The house was again safely Colonna, propertied, and Minoress; and it faced a long, secure, if not particularly spiritually brilliant career. In the early fifteenth century, in 1417, the convent could count in itself at least twenty-one nuns (and say that they were at least two-thirds of the convent), and in the mid-sixteenth century, in 1558, at least thirty-six nuns. The Catalog of Turin described San Silvestro as a house with thirty-six nuns and two friars. And the 1323 will of the rich sister (concerned with her rings and desirous of being buried in San Silvestro) of a San Silvestro nun, Sister Agnese, which remembered both sister (twenty florins to be kept by her mother and given at Agnese's will for her needs) and house with lavish bequests (including red velvet for a chasuble for Mass), left ten florins, two for each friar, and forty florins, one for each nun. The problem seems for a while at least to have been the danger of too many rather than too few nuns. In a clause dated November 11, 1322, and appended to Giacomo Colonna's rules, the abbess and nuns swore upon the altar of Our Lady never to give assent to the admission of any woman, regardless of her condition, to the house, if she raised the number over forty. The clause ends by restricting rather harshly the meaning of its phrase "of any condition whatsoever" by saying that the restriction does not apply to "the daughter of an emperor or a king or a prince or a duke or a marquess, etc." [64]

On December 29, 1310, Abbess Egidia, with four nuns (Francesca of Sant'Eustachio, Francesca of Parioli, Francesca of Palestrina, and Margherita of Palestrina) consenting for the convent, conceded to a priest (Francesco Custanzi of Perugia—one notes that the name of Francis is not forgotten) the church of Santo Stefano de Arcionibus and all its possessions for a yearly nonmoney rent at Christmas and Easter and the promised invitation to the friars of San Silvestro to celebrate Mass at Santo Stefano on Saint Stephen's day. The priest promised to avoid causing trouble between the convent and the people of Santo Stefano, and he was invested with the church's keys. In acts like these, and in these nuns, behind their property and their grill and their friars, the Franciscan revolution had been domesticated. (And by the early eighteenth century their kitchen in which they made pastries—presumably largely for external consumption—used ten thousand eggs a month.) [65]

It was equally domesticated across the river at San Cosimato in Trastevere. The San Cosimato story is in fact very similar to the San Silvestro story, but it is the same melody in a different voice. San Cosimato's holdings were out different gates and in different towns; San Cosimato changed order at a different date and to a different branch of Franciscan nuns. Its social caste was at least slightly lower; and it was a convent of southwestern not northeastern Rome, of the right not the left bank. It is now in the center of Trastevere, still behind its handsome, columned, arched, gabled, brick (closed) gate, with its romantic, distorted cloister, its pretty garden, and, before it, its hideously butchered piazza continually disguised by the brightness of market days.

San Cosimato entered the thirteenth century already an old Benedictine house. Its part of Trastevere was relatively rustic; it seems to have been slightly more a country convent than San Silvestro. Vineyards as well as gardens came up to its monastery walls. In the thirteenth century the house was small but not pathetically so. Under Abbot Ugolino (1200–1219) it seems to have been healthy; at any rate its documents show no striking signs of disease. In 1209 there were at least seven monks and *conversi:* the abbot, the prior, the steward, two monks, and two *conversi.* In 1227 under Abbot Clemente there were still at least seven, five of the same seven. Under Abbot Massano in 1232 there were at least eight; but the proportion of *conversi* (lay brothers and, perhaps, in this case, novices) to choir monks complicates the picture: there were the abbot, the steward, one monk, and five *conversi.*[66]

The monastery's properties centered around itself in Trastevere with holdings in neighborhoods called after the churches of San Callisto, the Quaranta Martiri (San Pasquale Baylon), and San Giovanni

The Spiritual Family

Lombrica. Besides Trastevere vineyards, San Cosimato had houses and gardens, bake-ovens, and river mills. The monks were local proprietors and landlords; and they collected money from rents and permissions to alienate. When, for example, in a transaction we have already noticed, from October 1218, Barone di Paolo and his wife Regimina di Biagio Romani Mellini (with the consent of her sister Caracosa and the sister's husband Pietro di Gregorio) sold, for four and one-half *lire* of the senate, a house with a garden beside it and behind it (and all its appurtenances) in the *contrada* Quaranta Martiri, Abbot Ugolino (or Hobolino as the document calls him) got six *denari* of the senate for his consent. And the instrument (decorated with the scribe Angelo's scrinarial sign, appropriately a wing) notes that on two of the three tenanted sides of the tenement (the fourth was a public road), the tenements were held of the abbey of San Cosimato—it was a San Cosimato neighborhood.[67]

San Cosimato's gates were the Trastevere ones of Porta San Pancrazio and Porta Portese, and the monastery held considerable holdings outside both. Some of the holdings still come with little fragments of narrative plot attached to them. In 1203 Abbot Ugolino bought from Tebaldo and Pietro, the son of Leone di Francolino, and from Giovanni and Romana, their minor siblings, and Scotta, their mother, four *pezze* of vineyard, with the attached land, and one vat on one of them, and with a common vat at the foot of another, for sixteen *lire* of the senate. The vineyards were in the Marcellis, or Bravi, an area just outside the Porta San Pancrazio (near now the American Academy). One plot was said to have had no fourth side because it was a triangle; neighbors were named, among them a holding of San Lorenzo on the Gianicolo.[68]

The family of Leone di Francolino needed its sixteen *lire*—and San Cosimato (as one interprets it) helped them in, or took advantage of, their need. The family had to pay back the usuriously increased debts contracted by their father and husband when he was alive: to Teodora di Buonofilio and Guido Piperarola her husband (and a witness) eight *lire* capital, and twenty-four *soldi* interest (*pro usuris*); to Pietro Roscimanni, four *lire* capital and twelve *soldi* interest. The remainder (two *lire, four soldi*), they said, was to be used "for our very great and evident need." The continuing reality of that need (and San Cosimato's continuing involvement) is stressed by another alienation by Leone's heirs in the following February, in 1204. In it they gave as a pledge, and conceded, to the steward of San Cosimato a small piece (*petiolum*) of arable land in the Marcellis Minimis, again outside the

Porta San Pancrazio, for thirty *soldi* of the senate. The land was on the public road and was between San Cosimato land and other land of the heirs of Leone di Francolino—one is watching the growth of one neighbor at the expense of another, of the institutional neighbor at the expense of the familial one. The thirty *soldi* is described as "what we owe the monastery for the grain it sold us to feed us." Among the witnesses to the alienation was a man called Nicholas "Bread and Wine" (Nicola Panis et Vinum), recalling in his perhaps rather ironic presence the good saint who succored the poor.[69]

The Porta San Pancrazio was clearly a San Cosimato area. This is shown again in 1221 when Enrico, the son of Enrico of Sant'Eustachio, with the consent of Teodora his wife, sold to Abbot Clemente a vineyard in the Marcellis on three sides of which the tenements were held of San Cosimato. And from the Marcellis, San Cosimato got some at least of its wine.[70] But the Porta Portese was San Cosimato country too, and particularly that part called "the plain of the palms." Some of the Porta Portese land was arable; and some of it marched with the fields held of Santa Cecilia in Trastevere.[71]

The holdings of San Cosimato were by no means limited to the immediate area of Rome—like all major Roman families and all other observable Roman monasteries it was involved in the tenements of country towns. The area around Campagnano di Roma, about thirty kilometers north of Rome, just off the Via Cassia, can be thought of in the surprising detail of street and ditch and cave and gully, of vineyard and neighbor, because San Cosimato preserved a long and elaborate document of renunciation (to it) so describing parts of the place, with other places, drawn up at the time of Abbot Ugolino and his steward, Mauro. San Cosimato's was a complex holding connected with the church of Santa Maria de Prato, on the edge of the town, which a compromise had alienated to the priest Angelo in 1200, while retaining his obedience; and his obedience included providing hospitality to the monks. San Cosimato's verbal map of the Campagnano area is not really surprising because San Cosimato was a good keeper of records. San Cosimato had a cartulary (an unusual object in these parts), or at least it began a cartulary in the 1190s. In 1282, considerably after its change of order, it had a list of its lands in Sutri drawn up.[72]

Sutri was in the early thirteenth century the town of San Cosimato's interests. These interests were tied to a subject institution, the hospital of San Giacomo (Santi Filippo e Giacomo) in Sutri. The monastery bought land in Sutri, and other property was given to it. In 1205, Finaguerra, the son of Giovanni di Sebastiano, gave to the house of

The Spiritual Family

Santi Filippo e Giacomo, through the hands of Matteo its provost, himself and his goods—houses, vineyards, lands, gardens, arbors—for the redemption of his soul and the souls of all his relatives.[73]

In 1220 San Cosimato continued to seem, as it had earlier in the century, a worthy recipient of gifts. If those gifts were acts of pure piety, then Cosimato must have seemed a worthy vessel of religion. If they were financial investments, insurance, annuities, then it must have seemed financially trustworthy and promisingly solvent. If, as is generally likely, they were a mixture of the two, then San Cosimato must have seemed to observers worthy of the trust that their mixed investment implied. It must have seemed at least stable. In 1203 the widow Bona made her son and her two daughters her legatees: to one daughter, Jacoba, Bona left what she had given her for a dowry, plus ten *soldi;* to her second daughter, Teodora, she left forty *soldi* (two *lire*); to her son Pietro she left all her other goods and her lands which she had in Marcellis. But she left three *lire* (more than for her daughter) to San Cosimato for her soul.[74]

In 1212 another widow similarly concerned for her soul arranged the disposition of her property for her soul's protection differently: she left the monastery of San Cosimato all her lands in the Marcellis which lay between those of the monastery and those of a man named Giovanni di Leone and his nephews (and at least one suspects that poor Giovanni di Leone di Francolino faced again and anew his Benedictine neighbors); the widow's lands were divided into two *pezze.* In October 1220 Giovanni di Benedetto and Marsilia his wife offered, for the healing and redeeming of their souls, themselves and their goods to San Cosimato, with the reservation of certain bequests (particularly for Giovanni's sister) and of the use of their goods for their lives, and with the arrangement that should Giovanni die of his current illness his great wine vat, full of wine at the time, should go to the monastery, and also twenty *soldi* for funeral expenses. In May of 1220, a man named Bentivenga had offered to God, to the monastery, and to its more local branch, San Giacomo of Sutri, himself, his wife Teodora, his son Cesareo, and all of his goods in Sutri (saving forty *lire*), to take care of them and their bodies.[75]

It was also in 1220, in October, that Teodora, daughter of Lord Obizione di Giovanni Ionaci and wife of Pietro Bonifilii, gave to her son Giacomo (and all her male heirs) her closed garden with its trees located by the church of San Biagio (Saint Francis's supposed hostel by the later church of San Francesco a Ripa) on a corner of the public road and the Vicolo San Biagio (across it from the church) and next to her paternal uncle Pietro's tenement. She reserved its use for her life. She also provided the condition that should her sons die without children the tenement should go to the monastery of San Cosimato—for the remission of her sins and those of her dead—saving the tenement's use for life by her own mother, Maria. This arrangement of the property of a Trastevere-Ripa family, visible in its various actual and potential generations, was witnessed by Pietro, priest, and Girardo, clerk of another local ecclesiastical organization, the church of Santa Maria in Capella (now famous for its old campanile). What they watched is very interesting—the gift of the reversion of a tenement in one of Rome's future Franciscan centers, where Francis himself had supposedly stayed ten years before, to a house of Benedictine monks itself due within fifteen years to go over to a branch of the Franciscan order. In fact, in July 1229, Gregory IX ordered the monks, for the love of religion and the good of their souls, to give the church of San Biagio, which pertained to them but which, the pope said, was derelict, to the Franciscans who needed it—an ominous sign.[76]

By August 18, 1234, Gregory IX (whose biographer explains his act by his devotion) had changed San Cosimato's order. Gregory had, after an initial flirtation with the Camaldolese, made San Cosimato a house of Damianites, a house of the followers of Clare, not yet ordered as they would be by their later rules. On that day Gregory appointed the Franciscan friar Giacomo the abbey's steward. The appointing papal letter is preserved in an appendix to an instrument which records Fra Giacomo's own appointing of Fra Paolo as the abbey's steward, an act and instrument dated April 25, 1235. The scribe, Marsilio, copied the Gregory letter, he said, because it was difficult to carry it about everywhere.[77]

Marsilio's instruments are a clue to the nature of San Cosimato's change. Marsilio worked for the abbey both when it held Benedictines and when it held Damianites or Claresses. There is no significant change in his neat, lined, nicely written and decorated, trapezoidal instruments. They look the same, and they do the same sort of business. Except for the masculine names of the monks and the more plentiful feminine names of the nuns it would be impossible, for example, to tell which of two documents, one from 1232 and one from 1244, was Benedictine and which was Franciscan.[78] The important point about the change was that it was very little change at all.

Some change, of course, there was. When in 1246, at the mandate of the papal vicar Cardinal Stefano, the bishop of Ascoli consecrated and gave indulgence to a new altar within the church dedicated

The Spiritual Family

to the ancient patrons Cosmas and Damian, an inventory of relics was compiled.[79] It included great relics (wood from the cross, part of the sponge from the Crucifixion, part of Christ's sepulcher and of the stone from which He ascended into heaven, clothing of the Virgin), pertinent relics (of Saints Cosmas and Damian and Cornelius), relics of the great saints (from Andrew's cross, Luke, Peter, and Paul), conventional Roman relics (Lawrence, Blaise, Sebastian, the Quattro Coronati, Linus, Valentine, many others), many virgins, Roman and distant, single and multiple. But the San Cosimato relic hoard also included by this date, by 1246, a relic of Saint Francis's stigmata. Similarly, by 1277, when Bartolomea di Gentile di Gentile di Pietro di Leone (just possibly a great-great-granddaughter of poor Leone di Francolino) became a nun of the house and offered all her goods to it through the hands of Abbess Giacoma, she explained the act as being out of reverence for Almighty God, the blessed martyrs Cosmas and Damian and of blessed Clare, and for the remission of all her sins. And as Francis and Clare had come to San Cosimato so, at least rather indirectly, had interests in Assisi tenements—in an elaborate papal transfer in 1238 San Cosimato got a church in the diocese of Porto from the monastery of Santa Maria Farneto in the diocese of Arezzo, while in exchange Farneto got a "monastery" in the diocese of Assisi.[80]

The change of order also seems, as it must have been hoped that it would, to have stimulated the giving of gifts to the convent (and perhaps too, some acceleration in the transfer from directly held vineyards with income in kind to money rents—the progress of economic history may have been slightly prodded by the nuns' drinking, if they did, less than the monks). Certainly the change of order brought the abbey more professions. The numbers of the nuns are immediately higher than those of the monks were, although of course initially this merely means a larger plantation. A list, from 1244, names Abbess Giacoma and twenty-eight nuns, their names a litany of Roman saints with appropriate additions. One from 1261 names Abbess Francesca and thirteen nuns. One from 1273 names Humelia and six nuns. One from 1280 names Abbess Giacoma and twenty nuns. Some of the nuns stayed in the house a long time; the Donna Clarastella of 1244 still appears in 1273. Some of the nuns, like Bartolomea, in her profession, or Abbess Giacoma II, the daughter of Pandulfo of the Suburra, identify themselves. Those who are identifiable were from prominent Roman families (the Ponti, the del Giudice), not Colonnas admittedly, although the early-century gigantic Pandulfo of the Suburra was in some ways a more prominent and impressive man than any single Colonna; there is, at least, nothing to suggest that San Cosimato was either modest or squalid. (But of course evidence of family is almost always evidence of prominent family, or it would be unrecognizable.) The house to which the nuns had come, moreover, already had respectable connections; the rich monk Oddone Benencasa (or Benencase) had brought, before 1205, half the valuable house called the Arco di Benencasa to the convent. San Cosimato's location in Trastevere should not in the thirteenth century indicate that the convent was in any way a particularly popular or vulgar one—nor unfortunately, as later Trastevere connections with the university might suggest, a learned one. In 1317 San Cosimato could still count among its something over twelve nuns (although, and one should not underestimate, the Catalog of Turin lists thirty-six nuns and two friars), both a Frangipane and a Scotti.[81]

At about the time of the change in order a different and in some ways as significant change in the convent's connections occurred. The convent became increasingly interested and involved in tenements in and near Tivoli, a papal summer capital and so a sensible place to invest. (Rents, it will be remembered, doubled in Viterbo when the curia was there; and the market in Tivoli wine, grain, oil, meat, and figs, as well as houses, was undoubtedly very profitably stimulated by the curia's being in Tivoli.) The convent's new connection with an order, one that was increasingly well organized and one that had captured the sympathies of some of the keenest contemporary minds, including investing minds, probably helped it to rearrange its portfolio profitably (although the Tivoli profit was one to be unexpectedly stolen by the Babylonian captivity).

A particular Tivoli involvement at about the time of the change of order was with the murderous Benencasa (or Benencase) family. In June 1237, five years after the documents (preserved in San Cosimato archives) dealing with the transfer and confiscation of Benencasa properties after Oddone di Giovanni Benencase had murdered Bartholomeo di Benedetto di Bartolomeo, San Cosimato bought, for one hundred *lire* of the senate, through a mediating Tivoli neighbor, from Oddone and Giovanni his son, a Tivoli tenement at a place called Cassano. This may well have been part of the tenement which Oddone saved from confiscation by his rather shady, but not unconventional, use-transfer. The Oddone and Giovanni sale in 1237 was consented to by their wives Tyburtina and Maria. These witnessing, consenting wives, brought forward to avoid future legal snags, are interesting women, and interesting in their actions. In 1283 six Papareschi wives

The Spiritual Family

and widows met in San Cosimato's cloisters to consent to a great Papa-reschi sale of country tenements north of Rome to the hospital of Santo Spirito in Sassia for four thousand *lire*. These sets of women thus not only witnessed the dismantling of great and less great Trastevere inher-itances, they also witnessed the swelling properties of Trastevere (or at least *trans*-Tevere) religious houses with which their families had con-tinued attachments.[82]

San Cosimato's interest in Tivoli, once begun, was maintained through the century. In 1272 the convent was again involved with the Benencase. It then bought, for twenty-one *lire,* a vineyard with olive and fig trees at Cassano through Oddone's son Giovanni, acting as the monastery's proctor. One of San Cosimato's new neighbors was Gio-vanni di Pietro Pazzi, as his father had been a neighbor in the Oddone transfer of 1237. The generations of Benencase and Pazzi progressed around the properties held by the dead hand of never-dying San Cosi-mato. In 1278 Bibiano, or Viviano, Bartolomei, a frequently employed convent proctor, bought two Tivoli tenements with vines and olives for the convent. The tenements, which cost sixty and ninety *lire,* were pro-cured with part of the five hundred *lire* which the monastery had from a major sale elsewhere of tenements with lands, rights, and men to Cardinal Matteo Rosso (who makes an Orsini connection for these Franciscans) through his proctor Pietro da Pofi—money which had been deposited with a Florentine merchant. In 1270 the convent bought a tenement near Tivoli from the Lady Bena, with her husband Bartolomeo di Giovanni consenting, for 160 *lire,* and paid 105 *lire* to the wife and the rest to her husband. In 1276 the convent (with its members listed) is seen granting a Tivoli tenement, an old vineyard with olives, through Viviano, to two Tivolese, Guglielmo di Guidone and his wife Resa, in perpetuity, on the condition that they make the tenement a productive vineyard and farm, and that their yearly return to the monastery include one-third of the new wine and of all fruits and grain and a canister of grapes.[83]

Although San Cosimato turned its attentions to Tivoli, it did not forget Sutri. In December 1235 Famiano, "abbot, provost, and rector" of San Giacomo, was engaged in trading tenements in a quite accus-tomed way, as if nothing had changed in the home monastery. In 1239, however, the church rented out the church and hospital to two men with local Sutri interests, Cencio, the archpriest of Capranica, and Pietro Zilla, canon of Sutri, for eight years, for improving the place. The renters were to pay a yearly rent of thirty-four *lire* of Siena, if they could get to Rome safely. The days of payment in this instance rather hilariously recall two of the basic reasons for the church's existence; the days were the feast of All Saints and the Sunday on which is sung, "I am the Good Shepherd." [84]

Continuity, renovation, and rent, in themselves, were insufficient at Sutri. Sutri had a bishop; and against him the nuns wanted protection for their holdings. They got it from the pope. On January 12, 1260, Alexander IV ordered his "venerable brother" the bishop of Sutri to stop trying to make collections and exactions from San Giacomo and other San Cosimato churches and their properties. On April 26, in another letter which survives in two copies in San Cosimato's archives, Alexander ordered the bishop to stop molesting the possessions of San Cosimato, and he declared null and void any pertinent episcopal excommunications. In December Alexander had already asked for the convent the protection of Guido di Enrico, "citizen of Rome." These actions formed part of a pattern of papal privilege and protection, into which also fitted the way in which the convent was shielded from Saint Peter's in the Vatican, as in a letter for them from Innocent IV to the vicar cardinal of Santa Maria in Trastevere, which had, a surviving copy says, to be gotten again because of faults in its phrasing, and another in which it is harangued that Saint Peter's acted as if it had been drinking the accursed water. For procuring papal letters the convent used the Franciscan order's proctor, Bonaspes.[85] As a member congregation in a new and privileged order and a Roman house directly subject to the pope, the convent was in a doubly advantageous position. It worked industriously to consolidate that position. That, in its quest for papal privilege and in its greed, it was violently violating the express and implicit desires of Clare and Francis seems to have been of no consequence to it. At San Cosimato the words of Clare and Francis could not be heard and understood, in this way, by 1260. Between that past and this present there was not that sort of communication, although Clare was only six years dead, although San Cosimato was still called Damianite and had not yet been sheltered by the protecting perversion of Urban IV's legislation, and although another memory of Francis was still alive in Greccio and the Marches.

In 1273 the nuns consented to a tenant's sale of his *pezza* of vineyard and as a compensation they accepted five *soldi* and a promise of one-fourth of the wine and a canister of grapes each year at the vintage. This is the sentence that describes in act the thirteenth-century Roman convent. In 1258 the convent consented to the sale by one tenant (Nicola Alexii) to another (Matteo Acti) of a tenement of vine and trees on a plot of about one *pezza* of vineyard outside the Porta Portese at

The Spiritual Family

Tertium and adjoining another convent tenement. The sale brought to the seller three and one-half *lire* of the senate and to the convent five *soldi* for its consent and also the continued promise of one-fourth of the new wine which came from the place at each vintage, and everything else that had been promised in the old charter. In 1262 another tenant sold similarly another similar tenement, but within the walls of Trastevere and for seventeen *lire*; the convent received five *soldi* and one-fourth the wine at vintage. In 1261 the convent itself leased a *pezza* of vineyard to be improved and one-third of a vat; it demanded nothing the first year, but one-fourth of future vintages, a canister of future grapes, and hospitality. The annals of thirteenth-century Roman monasticism (the preserved annals, one must note the possible distinction) are a list of vintages and saints' days with rents in wine and salt and grapes marked next to them.[86]

The records of behavior in property are not all leases for three generations, leases for limited numbers of years or in perpetuity, exchanges, consents and approvals of tenants' leases, and loans; there are also records of disputes with other landlords, including other churches, with tenants, sometimes, as early as 1244, with large numbers of them who perhaps questioned the Claress succession, or who perhaps recognized it fully and found the Clares heirs to the old fights. Amid alienation and dispute, too, the acquisition of property went on; and some of it was connected with continuing religious sentiment. When in 1246 the convent bought four *pezze* of vineyard from the four daughters (Paola, Constantia, Theodatia, Agelia) of a Trastevere notary, Rainerio, they bought it with the fifty *lire* that a man named Graziano (probably a Frangipane) had left them for the salvation of his soul. In 1260, Donna Beatrice, wife of Jaconello, in his house, before witnesses including a canon of Santa Maria in Capranica, assigned to the proctor of San Giacomo in Sutri all the things which Janconello had in his house, one vat of wine in the house, and three vats, two empty and one full, in another house. (The possibility that the nuns drank less wine than the monks may not, by this point, seem a strong one.) [87]

In 1256 Maria di Gennaro granted herself, the house in which she lived, and six *lire* from her goods to two friars, the abbess and three nuns, representing San Cosimato, for the salvation of her soul; and she reserved the control of her other goods and the use of her property for life, and she specified that, if her niece Clara should survive her, Clara should have the house to live in, but that any goods left when Clara died should go to the Clares' monastery.[88] Maria's "house in which I live" reminds one forcefully that in the everyday life of ordinary Romans these religious houses, which seem on the grand religious scale such horrifying failures, seemed nevertheless to offer security and promise of spiritual (if also perhaps animal) survival. In this, papal reformation had been successful, or at least not destructive; the old totems as well as the old aggregates of property still stood.

The problems of the two Franciscan nunneries, of San Cosimato and San Silvestro in Capite, are echoed in other houses, sometimes more grossly and litigiously, as at Sant'Erasmo sul Celio at the end of the century, where grave discord separated abbess, nuns, and the patron abbey of Subiaco. And the two movements of San Cosimato and San Silvestro were characteristic of their Rome—the dramatic movement of change of order; the repeated life-movement of dealing in property. In 1231, for example, Gregory IX gave Sant' Alessio on the Aventine to the supposedly relatively rigorous Premonstratensian canons. In April 1273 for a yearly rent Sant' Alessio leased in perpetuity to the nuns of Santa Maria Massima four *pezze* of vineyard outside the Porta Appia by Santa Maria "where the Lord appeared"—the *Domine quo vadis?* which touched the apostolic miraculous Roman past. Each of these two movements, that of 1231 and that of 1273, is a typical gesture.[89] But the second is a gesture not merely typical in the life of the thirteenth-century Roman monastery but in that of thirteenth-century Roman landholders in general. The failure of the reformed monasteries to rise above the level of more general society allows them, as historical sources, to represent a very broad segment of society. In some ways, moreover, their representation and failure are infinitely more broadly representative. The disappearance of spirit into the wearing but absorbing tedium of daily economic life—the shuffling of papers and fixing of rents—is the human tragedy fixed in the *exemplum* of these hapless but long-lived institutions—pretentious in their ambitions but not more pretentious than the Christian soul.

More violent in their rejection of convention than the Claresses, more difficult at first to control and consolidate, and to fit into (from the official church's point of view) appropriate houses were the Augustinian friars. Their naked poverty and ecstatic violence were at home in Umbrian and Tuscan hills, but, at least as they were calmed and caught in administrative nets, they came too to Rome. They found a protector and tamer in Riccardo Annibaldi, cardinal and Roman, who fitted their pieces together into an order, and then saved the new complex from the dissolution of new orders. He also engineered the order's new involvement in the topography of Rome in the middle of the century. In or about 1250 Riccardo managed a complicated and important

The Spiritual Family

settling for the Augustinians. He (according to their historian) got them the Franciscan church of Santa Maria del Popolo, at the rustic northern edge of Rome, just within the wall. The Franciscans got the senatorial church of Santa Maria in Aracoeli. The Aracoeli's Benedictines were dispossessed and sent to other houses. Few other monastic movements involved such prominent churches and such diversity of order. None makes more clearly the point of what was being done: the removal of relatively dead orders and the domestication of lively but difficult ones.

The partly domesticated, but even then relatively untamed or unhoused, black Augustinians were still swooping about the city late in the century when they were deposited also in the church of San Trifone (by Sant'Agostino). In May 1287 Honorius IV's plan for giving San Trifone to Santa Maria del Popolo and the Augustinians was still being worked out so that the clerks of the old secular establishment (whom Alexander IV had firmly limited to the number eight in 1255) should be financially compensated for their loss of church—and particularly so that Don Giovanni di Don Paolo Pietro di Jaquinto should. The Augustinians were to provide him for life with an appropriate house and twenty-six florins of gold a year (or a benefice that provided these things), the twenty-six florins to be paid in three installments a year, on the first of June, October, and February. For necessities like this the friar-hermits had gotten seven hundred *lire* of the senate for granting to Don Egidio di Paolo Roffridi, for forty years, the fruits of vineyards and other holdings of the church of San Trifone, including those outside the Porta San Valentino by the Milvian Bridge. These Roman Augustinians make an interesting assembly: lists from 1287 include thirty-two or thirty-three of them, of whom one, Walter, is English and one Florentine; two are from Gubbio, two from Orvieto, one from Siena, one Ascoli, one Lucca, one Narni, one Viterbo, one Anagni, one Genazzano; two on the other hand are called Roman, and several are from (if we can trust these name-identifications) specific Roman districts, two from the Mercato, one from the Monti, one from the Posterula, and one, Nicola, from the Calcarario. One of these Hermit gatherings was held in May in the complex of Augustinian houses which had been the houses of Don Egidio di Paolo Roffridi.[90]

One must turn again to the Aracoeli—one must always return to the Aracoeli in thinking of medieval Rome; everything is gathered there: the altar itself and its memory, the senate, the mosaic, the floors full of senators and lined with cosmatesque work, the old brick of the unfinished façade, the Franciscans, and the chapels, from various times, of a great collection of Roman families, the great and the senatorial names—Savelli, Orsini, Colonna, Capocci, Conti—and the names of neighbors—Margani, Delfini, Astalli, Mattei. In this very Roman place one of the century's great scenes of renunciation and conversion took place—in some ways the century's greatest scene in Rome.

Charles II of Naples had found out that one of his sons, Louis, that son in fact who had, it was thought, been destined to become King of Naples, had been moved by disposition and persuasion to reject all earthly blandishments and to desire only to follow the path of religion.[91] A sense of him and his act of rejecting the crown are caught in one of the greatest, from any point of view, of fourteenth-century paintings, the Simone Martini in Naples.

Charles II himself was in his way both a responsible monarch and a responsible father. He eventually allowed his son considerable freedom. He allowed him to retire from the royal court at Castel Nuovo in Naples to the austere sea-bound whiteness of Castel dell'Ovo, there to read the *Summa Theologica* of Thomas Aquinas which his father had given him and to cultivate the seed of Franciscanism planted in him by spiritual tutors in his youth. Charles insisted, however, that were Louis to enter the church he must enter it grandly and helpfully, and be a bishop—bishop, in fact, of Toulouse, a gift from Boniface VIII.

To accept office, Louis was allowed to go from Naples to Rome. There, meeting the pope, he demanded to be allowed to become a Franciscan as a price for accepting the wretched bishopric. Louis was privately received into the Franciscan order on Christmas Eve, 1296. But the secret was carefully kept from his father. Louis was to wear the habit only in private. On December 29 or 30 he was consecrated bishop of Toulouse with his Franciscan habit hidden under his robes. In early January he went back to Naples to see his family; he arrived on a mule. In late January he left his family to go first toward Paris then to Toulouse. By the end of January he had arrived in Rome. He got permission from Boniface VIII to be a Franciscan in public— Boniface's attitude toward Louis must have been extremely interesting. At the Aracoeli, on the feast of Saint Agatha, on February 5, 1297, he celebrated Mass, before a great crowd, of over one thousand people, it was said, including two important cardinals, the great Franciscan Matthew of Aquasparta and Giacomo of San Clemente, nephew of the pope. When the Mass was finished the cardinals came to Louis and said that he might then show himself a Franciscan. And immediately he stepped out of his robes and showed himself a Franciscan.

The Spiritual Family

This step at Aracoeli, in a huge crowd, above the city of Rome, was taken into the fragile spring of a Roman February; and Louis's period of conversion was almost as short as is that first spring. He died in August 1297, involved in celebrating the Provençal cult (of the southern France of his boyhood not of his bishopric) of the Magdalen. But Louis's moment of glory, blazing in his plainness above the February Rome, was really glorious. What his episcopal life would have been no one can know. But he had, at all that distance, made a real renunciation, made Aracoeli in a way really Franciscan, carried, in a way, Assisi to Naples and to Rome, and to 1297. And he caught forever the attention at least of his brother and successor to the succession, King Robert the Wise, who, as a royal lay-preacher, would tell his subjects of Louis. Still, even with its beauty and clarity of scene if not of meaning, the gesture at the Aracoeli cannot make disappear the opposite, heavy gesture of accumulation at San Cosimato and San Silvestro. Theirs is the main plot line of this thirteenth-century sort of Franciscanization; and it is tragedy, at Rome as elsewhere.

CHAPTER
VII

Last Wills and Testaments

MEN IN ROME KNEW THEY must die. Across the city, through the *rioni* and neighborhoods, fishmongers in Sant'Angelo, sailors in the Ripa of Trastevere, roe-eating Premonstratensians in Sant'Alessio, small farmers in their low houses north of San Silvestro, Orsini in their high, rebuilt, Theodoric-recalling ruins in the Theater of Pompey, all of them, and their wives and daughters, knew that death was the end, for each of them, of his earthly city.[1] They feared death. They avoided what seemed its causes, and they sought what seemed its preventives. As they felt, in their age and sickness, and sometimes even in youth and health, the stealthy movement towards their faces of those angels' wings which, particularly by end century, they could see in their churches, and could see in Santa Maria Maggiore almost, in the mosaic itself, moving across the faces of Giacomo Colonna and Nicholas IV—like Maitland "in 1906 when, during his last illness, the mosquito nets around his bed were translated," except that Romans with mosquito nets would have gone more slowly to their deaths—they called to themselves seven witnesses and a notary; and they made their wills.[2]

In his will, each articulate Roman testator tried to complete the structure of his life, to close its narrative with a proper ending, to achieve what had not been achieved and do what had been left undone, to secure his house, pay his debts, and do penance for his sins. He tried to find for himself (and his family) as much immortality in each world as the substance of his life (and his possessions) could afford. In this he moved in conventional paths to conventional acts—although not without that occasional rough originality which shows the individual beneath the type. His convention was informed by traditions, although not static ones, which he and his family and advisors and, perhaps particularly, his notary knew or assumed. Of his will's bequests and provisions, some, both conventional and individual, were specifically religious, the final (or so meant to be) physical expressions of a lifetime's complex of belief.

Last Wills and Testaments

263

Religion, as Acton said, is the "first of human concerns"; and its expression in piety is probably the most consciously exalted, the most serious, of the ordinary man's intellectual activities, and the one in which his rational and irrational qualities of mind are most interestingly combined.[3] So too Romans, in the pious acts of their wills and bequests, acted most interestingly, even when they were in fact guided by convention and cliché, although even in convention their piety seldom seems stale. Romans breathed the air of streets along which Francis walked early in the century, and Margherita Colonna and Altruda of the Poor, late, streets on which houses of the new orders—Franciscans, Dominicans, Augustinians, Carmelites, Friars of the Sack, Celestinians, and Silvestrines—were forming themselves.

The piety of thirteenth-century Rome did not, of course, exist in geographical or chronological isolation. Countless waves and winds and filaments tied it to the past, the future, and the outside world. Long before, in the north, Norbert of Xanten, like some gigantic electric prototype, and with him Robert of Arbrissel, had predicted the outlines of thirteenth-century piety; and John of Matha of the Trinitarians in Innocent III's Rome, at once hiding hermit in the aqueduct and active disciple in the hospital for freed slaves, clearly echoes them and at the same time establishes the type of the thirteenth-century Roman, and Italian, at least, religious enthusiast—that strange, tense, recurring combination, the active hermit. Between Norbert and Robert, at the end of the eleventh and beginning of the twelfth century, and John, at the beginning of the thirteenth, there flourished an "evangelical revival" which stirred the twelfth century and which has stirred again the perception of modern historians. In this revival the romance of poverty and humility grew, and it grew, in part, around the narrative of the life of a Roman boy, Alessio, son of the senator Eufemian, remembered in Rome in painting at San Clemente, in his father's stairs under which he supposedly lived and died, in the church and monastery which Honorius III rededicated to him on the Aventine (and in which the Premonstratensians ate their roe). Remembered in Rome and Roman in supposed origin, Alessio's fame spread through Christendom on the wings of a rhymed life:

> Sainz Alexis est el ciel senz dotance
> Ensemble o Deu, en la compaigne as angeles

with God before Roland and with less military virtues. Alessio's aspirations joined those "common to all religious reforms" from the eleventh to the fifteenth century, "the desire to emulate the life and teachings of Christ and his apostles." [4]

But within the general history of piety, in Rome as elsewhere, the period around 1200 was an extraordinary one. Then, in a peculiar way, in a new style of seeing and being (the simple sharpness of Francis's parsley), the New Testament—Matthew and Acts—began to come to life; and a domestic Christ started to eat at Martha's table. It was Innocent III as well as Francis. But when the cicada which had disturbed the peace of Innocent's camp at Subiaco, or her cousin, came at the Portiuncula, to the branch of the fig tree next to "the furthest cell next to the hedge of the garden behind the house," and Francis stroked her and said, "Sing, my sister cicada," she had clearly flown into a new bright world with Christ at its center.[5] It was a world in which, at best, exhilarated men and women, with sharply observant eyes open to the physical present, tried to re-create the structure, the act, the idiom, the vocabulary of the gospel, to feed the poor, to visit the sick—to do what Innocent III tried to do at Santo Spirito, what Margherita Colonna would try to do in Zagarolo, to wipe Christ's face with Veronica and touch his side with Thomas. It was a world in which William of Rubruck, traveling to the Mongol East, would say in answer to a jeering question, that, like the soul in the body, "God is everywhere." [6] Here, but perhaps in the lightly applied Word of country rusticity, the boy who would be Celestine V grew up; and from a painted cross, of a sort still familiar to us, from the school for example of Berlinghiero Berlinghieri, the painted Mary and John, whom the boy Peter of Isernia was still too simple to recognize, came down to him and read with him his Psalter, which his earthly mother was having him learn to read, so that she might have a second, and this time serious, cleric among her sons, in spite of the opposition of his brothers who had said " 'it is enough that one of us doesn't work'—because in their *castro* no clerk worked." [7] Through this countryside were spread, more convincing in their wooden impermanence, those affecting, sentimental depositions of Christ from His cross—those nail-ridden, stigmata'd feet—Tivoli, Norcia, Bulzi, and, most particularly, Volterra.[8] Here Charles II of Naples, "inflamed by the Holy Ghost," would give a hundred beds in a hospital at Pozzuoli to Christ in His poor and dedicate a chapel to his century's model, the working sister Martha.[9] This is the world of Franciscan literalness: "What is evoked by the crib, the rosary, the crucifix, is the Gospel in its literal sense"—Francis, his stigmata and his real Christmas *presepio* at Greccio with ox and ass, the Franciscan "spirituality of tenderness" (oddly and beautifully mon-

Last Wills and Testaments

umentalized by Arnolfo at Santa Maria Maggiore); [10] it is Ubertino writing "I was first the ass, then the ox, then the crib, then the hay. . . . He took me with him into Egypt." [11] Nothing is more obvious in this century, in this world, than the friars; and they are cause, result, sign, emblem, victory, and defeat. The were propagandists for their kind of piety, but also they were popular because of it.

The lively piety, the piety of life, of the thirteenth century, and certainly of thirteenth-century Rome, turned itself toward the face of Christ. This glistening, Christocentric complexity moved itself, at least in Rome, I think, around a central trinity of things: the real face (sometimes more complexly real because iconic) of the living Christ the man, in all its sacred representations, but particularly in the Veronica; the real presence of the living Christ, God and man, in the Eucharist; and the real remembrance, the literal copying, of the living Christ's acts on earth—and the presence of Christ in the least of his brethren—in the corporal works of mercy. It was a piety, then, centered in the Veronica, the Eucharist, and the corporal works, the triad of Santo Spirito.

To live like Christ, in His presence, with His words, is very hard. From the beginning there was a subversion, a sublimation, a spiritualization, an institutionalization of the literal interpretation of the apostolic life. Few men can kiss lepers, or be absolutely poor, or even feed the hungry with their own hands, or love their neighbors in any direct and personal way; driven by Christian need they can instead found or support hospitals, join or support orders, live with Christ by contemplating His face or consuming His body. (And, of course, one could argue that it was the act, the leprous kiss, which was the real sublimation.) The sublimation was sometimes very ugly, as it was in the life of "Saint" Rainaldo, bishop of Nocera Umbra, who died in 1222, and who kept in his house a poor orphan boy who went about to the bishop's guests each day and said, "For the love of God and of the Blessed Virgin Mary give alms to me a poor little creature," so that all there present would be reminded of Christ in His poor.[12] More frequently it led instead to the alternate beauty of hospitals and mystical devotion and, very prominently, devotion to the Eucharist.

Points of chronology and causal connections may remain in doubt; but no one can avoid the crucial importance of the Eucharist in popular devotion in the thirteenth and fourteenth centuries. The disputants over detail converge in making terribly obvious the general point of the laity's desire to see the Host, to see the Man-God in the Mass —bells, candles, kneeling. "See the Body of Christ at the Consecration and be satisfied! In the cities people ran from church to church to see the elevated Host. . . ." [13] The new Elevation at the Mass, the feast of Corpus Christi, the miracle at Bolsena, the Franciscans' host-making machine in Mongolia, Francis himself worrying with pyxes in the provinces and with reverence for the Sacrament in his Testament, pyxes among the precious stones and golden strawberries of the papal treasury, Honorius IV crippled before the altar, the ambry clear to the nave of San Clemente, Innocent III's doctrine, and Boniface VIII's "heresy" conspire to press the point.[14] The path is plotted which will lead to Thomas à Kempis and to Margery Kempe—Margery Kempe, who in her wildly extravagant, expressionist way, is explaining to us the quieter reactions of earlier and more reticent observers to the Consecration and the Elevation, and explaining, too, why Consecration, Elevation, and Eucharist (the Host fluttering in the priest's hands) are so central to the popular, relatively unlearned, particularly lay piety of the later Middle Ages, because they obviated any interference or explanation between the observed God and the observer.[15] The Eucharist was direct, emotional, and divine. It was a center upon which fragmented lives and attentions could easily seem to focus. So in 1355 Margherita Colonna, the widow of the Magnifico Giovanni Conti, as she stood one day at Mass in the Augustinian church of San Trifone at Rome, was so moved at the Elevation of the Body of Christ, that she made a great vow of gifts (of lands and rights and *vassalli*) to God and to the paupers of Santo Spirito in Sassia.[16]

The Blood of Christ in the chalice, like the Host in the pyx, attracted too as its legend of the grail spread.[17] (And in his will Cardinal Jean Cholet had one hundred gilt chalices sent to the dioceses of Rouen and Beauvais.) Blood was not unnoticed. In 1199 Saint Pietro di Parenzio, scion of a senatorial Roman family, minister of Innocent III, was killed by heretics in Orvieto. His "life" by John of Orvieto, romantic with its image of the tunic, ripping, wiping, its women's premonitions and tears, has the blood of its martyr collected, on a tunic, and put into a pyx—we are in the swing of the popularity of Lancelot's bloody sheets as well as the grail. One hundred and fifty years later Catherine of Siena savors the blood of Niccolò Tuldò of Perugia: "The fragrance of the blood brought me such peace and quiet I could not bring myself to wash it away." The Blood of the Eucharist as well as, although inseparable from, the Body, was very potent, as Bolsena's as well as Pietro di Parenzio's Orvieto makes clear.[18]

But the potency of the Eucharist, although it might seem sublimation, did not necessarily detract from, and oppose, the other elements of

Last Wills and Testaments

Christocentric piety. It might complement and support them, at least in thirteenth-century Rome, as it did in the minds and acts of Innocent III and Margherita Colonna. Nor should it be believed that, in the new enthusiasm, the old cults of Christianity and Rome were lost. The doorsteps of the apostles continued to attract the lovers of the poor. In the mosaics at Santa Maria Maggiore and the Lateran, new Franciscan saints joined the old, but the old remained. At San Cosimato, Clare and the Franciscan relics joined old saints and old relics. At the newly Franciscan great church of the Virgin (the thirteenth century's mother and queen), the Aracoeli, the "altar of heaven," found itself in a transept, but still present and honored.[19] In 1316 the abbess of the convent of Santa Maria Rotunda in the diocese of Albano still found it wise to have certified (and translated from Greek into Latin) the presence in her church of old and conventional relics (and their indulgences) in connection with the consecration, or reconsecration, of the church's major altar, dedicated to the Virgin, and minor altars, conservatively dedicated to Saints John, Nicholas, Bartholomew, and Augustine.[20] So, in the city of Rome itself, old Christian names and dedications remained as the new appeared.[21] Still, in spite of this not unexpected conservatism, the shape of modernism in religion was very clear, as clear as the new emphasis in Franciscan parts of the Christian community on the indulgenced feasts of Mary, Francis, Anthony, and Clare.[22] It was clear, and it was constructive: even in its institutionalization— with the spread of hospitals and relatively enthusiastic orders; even in the clogged, pedestrian daily lives of only partially committed laymen and laywomen—in their spiritual moments. It is, in fact, certainly not true in thirteenth- and fourteenth-century Rome that religious enthusiasm could only end in heresy or the foundation of an order; it ended normally, as one would expect, in the sporadic acts of spiritually moved, but indecisively moved, men and women.[23]

Among these sporadic acts, the making of wills was very prominent. The pious offerings of these wills have left their gorgeous residue not only, in some part, in the churches to which they were offered, but, more fully, in the necrologies of those churches, the books of anniversaries of the donors' deaths, and particularly in the necrology of the city's apostle's own great church, as a mid-fourteenth century (1335) will calls it, the church of Saint Peter's of the City (*Ecclesia Sancti Petri de Urbe*).[24] Even Boniface VIII, perhaps particularly Boniface VIII, who in his trial-self mocked immortality and the Eucharist, returned some part of the wealth he had extracted from the church to this church of Saint Peter's, and to the preservation of the sacrament there. There, in the chapel which he caused to be built, or rebuilt, and which was dedicated to his papal namesake, Boniface chose to have his (admittedly effigied and self-remembering) sepulchre. He did not deny himself the pomp of death and burial. And angels of stone guarded his tomb.[25]

Boniface (who is described in the necrology as "by nation a Campanian, from the city of Anagni, from the house of the Caetani, of great learning and eloquence") enlarged the number of canons and choral clergy, and provided for the continual saying of Masses at the altar in his "remarkable chapel" which, with his tomb and that of his nephew Benedetto Caetani, cardinal deacon of Saints Cosmas and Damian, was enclosed within an iron grill. Boniface furnished the chapel richly: a silver basin; four chalices with patens; two crosses of silver, one of jasper, and one of crystal; three pairs of silver candelabra and one of jasper decorated with silver and precious stones; a pyx of silver gilt, and two silver boats; two perforated silver ladles, or colanders; four pairs of silver cruets, of which one was gilt; three censers of silver, of which one was gilt; an icon or relief in ivory (*cona de ebure*) "precious enough," with twelve stories from the New Testament; eleven chasubles of various colors in rich cloths; two copes; six dorsals (three of noblest Cypriot work); seventeen whole cloths of various colors and of Lucchese work; five gold embroideries of which one was of English and three of Cypriot work, and one had enamel decorations and whole figures of saints (*nobilissimum*); four surplices or shirts with pectorals and embroidery of Cypriot work; three stoles and three manuals with Cypriot work; one pretty missal; one pretty breviary, noted, in two volumes; one small gradual, noted; twenty silk towels of German work (and the reader has curious memories of the great Boniface and of towels in Germany); three over-furs; two *arcupanili;* and because the bells had fallen down and broken, quickly a new and better wooden campanile had been built, and seven new bells been made, of the best, and double weight; and Boniface gave more not specified, besides the very lavish gifts and benefices, and properties to a value of almost fifty thousand florins, which he gave to the basilica as a whole. Clearly his priests need not sing naked the Masses for his soul (above his pretty missal), nor in a poor or little church; nor need, there, the sacred Host be touched by simple metals.[26]

Obviously, not all even of Saint Peter's thirteenth- and fourteenth-century gifts were so lavish as those from Boniface, nor are all the names on its calendar of obits so important as his. On August 22 the canons remembered Romana, the widow of a scribe who had left

Last Wills and Testaments

them four houses, one with the sign of a woman, one with a lion, one with an altar, and one with a priest, in Trastevere. On July 8 they remembered Rita, the daughter of Per Giovanni of the *rione* Ponte, a woman whose little nuclear family is recoverable in the necrology (her mother, Fasana, her brothers, Ciccho and Semprevivo—a curious necrology name—her sister, Perna) whose gifts included property in the piazza "Armenorum" by San Giacomo de Armenis, near the basilica, a house marked with a bow.[27]

But some gifts, if not names, were even more distinguished than Boniface's, most notably those of his own cardinal, the rich Jacopo Stefaneschi, who was remembered on June 22. Among other munificent gifts—some drawn from incomes as specific as that from the house of Ciccha in the parish of San Lorenzo de Piscibus—Jacopo gave Saint Peter's the painting to be placed above the high altar of the basilica, painted by the hand of Giotto (*de manu Iocti*) which cost 800 florins of gold, and also for the *paradiso,* or atrium, of the same basilica, the mosaic which told the story of Christ walking upon the waves and holding the hand of Peter so that he would not go under the water (the "Navicella"), done "by the hands of this same most extraordinary painter" (*per manus eiusdem singularissimi pictoris*) for which Jacopo paid 2,200 florins. The Stefaneschi were a giving family; certainly Perna Stefaneschi, the widow of Stefano Normanni and the daughter of Stefano and Perna Stefaneschi, gave generously to Saint Peter's. And Bertoldo, Jacopo's brother, is recorded in the necrology of Santa Maria in Trastevere, "because he caused to be made the whole mosaic work of the Blessed Virgin in our tribune." [28]

The Boccamazzi swarm in the necrology of Saint Peter's, as do, of course, the Orsini, with the Orsini nephew Cardinal Latino Malabranca, "who much loved our basilica and its servants and gave to us his palazzo with its cloister, houses, and vines placed next to the church of San Michele" and asked particularly that the canons celebrate the feasts of the great (Roman pope) Saint Gregory and of Dominic (the founder of Latino's order). And Boccamazzi and Orsini, like other people who knew the pleasures of family on earth, remembered their families in their prayers—"his daughter Letitia," "Masses for father, mother, brothers, nephews at the altars of San Biagio and Saint Mary Pregnant in the basilica." [29]

An Orsini daughter gave the basilica holdings including a slaughterhouse and a bake-oven, and three great silk draperies to be hung each year in the basilica on the feast of Corpus Christi. The great Orsini pope, Giangaetano, Nicholas III "by nation a Roman, from the family of the Orsini," as Boniface VIII would after him, erected an altar in honor of his papal namesake, consecrated his altar with his own hands, and chose, as Boniface would, to be buried beside his altar; as Boniface would too, but perhaps less obviously extravagantly, he furnished his altar with silver chalices and colored chasubles and other splendors. And he gave properties, including an orchard which had belonged to Compagio di Giovanni Lucidi, in the Prati—and for the church itself he gave much more: a golden cloth to make into a dorsal for the major altar, a most precious cope with images of the saints in English work, a tabernacle of silver with a pyx of gold for conserving the Body of Christ in the Supper of our Lord, and another pyx of silver for conserving the sacred Hosts, and a silver tube for the pope's *obserbendum* of the Body of Christ, and a silver container for the Hosts, and a jeweled miter and a gold ring and much else besides. Nicholas's translation to his new tomb was also remembered at the basilica; and his death, also, was remembered at Santa Maria in Trastevere to whom he left fifty florins and two images, one of ivory and the other of silver "which now are in our sacristy." [30]

Although the Orsini are very much present at Saint Peter's, the church was, again, Saint Peter's of the City, as well as Saint Peter's of the Orsini. Colonna, Ponte, Savelli, Sarraceni, Astalli, Frangipane are there. So are de Tartaris, with their gardens and buildings marked with their sign of the red cross, as in the *contrada* Satiri, and the parish of Santa Barbara—houses which had in the thirteenth century been normally identified by the people who lived in them, come increasingly in the early fourteenth century to have identifying signs on them like the house "with the sign of the man with the caraffe in his hand" (*cum signo hominis cum carrafa in mannu*)—a slight movement from people to place, which may seem to echo those major movements like "English" to "England." [31]

In this warehouse of silver and silk (the draperies of this drapperied and drapery-dealing—and *sudarium*—world) certain patterns do appear. In the first place, although it is true that families remain families dead or alive, in grants of land and in prayers for souls, it seems equally true that a good part of that wealth which wily Roman noble families extracted from ecclesiastical office, from the "church," went in fact back to churches in the form of lavish legacies. This was of course a sort of sophisticated investment for the family, but it was also a real return of "church" property. One observes a circular sort of movement. But the word "church" is not really an effective common noun. The action might in fact not be circular but rather a complicated

Last Wills and Testaments

way of moving property from one ecclesiastical institution, one sort of cult or stimulant to piety, one place to another (as, in emblem, Fieschi carried north their titular churches). It could effect a change in the control of property, which would correspond to a change in religious taste. One cannot, however, successfully see or question this movement, define its arc, in necrology evidence; one can only see where the money, the silk and silver, the orchards and vineyards went. One can, however, if one cares to, see other sorts of pattern more successfully defined in necrologies. At Saint Peter's one must notice the endless emphasis on the Mass and the Eucharist. One cannot help noticing that what Cecco di Cola Gabose, a denizen of the portico of Saint Peter's, gave for his soul and his relatives' souls was a silver gilt chalice with a paten, of eight-ounce weight, and that what Ceccharello Cecholi gave for his soul and his relatives' was a chalice worth thirteen florins. One cannot help noticing the celebration of the consecration of the altar of the *sudarium* (November 23). One certainly cannot avoid noticing, although it is late (1350), the gift from three Venetians "because of their special devotion to the sacred *sudarium*, which is kept in the basilica, and shown, much to the consolation of sinners and the remission of sin's punishment, of a very beautiful and wonderful table of crystal," the wood-framed guard, which was in fact used for centuries, both to protect and display the Veronica. One cannot, that is, avoid noticing the attention paid to, the devotion to, the Real Presence of Christ in the Sacrament, and the real face of Christ on the towel.[32]

The patterns of testators' attentions are much more fully apparent in their wills, and so are, if less clearly so, their transfers of property from one sort of insitution or cult center to another. Wills are formal documents; informality would have meant invalidity. They are often, almost surely, inexact descriptions of their testators' personal sentiments. How often did testators relatively carelessly follow contemporary fashions in legacy? How often did redacting notaries suggest to confused testators appropriate methods for effectively expressing piety? This sort of question must always be kept in mind (but its effect is not only negative—currents of conventional piety caught in a notary's cartulary are, very much, things historians want to know); sometimes the pattern of a will, Hugh of Evesham's, for example, gives these questions a partial answer.[33] And, even when the waters have been appropriately muddied, wills are wonderfully expressive and informative documents. It is hard to think of any evidence (except a penitentiary's) which could so quickly expose men's souls, so neatly define their final duties and affections.

Sometimes, particularly early in the thirteenth century, Roman wills are short and reticent. Such a will is that of the widow of a man named Astone, which was written in 1212 by the scribe Angelo, whose angel's wings decorate the document both at its invocational *incipit* and at the terminal place of its scribal sign. In the surviving document, the widow simply named her heirs and gave to the monastery of San Cosimato all the land (divided into two pieces, or *pezze*) which she had in the Marcellis, between the monastery's own land and the land of Giovanni di Leone and his nephew. The will shows the connection between a widow with a suburban farm and the monastery which owned a neighboring farm—that is essentially all.[34]

In the wills of Pietro Lombardo di don Giacomo di Enrico Lombardo (Peter the Lombard), from 1281, and of Pietro Saxonis, or Sassone, merchant of the *rione* San Marco, from 1295, there is a drastically different sort of articulateness. It would be strange if this change were not connected with chronology, if one could not assume that, with exceptions, wills (and testators) were becoming more articulate. The assumption is buttressed by the increasing pace and volume of self-revelation by testators as wills move into the fourteenth century —but, of course, the number of preserved wills also increases.

Peter the Lombard's will was redacted by the judge and scribe Pietro Piperis, a man whose professional family ties together seemingly disjointed pieces of later thirteenth-century Rome.[35] Peter the Lombard himself was not unwell, or at least he was "of sound body," when he wrote (although the presence in the will of his doctor, Pietro Romani, as a recipient of twenty-five *soldi,* in reward for his service to Pietro in his illnesses, suggests illnesses past, real or imaginary, as well as future); but the peculiar pattern of Pietro's familial heirs helps one to understand why he particularly would not want to die intestate. Pietro instituted as his heirs his brother Giacomo and his nephew Paolo, the son of his brother Giovanni, should they be alive at the time of his own death, and, if not, their male heirs. As for the rest of his family, he left money and/or property or life tenancies to an unidentified Giacomino Lombardo; to Teodora, the natural daughter of his brother Giovanni; to Margherita, his sister-in-law and the wife of Giacomo (forty *soldi*); to Contissa, his aunt; to Constantia his sister-in-law, the widow of Giovanni (six *lire*); to Gayta and Contissa, his own sisters; and to Ricka, his foster daughter. Although Pietro's own formal phrases clearly place potential male heirs before female ones, the presence of women in the will of this bachelor, or perhaps widower, is very noticeable; and his body of executors is composed of his sister Gayta

Last Wills and Testaments

and two friars (one the guardian of the Franciscans at Aracoeli) and a priest. Natural earthly family is important in almost all wills; and it is in this one. (It is a family which had appeared many years earlier in rather different dimensions and extensions, with emphasis upon its female members, and with already a Constantia, widow of Giovanni, in a lease of property out beyond the Milvian bridge.) The difference in the size and nature of the bequests to the various members of the family is shaped by a group of not entirely recovered variables: convention, affection, guilt, need, duty.

Peter the Lombard's will shows no wives or children. He had to think harder about leaving his property. He left much to his family. But, although the natural family and this earth are heavily present in wills, it is here rather the provision for the afterlife and for the spirit that is particularly provocatively interesting. First, for his soul, and before his detailed family bequests, Pietro left one hundred *lire,* fifty of which were to be collected by his executors and invested by them in an olive grove with its appurtenances in Tivoli, or some other place which they would select, with the profits from it going to two churches which were once each other's Roman neighbors near the present site of San Luigi dei Francesi and the Palazzo Madama, a short distance from the Pantheon, the two churches of San Salvatore de Termis and San Benedetto de Termis Lombardorum ("of the Lombards"), in the second of which, before its altar, he chose to be buried. Both of these churches were named from the ancient Alexandrine baths, and both had tangled, obscure histories connected with the great abbey of Farfa, and one, San Benedetto, with both San Lorenzo in Damaso and Sant'Eustachio.[36] The property's income was to go to the use of the churches for candles for the soul of Pietro and his relatives, and it was not to be alienated for other purposes. Pietro then dealt conventionally with debts and faults he might have had or committed (in an era when gained wealth was often a source of scrupulous worry, and in which financial involvements were intricate) [37] and to a small group of personal bequests: twenty *soldi* to Giacomino, ten *soldi* to a mason (Florio Muratore); ten to a man named Giovanni Neke. He then moved to money for pilgrimage for his soul: forty *soldi* in subsidy of anyone going to the Holy Land within five years or for poor pilgrims going to the sepulcher of our Lord Jesus Christ; twenty *soldi* for anyone willing to go to Sant'Angelo Maggiore (Gargano); five *soldi* for anyone willing to go to Farfa (geographically a great falling off from the Holy Sepulcher, perhaps, but a place connected with the Termis churches).

Then Pietro turned, most interestingly, to his bequests to religious people and institutions within the city: five *soldi* to the recluse of San Salvatore in Termis; fifty *soldi* to the neighboring church of Sant'Eustachio for buying a pyx with a silver cover, in which the sacred Hosts could be reserved; and money for a similar pyx for San Lorenzo in Damaso; thirty *soldi* for restoring the image of Saint Gregory (again Saint Gregory) on the wall of Santa Maria Rotunda (the Pantheon) opposite the house of the by now surely familiar Don Compagio di Giovanni Lucidi.[38] Then Pietro moved to his medical bequest and that for his foster daughter (thirty-five *soldi* but connected with other provisions), perhaps significantly placed within the spiritual part of the will, and then to provision for his funeral should he die in the city (ten *lire*). Again a pattern emerges: pilgrimage—and specifically to Christ's sepulcher; the pyxes for Christ in the sacrament; local and Lombard churches (and Gregory and a poor recluse)—Jesus Christ and the neighborhood. But three times as much for his widowed sister-in-law as for the Holy Land pilgrim.

The spread of Pietro Sassone's spiritual bequests is broader, and they are informative in an additional way.[39] In fact, Pietro Sassone's life seems in several ways to have been broader than Pietro Lombardo's: he had had two wives, to the living one of whom he left the bed in which he lay, its two mattresses, and some sheets, and a home to live in as long as she lived there honestly with his son Edward, or until she remarried; this Pietro's name, familiar as it is in Rome, combined with the Christian name of his son Edward (Adwardus) may suggest a foreignness more distant than Lombardy. Pietro was, though, as he states in his will, a merchant of the *rione* San Marco (not itself very far from the Pantheon), the *rione* of merchants, and his bequests make clear his attachment to that local merchandising place. And Pietro Sassone's will, unlike Pietro Lombardo's, assumes death in the city; but it was probably the will of a man critically ill, within sight of death, "fearing," as he said, "the danger of death."

The pattern of Pietro's ecclesiastical attachments is perhaps clearest if his pertinent bequests are repeated in the order in which they fall within the will, after certain initial settings and comments. The will was redacted in the garden of the Aracoeli, with eight witnesses (seven and an additional one), including a clerk from Tagliacozzo, a man from Penne, and five friars, of whom one is called of Colonna and one of Santa Maria Rotunda, one of Bolsena, and one of Orvieto; the fifth friar, "Thomasso de Alto Sancte Marie," is in fact the recipient of three *lire* within the body of the will. Tommaso is one of five recipients of bequests within the "spiritual" area of the will; although he is the only

Last Wills and Testaments

religious, two of the others, brothers, are specifically said to receive their bequest for the sake of Pietro's soul. One of these five recipients, Caradomna, is said to be Pietro's maidservant, and she is to get forty *soldi*. The heirs in the will are Edward and Sermonetta's womb if it should contain a son or sons (daughters were only to have 350 gold florins for dower and custody). The executors are Sermonetta and (as in the case of Pietro Lombardo) the guardian of the Franciscans at Aracoeli at the time of Pietro's death. The reserved gifts specified for Sermonetta are the stuff (silver, gold, scarlets, furs) of a rich (but not rich like the Stefaneschi) merchant's household, as well as land and income for life, and a measure of good oil each year (until she remarried).

Pietro left to the recluses of the city, ten *lire* (one should remember the equation of one *lira* to twenty *soldi*); to the church of San Marco, where he wished to be buried, fifteen *lire* for its *opere*, or for the fabric; to the church of Santa Maria de Capitolio (that is, the Aracoeli), twenty-five *lire* for the *opere*; to the church of Santa Maria de Minerva (that is, sopra Minerva), ten *lire* for the fabric; for the church of San Trifone, one hundred *soldi* (that is, five *lire*) for the fabric; for the church of San Giacomo in Septimiano (on the Lungara), forty *soldi* for the fabric; for the hospital of San Matteo de Merulana, ten *lire,* for paupers there; for the friars of San Cesario de Palatio, twenty *soldi* (that is, one *lire*); for the friars of San Giuliano, twenty *soldi*; for Santa Maria Grottaferrata, ten *lire* for the fabric; for San Salvatore in Santa Balbina, forty *soldi* for the *opere*; for San Giovanni di Mercato, twenty *soldi* for the *opere*; for San Salvatore in Pensilis "de apothecis" (*bottheghe*—"shops"), twenty *soldi* for the *opere*; for Sant'Andrea de Paracera (Paracenis) "de apothecis," for the *opere*, twenty *soldi*; to the hospitals of the Termine and Santa Maria Rotunda, forty *soldi*, that is twenty each, for the paupers. Then after personal bequests, Pietro's will makes another provision of another order. If his heirs fail, they shall be replaced by San Matteo in Merulana, which shall have half, and Sant'Eusebio, "where are the friars of Fra Pietro da Morrone," which shall have the other half of his residual inheritance. These last two legacies, to the church of the hospital on the Merulana and the church with the new enthusiastic friars, are on a different scale and at a different level of contingency from the other bequests; they are obviously significant. For the other bequests, it has been important, I think, to observe their order in the will, but their size should be looked at, too.

If one considers the bequests in units of one *lira*, one can line them up this way:

Santa Maria Aracoeli (Franciscans)	25
San Marco (neighborhood, merchant, burial)	15
Santa Maria sopra Minerva (Dominicans)	10
San Matteo Merulana (hospital, order of Crucifers)	10
Santa Maria Grottaferrata (country neighbor, [?] other connections)	10
The recluses of the city	10
San Trifone (Augustinian Hermits)	5
San Salvatore in Santa Balbina (portrait of Christ)	2
San Giacomo in Septimiano (Silvestrines)	2
Santa Maria in Rotunda (hospital)	1
Termine Hospital	1
San Giuliano (Carmelites)	1
San Cesareo de Palatio (Friars of the Sack)	1
San Giovanni de Mercato (neighborhood, merchant)	1
San Salvatore in Pensilis (neighborhood, important merchant center)	1
Sant'Andrea de Paracera (neighborhood, merchant)	1

The pattern of Pietro's selection is clarified. He is giving the Franciscans at their great city, senatorial church, in whose garden he is making his will and some of whose friars are witnessing it, lots more than he is giving any other institution (if his heirs survive him). But the Dominicans, the Augustinian Hermits, and the Silvestrines (the Benedictine order from the Marches recently shaped to a friarlike pattern) [40] appear, respectively, in the third, fourth, and fifth order of his bequests; and Pietro also includes the Carmelites and the Friars of the Sack. Pietro obviously believed in the relative spiritual efficacy of friars and new and enthusiastic orders (although he obviously did not find them equally persuasive); and this belief is heavily underlined by his selection of the Celestinians (the order of Peter Celestine) in Sant'Eusebio as one of his co-heirs should his natural heirs fail and confirmed by his selection as his other co-heir of a hospital run by the friar-related Crucifers. Probably a (to Pietro) similar pattern is reflected in his scattering of *lire* to the recluses of the city.

San Marco, Pietro's selected burial church, the church of his *rione,* the recipient of his second largest grant, establishes his second, but not in strength of attachment, principle of selection. It and San Giovanni, San Salvatore, and Sant'Andrea are churches of Pietro's neighborhood and of his mercantile profession, of the Mercato and the Botteghe Oscure. The selection of Santa Maria in Grottaferrata seems to have been

Last Wills and Testaments

another neighborhood selection, but the neighborhood of Pietro in another guise, as a landholder and controller of "my *casale*," with neighbors including Angelo Saxonis or Sassone, Pietro Capocci de Papazurri, from whom he had bought land, the church of Cosmas and Damian, as well as the monastery of Grottaferrata, land perhaps connected with an Oddone Benencase to whose sons he left a bequest for his soul— lands by the road to Frascati. Santa Maria was Pietro's country ecclesiastical charity; but the heaviness of his bequest is not completely explained by the surviving evidence, although that evidence includes our knowledge of extensive contemporary redecoration at Grottaferrata. Perhaps Pietro had intended to be buried in its church had his illness come upon him in the country.

The next act of Pietro Sassone's spiritual bequests is to hospitals and their paupers, to institutionalized corporal works of mercy. His gift to San Matteo in Merulana is large, and it is emphasized by his making San Matteo his co-heir should his natural heirs fail. The principle directing his gift to San Salvatore in Santa Balbina is almost surely the presence there of one of Rome's most famous portraits of the living Christ, a relative, in significance, of the Veronica.

Pietro, then, endowed the friars and the new orders, hospitals for paupers, the portrait of the living Christ, his own neighborhood, and the churches of his profession. It is also noteworthy that, although he specified that his gifts for San Cesareo and San Giuliano were for the friars there and that the hospital gifts were for the paupers, all the other institutional gifts were specifically *pro fabrica* or *pro opere*. San Marco's money was *pro opere sive fabrica,* but all the others were one or the other, with the Aracoeli, San Salvatore in Santa Balbina, and the merchant churches getting *pro opere*. The distinction may be meaningless. But it could possibly imply a distinction between building and decorating, or furnishing, and it could imply that some of the churches were known to be actively involved in building campaigns, and thus even explain the size of Grottaferrata's gift. More surely, perhaps, both sorts of specification help reveal the cast of mind of a merchant, not without investment in land bought by himself, but attracted to things that were active and new.

Surviving Roman wills do not repeat one another as if they were written from a stenciled form, but lumped together they reenforce one another, substantiate one another's testimony about the spiritual affections and attractions of the denizens of thirteenth- and fourteenth-century Rome. Hugh of Evesham, the English medical cardinal at San Lorenzo in Lucina, made a will which left money for the fabric of the great Roman churches of Saint Peter and Saint John Lateran and for *ornamentis* for San Lorenzo and which stretched out to the places of Hugh's life—Oxford, Paris, and the places where he had or had had livings in the north and west of England. Besides fabric, and much more frequently, Hugh spoke of bread, or money and grain for bread, for paupers; he provided for poor girls who needed dowries and for bridges that needed to be built, and for hospitals and *leprosaria*—corporal, and active, works of mercy. He provided, too, for scholars, nuns, and recluses; and he distributed his books.

But Hugh's will is literally filled with gifts to the friars of three countries, of England, France (Paris), and Italy (Rome). To the friars of the place where he would die, an unknown town and so one in which Hugh would not have had personal favorites among the houses of friars (and in which he could assume the presence only of omnipresent orders), he left fifty pounds Tournois to be divided equally between the minors and the preachers, that is, the Franciscans and the Dominicans, and ten pounds for the Augustinian (Hermit) friars, as well as forty pounds to be divided between recluses and hospitals (the last a combination, repeatedly apparent, which recalls John of Matha and San Tommaso in Formis). But in Rome itself Hugh's division among the friars was different, forty pounds to be divided equally between the Franciscans and Dominicans, that is, twenty pounds each, but thirty pounds for the Augustinans, and fifty for the recluses; in the city he knew, he favored the Augustinians (because they were closer neighbors to San Lorenzo?) and recluses (because they were less caught in a hardening shell of property?). In Oxford, which he had known, he remembered the Carmelites as well as the Augustinians, but gave each only one-fifth of what he gave the Dominicans, and he gave the Franciscans only two-thirds of what he gave the Dominicans. And so his will moves from place to place, Paris, York, Beverly, Grimsby, Worcester, adjusting his gifts to the friars.[41]

In contrast with Hugh's long and articulate will from 1286, is the short, reticent will from 1270 of "the noble woman Risabella (of Tagliacozzo), wife of the noble man Napoleone di don Giacomo di Napoleone Orsini," who left something to her mother, but who essentially put herself and the charity to be administered for her soul's sake in her husband's hands. But she did make one specific bequest for her soul's sake: "I will that my husband should give my breviary to Santa Maria de Minerva for my soul."[42] The friars attracted the reticent and the garrulous alike.

As early as 1232, when Giangaetano Orsini, whose greatest gift

Last Wills and Testaments

to religion was probably his daughter Margherita, the mother of Margherita Colonna, made his will, he had friar witnesses to it and was counseled in making it by the prior of the Dominicans at Santa Sabina.[43] It is a will, rather nervously codiciled in 1233 and 1234, which suggests a life entangled in remembered sexual problems. It grants large sums of money for the souls of specific people; it grants money to paupers and money for repairing damage done to churches like Saint Paul's. As Hugh of Evesham would send money back to the places of his livings, Giangaetano left money to the men of his fiefs. Giangaetano's soul was intricately insured, even with two featherbeds to a Franciscan friary. The friars are omnipresent, but of course not alone: in 1285 Giovanni, son of the noble lord Pietro Romani, included Cistercians with his Dominicans, Franciscans, and Augustinians; and in 1290, in the will with which he enriched the Minoresses of Margherita Colonna's San Silvestro in Capite and gave the Franciscans of the Aracoeli, if he should die in the city, besides his body one hundred florins, Pietro Colonna gave money to other diverse institutions.[44]

Pietro Colonna also granted money, fifty *lire,* and a house worth ten *lire,* to a woman named Perseta, who is described as the daughter of the then dead pauper of Gallicano named Chiara. In a will that at other points must deal with natural daughters, Perseta's sudden riches are suspect, but the description of her mother as a pauper, and even more the description within the will of the nuns at San Silvestro's as *pauperes . . . spiritu* trills in, in its slightly off tone, one of the recurring themes of this period's wills and piety, one obviously connected with the friars (and perhaps in Rome as in other places with the diocesan bishop), "the poor of Christ." So begins, essentially, the will of Crescenzio di don Leone di Giovanni de Giudice in 1271: "In the first place I institute as my heirs the poor of Christ" (*pauperes Christi*); and so provides the will of Giovanni Frangipane de Septizonio, in 1266, that should he die without heirs the poor of Christ (*pauperes Christi*) should become an heir. So, too, in 1272 Angelo di Paolo Bobone made his heirs the poor of Christ (*pauperes Christi*), although this inheritance did not need to imply the bulk of an estate, and in Angelo's case the implication is that it might be only ten *soldi,* and that the *pauperes'* becoming heir was essentially a convenience to the structure of the will.[45]

Wills often imply that the executors would know or select "the poor of Christ," so the reader in a distant century need not be told whom the testator has in mind. The identification of *pauper* may of course come to us from the other side, if sometimes a little oddly, as when Margherita Colonna, as a *"pauper Christi,"* declined to make a will. But the need for a convenient institutional receptacle for bequests to paupers had of course been more than adequately supplied, not only by the orders of poor friars, but even more specifically and exactly by hospitals for the poor. So, in disposing of two thousand florins in 1301, Filippo di Matteo di Pandulfo de Suburra could leave five hundred to the paupers at the hospital of San Matteo on the Merulana (and honor his father as well). Thus the wealth of Santo Spirito grew, and as its necrology makes clear, testators who wanted a closer connection with the poor and Christ than mere money could bring might leave pertinent specifics, as did a bishop who left his bed and bedclothes to the hospital, along with a silver chalice. The hospitals which men who lived in Rome remembered, moreover, could be far away, in Genoa, for example; and the taste for hospital bequests was not locally Roman. Hospitals in Pistoia, for example, were already beginning their "amazing growth." [46]

The themes of piety that can be collected and observed in thirteenth-century Roman wills are blatantly exposed, their message intensified, and the number of wills which exposed them stunningly increased in the notarial cartularies of the 1360s from the collection of Sant'Angelo in Pescheria. One can focus one's attention first on the wills of two specific women, which are immersed in the cartularies' vast richness. Both of these wills are from 1364, and both from women then living, and thinking of dying, in central, urban Rome.

The first of these wills, written in April at the Aracoeli, is that of an apparently childless noblewoman, Donna Paola Savelli of the Ripa, who had taken as her second, and, one might guess, not entirely pleasing, husband, one of the Savelli. (And one should note again the repeated unevenness of life—not only the seemingly omnipresent widows and women taking second husbands, proclaiming, as they do, the hardiness of the female, but also the men with second wives. Life was not simply short in thirteenth-century Rome—that is not the point; it was short for some people, perhaps the great majority, but long for others, very uneven.) [47] Donna Paola's will is obviously interested in protecting the spiritual memory of her first husband, Massaretto, who himself had had built a chapel in the monastery of San Gregorio. It is a rich will, with dresses pleated and sequined from head to foot, scarlet and satin, with a furred mantle and pleated belts, with pearls and sequins and silver gilt, tunics with red and ribbons with silver. It is also, and clearly connectedly, a will that worries very seriously over spiritual

Last Wills and Testaments

matters. Donna Paola, who made her mother Perna her heiress, left for three roofless paupers of Christ, twelve *denari* each; for the repair of the Lateran (recently damaged), thirty florins; to the image of the Virgin at Aracoeli, her precious sapphire ring worth twenty florins and also twenty florins to the Franciscans there for singing Masses; to the brothers of Santa Maria Nova, where a sacred image was, forty florins and ten florins more for buying a silver chalice; to Santa Balbina, where a sacred image was, eight florins for decorating an altar; to the paupers of the hospital of Sant'Angelo of the *Recommendatorum* of our Lord Lesus Christ, one hundred florins of gold; to her husband's chapel in San Gregorio, a monastery popular in wills, one hundred florins for building and a weekly Mass; to nearby Santa Maria Guinizio, ten florins for an altar's decoration; to Santa Cecilia in her in-law's Monte Savelli, six florins; to a fishmonger's wife and a miller's wife, six florins, for which they were to make the devotion of visiting the *limina sanctorum* for the souls of the testatrix and her first husband, and more money for the souls of the two, to Altruda, a butcher's mother, and to a nurse and to a woman named Olive. It is a will which constantly insists that the ecclesiastical bequests insure future Masses for the souls of Paola and her first husband. And in order to finance her bequests, Donna Paola instructs her executors (who include the prior of Santa Maria Nova, the abbot of San Gregorio, the guardian of the Aracoeli, and the guardian of the society of the *Recommendatorum*) to sell her sequined, pearled, pleated, scarlet, furred, red, silk with silver-barred wardrobe. Donna Paola's seven witnesses were seven friars (with names from places far from Rome, except one from Trastevere).

Less than a month before, Donna Agnese, the wife of the *pezzo grosso* of the *rione* Sant'Angelo, Matteo de Baccariis, made her will.[48] Donna Agnese was Matteo's second wife; he had earlier been married to a Capudzucche. And Matteo had had a daughter by each wife. Matteo was also Agnese's second husband. She too had had a daughter by each spouse; and she made Andreotia, the daughter of her first husband, Ceccho Tatiotii, her universal heir (although much in the will depended upon Agnese's not producing male heirs in the future). The will suggests that Andreotia become a nun of San Lorenzo in Panisperna and take with her her cloth and furniture. Donna Agnese herself was a woman from a prominent family in the Campo Marzio, from the family and the piazza de Riciis; and she obviously maintained her attachment to both. Agnese left to San Nicola de Riciis (San Nicola dei Prefetti), a house; to the church and monastery of Santa Maria in Campo Marzio, a house and three *lire* for her soul; to the church of

Sant'Agostino (San Trifone's successor, which had been talked of in the previous year as the new church), ten florins of gold with the stipulation that Mass be celebrated every Saturday for a year in the chapel of Saint Catherine in Sant'Agostino where Agnese's family (*progenie*) was buried; to Saint Peter's, the reversion of a house in the Vasca San Pietro; for the repair of the Lateran, twenty-five florins; to the hospital of Sant'Angelo of the society of the *Recommendatorum* of Jesus Christ, fifty *lire*, with the stipulation of an annual anniversary Mass by the society; twenty-five florins, too, were left for the poor, and twenty-five for the singing of Masses, and ten for the recluses (to the *carceratis*) of the city, four *soldi* to each.

The patterns of these wills are repeated more modestly in the wills of wives of butchers and fishmongers and, perhaps, a little less effusively (and less in silk and sapphire) in the wills of their husbands, and a little more pompously in the roughly contemporary will of a canon of Saint Peter's who worried about the status of men who should carry him to his grave.[49] These wills tend to be more specific than their thirteenth-century predecessors, to talk more of chapels and parishes, of institutional hospitals, of the decayed and burned Lateran, of detailed pieces of furniture and cloth. They repeat again and again their talk of gifts to paupers of Christ, to the friars' churches, to neighborhood churches, to specific societies and hospitals like San Matteo or Santa Maddalena at Santa Maria Rotunda, and particularly those of the confraternal societies of *Recommendatorum* (including that of the Blessed Virgin in the church of Quaranta Martiri in Trastevere), to specific images, or in honor of them, like the face of the Savior in the tribune of the Lateran, or the faces of the Savior at Farfa, or the Blessed Virgin on the island, in honor of saints like Gregory and Nicholas. They talk of candles to be lighted when the Blessed Sacrament is shown or elevated, and they talk of precious metals to touch the Body and Blood of Christ. The message of the wills from Sant'Angelo is amplified but not contradicted by wills from Santo Spirito, San Silvestro, and Sant'Agostino.

A Ponziano fishmonger leaves money (a florin) for oil for the lamp of the image of the "glorious Virgin Mary" (now, in our own times, restored) at Sant'Angelo in Pescheria, and he asks that when Masses are being said for his soul, twelve paupers should be fed from the money of his estate.[50] In Sant'Angelo, too, a more pretentious neighbor worries with the arrangement of the new chapel, partly mosaicked, by the door next to the stand or counter of Giovanni della Piazza. An English woman in the Biberatica, Rosa Casarola (or Rosa Anglica Casarola, "English Rose"), includes in elaborate provisions for

Last Wills and Testaments

her soul's safety, four parts of two houses for San Gregorio and remembers three paupers of Christ. A butcher's wife named Lella, from the Arenula, leaves her empty painted chest to her sister, a miller's wife, but she wants her new dress given to San Giacomo Septimiano (whose order seems to have kept it attractive) for a light to be lighted when Mass is said. Francesco de Tartaris gives the paupers of Santo Spirito rich tenements in Monte Mario, providing that they should keep there in the chapel of Sant'Angelo a priest to pray for his soul and those of his parents. One woman demands that a ring be sold immediately after her death, and another a dress; and one of them, a woman named Perna, asks too that a tablecloth of gold and four tablecloths of silk "which I have in my chest" be sold, and that the money be used for their souls' sakes. In 1323 a woman named Thedallina from the *rione* Monti and closely connected with San Silvestro, in which she chose to be buried, had left San Silvestro her red silk from which she wanted a chasuble made to be used at the Sacrifice of the Mass. And a woman named Ceccha from Sant'Angelo, who makes paupers her heirs and leaves a mattress with a pair of sheets to San Matteo Merulana, and another to Santa Maria Massima where she chooses to be buried, leaves to that monastery's figure of "the glorious Virgin" a striped tablecloth for her soul.[51]

It is difficult to drag oneself away from these evocative, fourteenth-century pictures of soul and furniture and life which reveal the direction in which the more reticent wills of the thirteenth century are saying that they are moving, but let the consideration of the spiritual contents of one last will, that of a woman named Baccha from the *rione* Arenula drawn up in June 1364, sum them up.[52] Baccha wished to be buried in Santa Maria sopra Minerva, and to the Dominicans there she left fifteen *lire.* She left much less, a single florin, to her parish church of San Benedetto de Clausura, for a candle to be lighted when the Body of Christ was exposed. She left a florin for someone who would climb for her the sacred stairs of the Lateran (to the life of Christ), and to someone who would make the customary devotions at San Lorenzo outside the Walls. She left money to a specific nun of Santa Maria in Giulia, a nunnery in her neighborhood, and money for the image of the virgin on the island in the Tiber, again, very close. She left money for paupers and orphans and Masses. And she left her tablecloth of gold and silk for Santa Maria in Giulia. The paupers, the friars, the neighbors, the image, the Body of Christ and the life of Christ are here. And, perhaps most strikingly of all, that splendid recurring theme of tablecloths turning into altar cloths.

These chests of silk and cloth of gold were for these Roman women, these givers of rings, obviously, treasure hoards (like the dragon's gold of an heroic age), stored away in part to give them celestial immortality. But they were stored by women with a quickened perception of Christ, the daughters or granddaughters of women who had been children when Altruda, their neighbor, walked among Christ's poor. It is unlikely that any of these women of the wills was another Margherita Colonna, but Margherita crystallizes and binds together more permanently in herself the perhaps fleeting aspirations which one finds in the testamentarily expressed souls of these self-confessing sinners, these hopeful donors of lighted candles and chalices and pyxes, these admirers of painted sanctity (gaining credit "only for recounting"), turning in mocking inversion of the Franciscan dream their silks and damasks into sackcloth.[53] Their domestic, dying world remembers a Christ, who might walk in the door, come to dinner (or be a beggar like Lazarus outside), a Christ who could have known their tablecloths even before they were turned, as so many of them were, into altar cloths. The women were in fact dying Marthas.

Wills are obviously not the only means of access to this or any recorded society's piety. Other means fly to mind: the paths of pilgrimage, the patterns of iconography, the structure of prayer, the names of chapels, the conglomeration of acts, artifacts, and ideas which shape a life like Margherita's or a dedication like Innocent's at Santo Spirito. There is now at the Palazzo Venezia in Rome a recently restored, painted crucifix, the "cross of San Tommaso dei Cenci," which was made at the end of the thirteenth century and which was originally at the Aracoeli. It still reflects remarkably affectingly the piety of the Romans who first looked at it. At its top on either side of an enthroned divine figure, blessing, are the busts of unidentified saints, possibly John and Nicholas, the baptismal and papal name saints of the great Orsini patron pope, Nicholas III; the "Nicholas" particularly ties the painting to a conventional religious past. Beneath is exposed the family (and it should be thought of as that) scene of the crucifixion. There, decorated with cloth and sandals, the new plastic naturalism and personal, sentimental religion are powerfully combined; the combination exposed itself in the Franciscans' and Rome's Aracoeli. The cross is very persuasive.[54] But wills, in their connection with death, have a special strength. They are a particularly nice token of that piety which hovers over a community like a guarding genius, a distillation of a community's feelings of guilt and inadequacy and fear, as well as of hope and wonder. Thirteenth-century Rome's piety seems to collect its spirit, to

Last Wills and Testaments

collect itself, like that figure of Rome in later map-drawings which weeps among her less personal and more secular ruins.

Rome's piety gives a curious unity to this weakly gilded, weakly governed city with its indecisive neighborhoods; it gives a curious luminosity to this "great solid fact of the past," through which Innocent III rode like Constantius II, a human statue, but a thirteenth-century statue with burning eyes and active mind.[55] Piety collects Rome's families in a new way and gives another dimension to their structure; Appian onions, Tiber roe, Savelli beasts, Santo Spirito settle themselves, in its light, into a sort of related significance which is an understandable and believable, not just fragmented and incoherent, way of life. The pieces and parts find within themselves (for us) a common formal distinction; they exist within a more comprehensible, imaginable historical space, because the people with whom they are connected say prayers that we can understand. We can allow these people to sit down and eat their *fritate* of sambuco flowers. We can savor, in mocking it, Gregorovius's mockery when he writes: "There is no better satire on all the most exalted things of earth, than the fact that Rome knew a time when her Capitol was given into the possession of monks who prayed, sang psalms, scourged their backs with whips and planted cabbages upon its ruins." [56]

On a different level, with a different sort of seriousness, Romans sold replica images of the Veronica to pilgrims who had come to see their city's ruins and relics,[57] to feel the pleasure of discriminating among the three names given to the horsed Marcus Aurelius at the Lateran, or to sense the exhilaration of the danger of poison and cheating as they did business at the curia. The petty Roman businessmen who dealt with these pilgrims seem to have been sharp, self-observant, and sharply observed by their obscurely selected rulers, like that Angelo Malabranca who, in 1235, regulated their greed, to its own profit. Rome, for all its romance, necessary romance, of jumbled, buried relic and treasure and for all the surface incoherence of its urban institutions, seems to have been a precociously self-regulating community. The quick smallness of its repeated and observable financial transactions, when they were small, equipped the men resident within its walls, perhaps, with the sophistication requisite for this behavior. In all probability the heavy burden of, and heavy profit from, bureaucracy and court quickened the Romans' understanding of the tourist trade, and the tourist trade formed at least a partial pattern to the Romans for making profit from bureaucracy and court.

Among the denizens and citizens of Rome, within this profit-making machine, the most apparent mold of order was the family (and related institutional *consortia*), and the most imposing mold of order was the great family. In its idiom, Roman organization, Roman politics, and even Roman neighborhood are most convincingly discussed. The great families, the Colonna, the Orsini, their fellows, controlled areas, but not solid areas, of Rome and of the *campagna*. Groups within families bought and sold, they mated their young and sent them to the church. They did this with a knowledgeable self-interest and a relative lack of concern about hierarchical dignity and above all a quick metropolitan tempo which was perhaps peculiar to them and their situation. But their power was based on wealth in land in the country, with its dependent laborers and seignorial income, and land in the city, with its profit in rents (and on both sorts of land the families had fortresses with which they expressed a source of power more primitive than, and as basic as, wealth). The source of Roman family wealth and power was thus ordinary, conservative, conventional; but the extension of family wealth, the sort of extension which led to greatness, was local and peculiar and was dependent upon office, particularly the office of pope and cardinals, but also all the smaller connected offices of the papal curia. The families themselves were loosely and difficultly organized. Neither did they include in their nexus of power and influence and friendship all their own genetic members, nor did they exclude men of different name or even different blood. Still the heart of the family, its nucleus, was a domestic family of parents, children, siblings, cousins, participants in the life of the same domestic household with jointly held property. It was a relationship as central to this organization as the relationship of lord to man was to conventional feudalism.

The revolutions of thirteenth-century Rome can be described in family terms. But although the great families changed position within the century, and the relation of great family to community changed too, the family itself and its way of getting and maintaining wealth and power changed very little—except perhaps in the brilliance of individual performances: Orsini acquisitiveness, Colonna eccentricity, and reckless Caetani speed.

For reasons which will have become apparent of this bazaar and vineyard, court and cult centered city, the interests of great family, of Colonna and Orsini, often and in many ways coincided with the interests of members of those similarly but more modestly constructed families of shoemakers, notaries, vine-growers, shopkeepers, by whose houses their own were surrounded and even infiltrated. As they shared the same impulses to piety and the same ages and senses (those painted

Last Wills and Testaments

at Tre Fontane), they shared, eventually, the same sources of satisfaction for their greed (and the same victims). "Gold and silver" were similarly "not for" them. Although they played different roles in militia and council, they were both represented by militia and council within the same walls, around the same ruins. Nobles, citizens, and denizens shared interest as Romans. They shared, too, with their predecessors and successors, admitting appropriate change, the city seen from Montorio, a place that could seem another Cana in Galilee (at least at Mass and siesta time), the place of Saint Peter's men, but also a militia'd, coined, notaried, traditioned, always dying, liturgied, mosaicked, lizarded, walled, physical place:

> the tile-clad streets, the
> Cupolas, crosses, and domes, the bushes and kitchen-gardens
> Which, by the grace of the Tiber, proclaim themselves Rome of the
> Romans, . . .

Notes

Introduction

1. The quotations are from Nathaniel Hawthorne, *The Marble Faun,* chapter 12.

2. The quotations are from *The Marble Faun,* chapters 34 and 12.

3. *The Roman Journals of Ferdinand Gregorovius 1852–1874,* ed. Friedrich Althaus, trans. Annie (Mrs. Gustavus W.) Hamilton (London, 1907), 26. For Margaret Fuller, see *Memoirs of Margaret Fuller Ossoli* (Boston, 1852), II, 216; my attention was drawn to her remarks by Giuliana Artom Treves, *The Golden Ring, The Anglo-Florentines, 1847–1862,* trans. Sylvia Sprigge (London, 1956), 160. *Renan's Letters from the Holy Land,* trans. Lorenzo O'Rourke (New York, 1904), 33.

4. *Codice topografico della città di Roma,* ed. Roberto Valentini and Giuseppe Zucchetti (Rome, 1940–1953), III, 143–144.

5. The quotations are from Edward Gibbon, *Memoirs of My Life and Writing.*

6. *The Roman Journals,* 2.

7. Ferdinand Gregorovius, *History,* VIII, 725; V, 660 n.; VIII, 705. This preface was written sometime before the publication of Ferdinand Gregorovius, *Rome and Medieval Culture,* ed. K. F. Morrison (Chicago, 1971), to which Professor Leonard Krieger has prefixed three intensely enlightening pages (vii–ix). He speaks of Gregorovius as "an outstanding representative of that pungent mid-nineteenth-century historiography . . ." (vii). "Pungent" is exactly right for Gregorovius.

8. Gregorovius, *History,* VIII, 765; *The Marble Faun,* chapter 12; *Passages from the French and Italian Note-books of Nathaniel Hawthorne* (Boston, 1884), 221, 219, 496–497, 506; or Fuller, *Memoirs,* "The whole heart must be yielded up to it [Rome]."

9. The quotations are from Leopold Ranke, *Ecclesiastical and Political History of the Popes of Rome,* trans. Sarah Austin (London, 1840), III, 73; and from *History of the Popes,* trans. Mrs. Foster (London, 1913), I, 382.

10. Sigmund Freud, *Civilization and Its Discontents,* ed. and trans. James Strachey (New York: W. W. Norton, 1962), 382.

CHAPTER I
The Physical City

1. For population and size, see particularly J. C. Russell, *Late Ancient and Medieval Population* (Philadelphia, 1958), 109–110, and J. Beloch, *Bevölke-*

rungsgeschichte Italiens (Berlin and Leipzig, 1937–1961), II, 1–2, 19; but see Giovanni Mossa and Maurizio Baldassari, *La vita economica di Roma nel medioevo* (Rome, 1971), 15. Georgina Masson is particularly interested in, and interesting about, Roman flora—see, for example, *The Companion Guide to Rome,* 357. The best introduction to the type of thought represented by pseudo-Bede and the literary tradition concerning the Colosseum is probably Theodor E. Mommsen, "Augustine and the Christian Idea of Progress: The Background of the City of God," in *Medieval and Renaissance Studies,* ed. Eugene F. Rice (Ithaca, 1959), 265–298 (originally, *Journal of the History of Ideas,* XII [1951], 346–374), 271. See Riccardo's *Chronica,* ed. Carlo Alberto Garufi, in *Rerum italicarum scriptores,* ed. Ludovico Antonio Muratori (hereafter, Muratori, *RIS*), new series VII (Bologna, 1938), 174. In general, the history of Rome for the period is covered in two volumes, X and XI, of the *Storia di Roma,* published by the Istituto di Studi Romani; vol. X is Paolo Brezzi, *Roma e l'Impero medioevale* (Bologna, 1947), and vol. XI is Eugenio Dupré Theseider, *Roma dal Comune di popolo alla Signoria pontificia (1252–1377)* (Bologna, 1952).

2. *Le liber censuum de l'église romaine,* ed. Paul Fabre and L. Duchesne (Paris, 1889–1952), I, 566.

3. This account of John of Matha is based on an article in the *Bibliotheca Sanctorum,* VI, 828–838, by Ignazio del Santissimo Sacramento; for the medieval Claudia, see Pasquale Adinolfi, *Roma nell'età di mezzo* (Rome, 1881–1882), I, 159. For the necessary reservations about John's historicity, see Paul Deslandres, *L'Ordre des Trinitaires* (Toulouse and Paris, 1903), 1–19. See Vincenzo Forcella, *Iscrizioni delle chiese e d'altri edificii di Roma* (Rome, 1869–1884), VII, 193, nos. 396, 397.

4. *Codice topografico,* III, 303 (Catalog of Turin, for San Salvatore); "Regesto del Monastero di S. Silvestro in Capite," *Archivio della Società romana di storia patria* (hereafter, *ASRSP*), XXIII (1900), 97–98, no. 124 (Rome, Archivio di stato, San Silvestro, 119), for Gregorio. The detailed description of action at Sant'Angelo is actually from the fourteenth century, see below. For monumental names, see Giuseppe Marchetti-Longhi, *I Boveschi e gli Orsini* (Rome, 1960), 57.

5. For papal vineyards, see Adinolfi, *Roma,* II, 215, and *Via del Corso* (Rome, 1961). For the wall, see Ian Richmond, *The City Wall of Imperial Rome* (Oxford, 1930), particularly 10, 43–52, 89.

6. *Le opere di Ferreto de' Ferreti Vicentini,* ed. Carlo Cipolla (Rome, 1908–1920) [Istituto storico italiano: Fonti per la storia d'Italia], I, 164.

7. Rome, Archivio di stato, Santo Spirito in Sassia, Coll. B, 5, for San Paolo. Carlo Cecchelli, in his extreme lamentations (and perfectly understandable ones) over the destruction of medieval Rome in his own time, suggests places like Anagni: *Topografia e urbanistica di Roma,* ed. Ferdinando Castagnoli, Carlo Cecchelli, Gustavo Giovannoni, Mario Zocca (vol. XXII of the *Storia di Roma* of the Istituto di Studi Romani), 326. For towers, see particularly Francesco Tomassetti, "Torri di Roma," *Capitolium* (1925), 266–277, and A. W. Lawrence, "Early Medieval Fortifications near Rome," *Papers of the British School at Rome,* XXXII (1964), 89–122.

8. For a sense of the country feudal holdings of the Roman nobility, see

Notes

G. Tomassetti, "Documenti feudali della provincia di Roma nel medio evo," *Studi e documenti di storia e diritto,* XIX, (1898), 291–320, particularly 296–304.

9. For Antonio (or Antonano) and Pietro, see "Regesto di Monastero di S. Silvestro in Capite," *ASRSP,* XXIII (1900), 101–102, 418, nos. 174, 129 (Rome, Archivio di stato, San Silvestro, 161, 125). For the *rione* Ripa or Sant'Angelo and the Campo Marzio, see Rome, Archivio di stato, Santo Spirito in Sassia, Coll. B, 72, 106 (for 1325 and 1340), and 2, for Malaspina. See Brezzi, *Roma,* 325, 334; Dupré Theseider, *Roma,* 28; and Cecchelli in *Topografia,* 198; particularly Dupré Theseider for the twenty-six *rioni* of 1267. See Rome, Archivio di stato, Santo Spirito, B, 31, and San Cosimato, 317, for a notary (in 1283) of the city *de dicta contrata Trastibere*; Gonippo Morelli, *Le Corporazioni Romane di arti e mestieri del XIII al XIX secolo* (Rome, 1937), 12; A. de Boüard, *Le régime politique et les institutions de Rome au moyen-âge, 1251–1347* (Paris, 1920), 47–48.

10. In general, for a rather limited introduction, see Ottorino Montenovesi, "L'Archiospedale di S. Spirito in Roma," *ASRSP,* LXII (1939), 17–229. A number of related works about the hospital and order have been written by Pietro de Angelis; among these the most significant is probably *L'Ospedale di Santo Spirito in Saxia* (Rome, 1960–62), but one should also know of the existence of *L'Ospedale apostolico di Santo Spirito in Saxia nella mente e sul cuore dei papi* (Rome, 1956), and *Regula sive Statuta hospitalis Sancti Spiritus* (Rome, 1954). See, here, particularly *L'Ospedale di Santo Spirito,* I, 373–384. For Thomas Anglicus, see Rome, Archivio di stato, Santo Spirito, B, 32 (1290).

11. Innocent III, *Opera omnia,* IV, in J.-P. Migne, *Patrologiae cursus completus . . . Latina* (hereafter, Migne, *Patrologia Latina*) CCXVII (Paris, 1855), cols. 345–350 (sermon for the Sunday after the octave of the Epiphany), particularly 348, 350. De Angelis, *L'Ospedale di Santo Spirito,* I, 375–377, 380. S. J. P. van Dijk and J. Hazelden Walker, *The Origins of the Modern Roman Liturgy* (Westminster, Maryland, and London, 1960), 95, 102–103, 460–461. Rome, Archivio di stato, Santo Spirito, A, 2; for expenses of the procession in the late thirteenth century, see Friedrich Baethgen, "Quellen und Untersuchungen zur Geschichte der päpstlichen Hof- und Finanzerwaltung unter Bonifaz VIII," *Quellen und Forschungen,* XX (1928–1929), 114–237, 209. See Carlo Bertelli, "Storia e vicende dell' immagine edessena," *Paragone,* no. 217, Arte (1968), 3–33, 3–4.

12. Vincenzo Monachino, et al., *La Carita Cristiana in Roma* (Bologna, 1968), 125–187; Mariano da Alatri, "Il Medio Evo," 140. The rule is, in my opinion, still most easily read in manuscript: Biblioteca Apostolica Vaticana, Codices Borghesiani 242, fos. 1–8, here particularly, fos. 3, 3^v, 4, 3, 1^v–2, 7–7^v, (and chapters 36, 41, 55, 35, 8). For a description of the manuscript, see Anneliese Maier, *Codices Burghesiani Bibliothecae Vaticanae,* Studi e Testi, CLXX (Città del Vaticano, 1952), 290–292, no. 242. See also Innocent III, *Opera omnia,* IV, cols. 1137–1158.

13. L. Schiaparelli, "Alcuni documenti dei magistri aedificiorum Urbis (secoli xiii e xiv)," *ASRSP,* XXV (1902), 5–60, 50–53, no. 10 (1306), 26–27, no. 1 (1233), 33–35, no. 5 (1279); Rome, Archivio di stato, Santo Spirito, B, 2, 4, 59, 61, 64, 108, for archival marks—2 and 4 are fourteenth-

Notes

291

century transcriptions of thirteenth-century business; the seal is described, in Santo Spirito, B, 119, as being on a 1348 document, and in green wax, with a dove above its cross: "on the middle of the seal is carved a great double cross like the cross of the Hospital of Santo Spirito in Sassia of the City"; and see a set of crosses marking property at Nepi in 1349 in Santo Spirito, B, 123. For the *crux gemina,* see Herbert Paulus, "Doppelkreuz," in Otto Schmitt, et al., eds. *Reallexicon zur deutschen Kunstgeschichte,* IV (Stuttgart, 1958), cols. 215–223, particularly col. 220. Roman monasteries other than Santo Spirito used monuments as markers, for example, a Santa Maria in Campo Marzio monument in the Trevi (Arcione), "Regesto del monastero di S. Silvestro in Capite," *ASRSP,* XXII (1899), 513, no. 47; Rome, Archivio di stato, San Silvestro, 44; Biblioteca Apostolica Vaticana, Codices Borghesiani 242, fo. 3^v (chapter 53). For business documents discussed here, see Rome, Archivio di stato, Santo Spirito, B, 7, 149, 38, 26, 36, 6, 5, 123; and see also de Angelis, *L'Ospedale di Santo Spirito,* I, 380–381.

14. Codices Borghesiani, 242, fos. 2–2^v, 3–4, 5^v–6, 7 (particularly chapters 8, 12, 35, 36, 40, 41, 47, 50, 54, 55, 57, 58, 59, 80, 81, 82, 83, 85, 107). For Eynsham, see Antonia Gransden, ed., *The Customary of the Benedictine Abbey of Eynsham:* Corpus Consuetudinum Monasticarum, II (Siegbury, 1963), 90–91.

15. A. Monaci, "Sant'Alessio all'Aventino," *ASRSP,* XXVIII (1905), 41–112, 151–200, 155–157, 173–174, 180.

16. Rome, Archivio di stato, Santo Spirito, B, 16; "Regesto del monastero di S. Silvestro in Capite," *ASRSP,* XXIII (1900), 113, no. 145; Rome, Archivio di stato, San Silvestro, 138, San Cosimato, 296.

17. Rome, Archivio di stato, Santo Spirito, B, 131. Recipes are hard to date and place, but see, for example, Ludovico Frati, *Libro di cucina del secolo XIV* (Leghorn, 1899).

18. Rome, Archivio di stato, San Cosimato, 244, (San Biagio), 188 (example of triangle); "Regesto del monastero di S. Silvestro in Capite," *ASRSP,* XXII (1899), 525, no. 61. For a feudal tenement in 1207, see Rome, Archivio di stato, San Silvestro, 59; for women and the law, in Brezzi, *Roma,* see P. S. Leicht's "Lineamenti del diritto a Roma dal IX al XII secolo," particularly 588–589, and Camillo Re, *Statuti della Città di Roma* (Rome, 1880), for example, 63.

19. See "Regesto del monastero di S. Silvestro in Capite," *ASRSP,* XXIII (1900), 95–96, no. 122; Rome, Archivio di stato, San Silvestro, 117; Santo Spirito, B, 10. See also Rome, Vallicelliana, Archivio storico capitolino, Archivio Orsini, II.A.I.33(ol.31).

20. Rome, Archivio di stato, Santo Spirito, B, 6. For a convenient and bright introduction to "consortial" family holdings, see David Herlihy, "Family Solidarity in Medieval Italian History," in *Economy, Society and Government in Medieval Italy,* ed. David Herlihy, Robert S. Lopez, and Vsevolod Slessarev (Kent, Ohio, 1969), 173–184, particularly 174–175, 176–178; also notes 6–7 and 10–13. For a convenient definition of various types of leases and alienations in medieval Italy, see Carlo Calisse, *A History of Italian Law* (Boston, 1928), 721–728.

21. "Regesto del monastero di S. Silvestro in Capite," *ASRSP,* XXIII

Notes

(1900), 111–112, no. 142; Rome, Archivio di stato, San Silvestro, 136. For Victor IV's family, see Brezzi, *Roma*, 350–351, Paul Kehr, "Zur Geschichte Victors IV," *Neues Archiv*, XLVI (1925–26), 53–85, and, particularly, Hansmartin Schwarzmaier, "Zur Familie Viktors IV in der Sabina," *Quellen und Forschungen*, XLVIII (1968), 64–79, 74.

22. "Regesto del monastero di San Silvestro in Capite," *ASRSP*, XXIII (1900), 111–112, 110–111, nos. 142, 141; Rome, Archivio di stato, San Silvestro, 136.

23. "Regesto del monastero di S. Silvestro in Capite," *ASRSP*, XXIII (1900), 109–110, no. 140; Rome, Archivio di stato, San Silvestro, 134.

24. "Regesto del monastero di San Silvestro in Capite," *ASRSP*, XXII (1899), 518, no. 54; Rome, Archivio di stato, San Silvestro, 51.

25. Rome, Archivio di stato, San Cosimato, 230; Archivio segreto vaticano, Instrumenta Miscellanea, 145, and cf. 137, and *Liber censuum*, II, 53–54.

26. Rome, Archivio di stato, San Cosimato, 266, 212. For *emphyteusis*, see Calisse, *History*, 726.

27. "Regesto del monastero di S. Silvestro in Capite," *ASRSP*, XXII (1899), 508, no. 38; XXIII (1900), 84–85, no. 106; Rome, Archivio di stato, San Silvestro, 35, 102; *Tabularium Ecclesiae S. Mariae in Via Lata*, ed. L. M. Hartmann and M. Mezores (Vienna, 1913), 99–100, no. 260.

28. Rome, Archivio di stato, Santo Spirito, B, 4; San Silvestro, 84; "Regesto del monastero di S. Silvestro in Capite," *ASRSP*, XXIII (1900), 69–70, no. 87.

29. Rome, Archivio di stato, Santo Spirito, B, 93; for a very long and detailed article on the Calcarario, see Giuseppe Marchetti-Longhi, "Le Contrade della zona 'in circo Flaminio': Il Calcarario," *ASRSP*, XLII (1919), 401–535.

30. Gregorovius, *History*, V, 632; *Studi e documenti per la storia del Palazzo Apostolico Vaticano*, ed. F. Ehrle and Hermann Egger (Vatican City, 1935), II, 37–52 (and, for an illustration of Nicholas's inscription, plate II); "Martini Poloni Continuationes," *MGH,SS*, XXII, 476; *Liber censuum*, II, 43–60, particularly, 49–54; Friedrich Baethgen, "Quellen und Untersuchungen zur Geschichte der päpstlichen Hof- und Finanzerwaltung unter Bonifaz VIII," *Quellen und Forschungen*, XX (1928–29), 114–237, 210; *Necrologi e libri affini della provincia Romana*, ed. Pietro Egidi (Rome, 1908–1914), I, 262–263.

31. "Regesto del monastero di S. Silvestro in Capite," *ASRSP*, XXIII (1900), 71–72, no. 90; XXII (1899), 531–532, no. 73; Rome, Archivio di stato, San Silvestro, 86, 70. For a three-generation alienation of a house "in the *vicolo* before the church," see "Tabularium S. Mariae Novae," ed. P. Fedele, *ASRSP*, XXVI (1903), 130–131, no. 166. For the Orsini, see particularly Rome, Vallicelliana, Archivio storico capitolino, Archivio Orsini, II.A.II.47(ol. 44), II.A.II.48(ol.45), II.A.I.25(ol.24), II.A.II.29(ol.27), II.A.II.36(ol.34), II.A.II.42(ol.40), II.A.I.50(ol.44), II.A.I.51(ol.49), II.A.II.2. I should like specifically to thank Mr. Richard Mather for having called my attention to this rich collection. It has an unusually helpful calendar-catalog: Cesare de Cupis (fu Natale), *Regesto degli Orsini*, I (Sulmona, 1903), a copy of which is conveniently kept on hand in the Vallicelliana, as is a copy of Giuseppe Marchetti-Longhi, *I*

Notes

Boveschi e gli Orsini (Rome, 1960), a small book which attempts, although sometimes more bravely than successfully, to sort out a number of Orsini problems like genealogy, and which makes helpful suggestions, like those echoed here about Donna Maralda and *arpacasella,* and which indicates the centers of Orsini urban properties and their connection with the Boveschi (particularly 41–58, 55, 59). For the Theater of Pompey and the Orsini area around it see also Cecilia Pericoli Ridolfini, *Guide rionali di Roma: Rione VI—Parione,* II (Rome, 1971), 147–174. Saba Malaspina, *Historia,* in Muratori, *RIS,* VIII (Milan, 1726), cols. 785–874, see particularly 864, but also 813–814, and 835.

32. Axel Boëthius, *Golden House of Nero* (Ann Arbor, 1960), throughout chapter IV and its plates. For the Boveschi tower (Turris Johannis Bovis), see Marchetti-Longhi, *I Boveschi,* 45, 48, plate 1.

33. Rieti, Archivio capitolare, III, D, 1, and the less important III, D, 2 and 3. For the neighborhood Tor di Nona, see Carlo Pietrangeli *Guide rionali di Roma: Rione V—Ponte,* I (Rome, 1968), 49, 51, 59–60.

34. For colonnaded houses, see Rome, Archivio di stato, San Silvestro, 186, 200—186 is from 1310 and the *rione* Sant'Eustachio; 200 is from 1352, *rione* Colonna. In the latter document the abbess of San Silvestro, Jacoba Annibaldi, rents to Petruccio di Silvestro, for ten *soldi* and two *denari* each year for a period of nineteen years, a house in the Colonna, between the public road and the old wall of the monastery, a house with a garden behind it and a colonnade in front of it—rented in the presence of a Franciscan witness, Hugone de Anglia (Hugh of England).

35. The notarial books in the Archivio di stato begin in the mid-fourteenth century. They are discussed in an article by Clara Gennaro, "Mercanti e bovattieri nella Roma della seconda metà del Trecento," *Bulletino dell'Istituto storico italiano,* LXXVIII (1965), 155–203. But they (even Archivio del Collegio de' notari capitolini, 475, 849, and 1703) are considerably less concentrated and interesting than the cartularies from Sant'Angelo of which nineteen (Biblioteca Apostolica Vaticana, Sant'Angelo in Pescheria, I, 1–19) are fourteenth-century. There is also an interesting notarial fond, partly Roman, from 1344, at Farfa; see I. Schuster, "Un protocollo di Notar Pietro di Gregorio nell'archivio di Farfa," *ASRSP,* XXXV (1912), 541–582.

36. For Cola, see particularly, Gregorovius, *History,* VI, 232, 240, 372–373.

37. Biblioteca Apostolica Vaticana, Cod. Vat. Lat. 10372; I should like to thank Mr. Anthony Luttrell, who is working with this manuscript, for having pointed it out to me.

38. The first cartulary is Biblioteca Apostolica Vaticana, Sant'Angelo in Pescheria, I, 1. For a drawing of another secondary church in the neighborhood, San Nicola in Carcere, only some forty years later, and for a discussion of the conflation of ancient memories and contemporary life, see Theodor E. Mommsen, "Petrarch and the Decoration of the Sala Virorum Illustrium in Padua," in *Medieval and Renaissance Studies,* ed. Eugene F. Rice, 30–174 (originally, *The Art Bulletin,* XXXIV [1952], 95–116), fig. 23, 160–162, 165. For drawings and photographs of the neighborhood see Carlo Pietrangeli, *Guide rionali di Roma: Rione XI—S.Angelo* (Rome, 1967).

Notes

39. Biblioteca Apostolica Vaticana, Sant'Angelo in Pescheria, I, 1, fos. 10, 17, 7ᵛ–8, 43–44ᵛ, 14ᵛ, 19ᵛ, 176–176ᵛ, 182ᵛ, 190–191; I, 3, fos. 230–230ᵛ.

40. Sant'Angelo in Pescheria, I, 1, fos. 180–182ᵛ, 5ᵛ, 125–128, 41ᵛ–42ᵛ, 16ᵛ, 192–195; I, 2, 10ᵛ–11.

41. Sant'Angelo in Pescheria, I, 1, fos. 105–106, 10–13ᵛ, 114ᵛ–115, 145–146ᵛ, 9–10, 58–59 (for the Soricara, see Tomasetti, "Torri," 273); Rome, Vallicelliana, Archivio Orsini, II.A.II.47(ol.44), II.A.II.48(ol.45).

42. Sant'Angelo in Pescheria, I, 1, fos. 44ᵛ–45, 158ᵛ–159, 113ᵛ–114, 107ᵛ–108ᵛ, 54–55, 112–113.

43. Sant'Angelo in Pescheria, I, 1, fos. 145–146ᵛ, 30ᵛ–32, 52ᵛ–53ᵛ, 81–83ᵛ, 16ᵛ, 118ᵛ–119.

44. Sant'Angelo in Pescheria, I, 1, fos. 84ᵛ–86, 87ᵛ–89, 164ᵛ–165ᵛ (and unnumbered folio), 75–77, 2ᵛ, 32ᵛ–33, 36ᵛ, 101–103ᵛ, 125–128, 161–164, 150ᵛ, 173ᵛ–174ᵛ, 99–100, 34ᵛ–35, 41ᵛ–42, 4ᵛ–5ᵛ, 54–55, 16ᵛ, 10, 115ᵛ, 111–111ᵛ, 165ᵛ–166.

45. See Emmanuel Rodocanachi, *Le Saint-Siège et les Juifs, Le Ghetto à Rome* (Paris, 1891), 321–323, 137–140, 163, 132, 244; Sant'Angelo in Pescheria, I, 1, fo. 104. See Marchetti-Longhi, "Calcarario," 488, for the name "de Judeis" (and, also, Rome, Archivio di stato, Santo Spirito, B, 93 (1332) with its three "de Judeis" witnesses). For a general discussion, see Hermann Vogelstein, *Rome,* trans. Moses Hadas, Jewish Communities Series (Philadelphia, 1940), particularly 168, 175, 181 (popes), 183 (Elijah de Pomis), 179–180 (Isaac ben Mordecai, Abu'lafia), 188, 201 (citizenship). For Abu'lafia, see also the original German edition of Vogelstein, *Geschichte der Juden in Rom* (Berlin, 1895–1896), I, 247–249. This matter has been discussed recently in Edward A. Synan, *The Popes and the Jews in the Middle Ages* (New York, 1967).

46. Rome, Vallicelliana, Archivio Orsini, II.A.I.51(ol.49), II.A.II.2, II.A.I.46(ol.44); Biblioteca Apostolica Vaticana, Sant'Angelo in Pescheria, I, 1, fos. 191ᵛ–192, 101–104; Rome, Archivio di stato, San Cosimato, 247; Schiaparelli, "Alcuni," 29–30. For *cloaca,* see Marchetti-Longhi, 492–494.

47. Rodocanachi, 28, 29, 31, 33; Cecchelli in *Topografia,* 200; Hawthorne, *The Marble Faun,* chapter 42; for the annual fast on the anniversary of the synagogue's burning, see Vogelstein, *Rome,* 173. For the 1300 evidence see Francesco Savini, "Il Cardinale Tommaso 'de Ocra o de Aprutio' e il suo testamento del 1300," *Archivio storico italiano,* series 5, XXII, 87–101, 95; for the 1404 evidence see Ermanno Loevinson, "Documenti del Monastero di S. Cecilia in Trastevere," *ASRSP,* XLIX (1926), 355–404, 374–375, and for the further use of *curia Judeorum,* in 1438, see also 377.

48. Cecchelli in *Topografia,* 312, 313; Marchetti-Longhi is interesting on churches in the Calcarario, 403–404, but also palazzi and towers, for example, 439, the house with stone in front of it. See also Giorgio Falco, "Il Catalogo di Torino delle chiese, degli ospedali, dei monasteri di Roma nel secolo XIV," *ASRSP,* XXXII (1909), 413–443 (and in *Codice topografico,* III). For the Ripa, see Theodor Hirschfeld, "Genuesische Dokumente zur Geschichte Roms und des Papsttums im XIII. Jahrhundert," *Quellen und Forschungen,* XVII (1914), 108–140, 136–140.

Notes

49. Rieti, Archivio di stato, Fondo Comunale, 26, calendared in Alessandro Bellucci, "Inventario dell'Archivio Comunale di Rieti," 6, no. 18.

50. Gregorovius, *History,* V, 660; see Rome, Archivio di stato, Sant'Agostino, 7, for Pietro di Bartolomeo, *macellario* of the Torre Conti, in 1308. For Conti involvement with Sant'Eufemia, see Sant'Angelo in Pescheria I, 1, fos. 112–113.

51. *Liber censuum* I, 566; Rome, Archivio di stato, Santo Spirito, B, 5; San Cosimato, 260.

52. Gregorovius, *History,* V, 366; Angelo Celli, *Storia della malaria nell'agro Romano* (Città di Castello, 1925), for example, 175; Schiaparelli, 37; Rome, Archivio di stato, San Cosimato, 264, 269; San Silvestro, 102. "Regesto del monastero di S. Silvestro in Capite," *ASRSP,* XXIII (1900), 84–85, no. 106—it must always be kept in mind that *medico* may be a name as well as an occupation. For the names of possible patients used here, see Rome, Archivio di stato, Santo Spirito, B, 27, 16; San Silvestro, 68, 133, 66, 73; "Regesto del monastero di S. Silvestro in Capite," *ASRSP,* XXII (1899), 530–531, no. 71, 529, no. 69, 532–533, no. 75; XXIII (1900), 108, no. 137. For the shrewd German, see "Annales Stadenses auctore Alberto," ed. I. M. Lappenberg, *MGH,SS,* XVI (Hanover, 1859), 271–379, 340; for communes and doctors, see William M. Bowsky, "Medieval Citizenship: the Individual and the State in the Commune of Siena, 1287–1355," *Studies in Medieval and Renaissance History,* IV (Lincoln, Nebraska, 1967), 195–243, 213–214; for the Talmudic sons, see Vogelstein, *Rome,* 194. For Nicola "One-Hundred Lire" acting as a proctor for the Magnifico Napoleone di Matteo Rosso Orsini in 1249, see Rome, Vallicelliana, Archivio Orsini, II.A.I.33(ol.31).

53. Rome, Archivio di stato, Santo Spirito, B, 5, 27, 41; San Cosimato, 261, 264, 269, 285, 295 Sant'Agostino, 8 (1314), for a combmaker and spicers; San Silvestro, 60, 68, 75, 86, 99, 103, 104, 107, 108, 123, 129, 133; Rieti, Archivio di stato, Fondo comunale, 4, 5 (the Giunta Pisano documents); *Liber censuum,* I, 566; II, 47 (Cosmato); "Regesto del monastero di S. Silvestro in Capite," *ASRSP,* XXII (1899), 526, no. 62, 530–531, no. 71, 534, no. 77; XXIII (1900), 71–72, no. 90, 82, no. 104, 85–86, nos. 107, 108, 87–88, no. 110, 89–90, no. 113, 100, no. 127, 106, no. 134, 108, no. 137. The reason that a man called "Smith" only *seems* to be a smith is suggested by the name of "Magister Thomasius Ferrarius" (Rome, Vallicelliana, Archivio Orsini, II.A.I.51 [ol.49]), who may have been an artisan with a title *Messer,* in a city in which that was certainly still unusual, but who may have been a more normal *magister,* a notary, named "Smith."

54. Morelli, *Le Corporazioni,* 22; Emmanuel Rodocanachi, *Les Corporations ouvrières à Rome depuis la chute de l'Empire Romain* (Paris, 1894), x, xi, xiv; Cesare De Cupis, *Per gli usi civici nell'agro Romano* (Rome, 1906), 3–6; Brezzi, *Roma,* 353, 363; Cecchelli in *Topografia,* 208; Dupré Theseider, *Roma,* 28; *Liber censuum,* I, 304–358.

55. Giuseppe Gatti, *Statuti dei Mercanti di Roma* (Rome, 1885), xli–xlii, 1–57.

56. Gatti, 44–45. For a photograph of an inscription from 1285 on a tablet once within S. Salvatore *in pensili* and referring to its reconstruction, see Pietrangeli, *S. Angelo,* 81.

Notes

57. For Saint Paul's, see Gulielmus Ventura, "Memoriale," *RIS,* XI, col. 192. The development of Vatican booths is explored in an article by Pio Pecchiai, "Banchi e botteghe dinanzi alla Basilica Vaticana nei secoli XIV, XV, e XVI," *Archivi,* ser. 2, XVIII (1951), 81–123, here especially 91–94, 95, 97, 99. For the license to the canons of Saint Peter's, see *Die Register Innocenz' III,* I, ed. Othmar Hageneder and Anton Haidacher (Graz and Cologne, 1964), 772–773.

58. *The Chronicle of St. Mary's Abbey, York,* ed. H. H. E. Craster and M. E. Thornton (Durham: Surtees Society, 1934), 31, 132, 30.

59. *Codice diplomatico del Senato Romano MCXLIV–MCCCXLVII,* Fonti per la storia d'Italia, 87 (Rome, 1948), 131–134, no. 81, 143–145, no. 86; "Regesto del monastero di S. Silvestro in Capite," *ASRSP,* XXIII (1900), 431, no. 186; Rome, Archivio di stato, San Silvestro, 173.

60. *Villani's Chronicle,* ed. Philip H. Wicksted and trans. Rose E. Selfe (London, 1906), 326 (Book VIII, 36); "Historie fiorentine di Giovanni Villani," *RIS,* XIII (Milan, 1728), col. 367. For Ventura, see *Chronicon Astense, RIS.,* XI (Milan, 1727), cols. 191–192; Giacomo Stefaneschi, *L'anno santo del 1300* (Rome, 1900), 17; Gregorovius in his *History* (V, 562) translates Villani: ". . . the Romans all grow wealthy by the sale of their goods." See also Ptolemy of Lucca, *Annales,* ed. Bernhard Schmeidler, *MGH,* New Series, VIII (Berlin, 1930), 236 (daily offerings of 1,000 *li. prov.,* and also see an editorial cross reference to *Historia Ecclesiastica*), and G. C. Bascapè, "Le vie dei pellegrinagi medioevali . . . ," *Archivio storico della Svizzera italiana,* XI (1936), 129–169.

61. Cesare Pinzi, *Storia della Città di Viterbo,* II (Rome, 1889), 59, n. 1.

62. Rose Graham, "Archbishop Winchelsey from His Election to His Enthronement," *Church Quarterly Review,* CXLVIII (1949), 161–175, particularly 168–172. For the number of members within foreign cardinals' households, see Agostino Paravicini Bagliani, "Un frammento del testamento del cardinale Stephanus Hungarus (†1270) nel codice C 95 dell'Archivo del capitolo di San Pietro," *Rivista di storia della chiesa in Italia,* XXV (1971), 168–182, 176–179; and now see the full study by Paravicini, *Cardinali di curia e 'familiae' cardinalizie dal 1227 al 1254,* Italia Sacra, XVIII (Padua, 1972), and particularly I, 158, 366–379, for the full households of Riccardo Annibaldi and Ottobuono Fieschi.

63. Canterbury Cathedral Archives, Ch. Ant. P 58, 59.

64. The account of the Trastevere procession is taken from Stephan Kuttner and Antonio García y García, "An Eyewitness Account of the Fourth Lateran Council," *Traditio,* XX (1964), 115–178, 125, 143–146. For the crowds and the resulting high cost of Roman living in 1215, see C. R. Cheney, "Gervase, Abbot of Prémontré: a Medieval Letter-Writer," *Bulletin of the John Rylands Library,* XXXIII (1950–51), 25–56, 40. For a recent discussion of the Cavallinis in Santa Maria, see John White, *Art and Architecture in Italy: 1250–1400* (Baltimore, 1966), 97–99. My discussion of the Cavallini Nativity is completely dependent upon Paul Hetherington, "The Mosaics of Pietro Cavallini in Santa Maria in Trastevere," *Journal of the Warburg and Courtauld Institutes,* XXXIII (1970), 84–106, quoted words on 97, particularly also 88, 95; one need not completely agree with the argument of this essay

Notes

to find it helpful. For excellent illustrations of Cavallini's work, see Pietro Toesca, *Pietro Cavallini* (Milan, 1959). The phrase "lu baron san Piero" comes from a delightful Orvieto Corpus Christi play, which has been dated c. 1325–1330 by Andrea Lazzarini and which argues the efficacy of the relics of the miracle of Bolsena; the phrase comes from line 3, the penance assigned to a doubting priest by his confessor: "che vada a Roma a lu baron san Piero"; it is repeated later at lines 44–45 when the returning priest tells an inn-keeper of his stay in Rome:

a San Gianni beato
ed a San Polu ed al baron San Piero;

the play is to be found in *Sacre Rappresentazioni per le fraternite d'Orvieto nel cod. Vittorio Emanuele 528,* Bollettino della r. deputazione di storia patria per l'Umbria, Appendice 5 (Perugia, 1916), 77–83, no. XXIV; Lazzarini's date is in his *Il Miracolo di Bolsena* (Rome, 1952), 33; Carroll Winslow Brentano pointed out these references to me.

65. Saba Malaspina, *Historia,* cols. 815, 842. It seems to me that Kantorowicz, in a wonderfully surprising turn, rather plays down Corradino's procession (Ernst Kantorowicz, *Frederick the Second* [New York, 1957], 675–676), although his explanation for the brilliant reception is neither played down nor convincing.

66. Florence, Biblioteca Medicea Laurenziana, Plut. 33, sin., fo. 32; Reinhard Elze, *Ordines Coronationis Imperialis* (Hanover, 1960); G. Rohault de Fleury, *Le Latran au Moyen Age* (Paris, 1877), 153–199; P. Lauer, *La Palais du Latran* (Paris, 1911); A. Cempanari and T. Amodei, *La Scala Santa* (Rome: Le Chiese di Roma illustrate, 1963); "Petri Mallii Descriptio Basilicae Vaticanae aucta atque emendata a Romano Presbitero," and "Descriptio Lateranensis Ecclesiae," in *Codice topografico,* III, 375–381 and 319–373, quotations from 424 and 342. For Innocent and Gregory, see *Gesta Innocentii III,* Migne, *Patrologia Latina,* CCXIV, col. xxi, and Pietro Maria Campi, *Dell'Historia Ecclesiastica di Piacenza* (Piacenza, 1651–1662), II, 343–349 (Gregory X), 346.

67. Brilliantly, if conflatedly, described by Gregorovius in *History,* V, 6–15; Michel Andrieu, *Le Pontifical Romain au Moyen-Age:* II, "Le Pontifical de la Curie Romaine au XIIIe siècle" (Studi e Testi, 88) (Vatican City, 1940), 665–669. For some of the places along the way, see Mariano Borgatti, *Castel Sant'Angelo* (Rome, 1930), 110–115; Gregorovius, *History,* V, 440–441 (Campo dei Fiori), and Rome, Archivio di stato, Santo Spirito, B, 145; Gian Filippo Carettoni, "Il Foro Romano nel medio evo e nel rinascimento," *Studi Romani,* XI (1963), 406–416; Ettore de Ruggiero, *Il Foro Romano* (Rome, 1913), 85–124; Richard Brilliant, *The Arch of Septimius Severus in the Roman Forum* (American Academy in Rome, Memoirs, 29, 1967), 254–257; P. Fedele, "Una chiesa del Palatino: S. Maria 'in Pallara,' " *ASRSP,* XXVI (1903), 343–380, 377–379. The quotation is a much used one from Acts 3:6. For a brief, recent consideration of "The Processional Character of Society," with helpful bibliography, see Samuel Berner, "Florentine Society in the Late Sixteenth and Early Seventeenth Centuries," *Studies in the Renaissance,* XVIII (1971), 203–246, 221–227.

Notes

68. "Regesto del monastero di S. Silvestro in Capite," *ASRSP*, XXIII (1900), 74–75, no. 94; Rome, Archivio di stato, San Silvestro, 90.

69. Francesco Fabi Montani, *Feste e spettacoli di Roma dal secolo X a tutto il XVI* (Rome, 1861), 33, 4, 5–7; Gregorovius, *History*, V, 418; Cecchelli in *Topografia*, 208; *Liber censuum*, II, 42–43, 172; Filippo Clementi, *Il Carnevale Romano*, I (Città di Castello, 1949), 31–69; Re, *Statuti*, 241–242. See also Luigi Fiorani, Giuseppe Mantovano, Pio Pecchiai, Antonio Martini, Giovanni Orioli, *Riti ceremonie feste e vita di popolo nella Roma dei Papi* (Bologna, 1970), 55–120. Two essays should press the reader to a more serious and interesting consideration of the significance of games, and particularly the games of elite youth in society: Natalie Zemon Davis, "The Reasons of Misrule: Youth Groups and Charivaris in Sixteenth-Century France," *Past and Present*, no. 50 (1971), 41–75, and Georges Duby, "Dans la France du Nord-Ouest. Au XII siècle: les 'jeunes' dans la société aristocratique," *Annales*, XIX (1964), 835–846. The Duby essay has been translated by Frederic L. Cheyette, in his *Lordship and Community in Medieval Europe* (New York, 1968), 198–209, as "In Northwestern France, The 'Youth' in Twelfth-Century Aristocratic Society."

70. *Liber censuum*, II, 172.

71. G. Tomassetti, *La campagna romana antica, medievale, e moderna* (Rome, 1910), I, 317–318. I should like to thank Mr. Randolph Starn for helping me to understand the charm of the hair and the grain in the glass of water. Since I wrote this paragraph, I have watched both Professor Richard Krautheimer of New York University and Professor Ronald Malmstrom of Williams College dating brick walls with tape measures, and I would now speak less negatively. I have also read Professor Ignazio Baldelli, sensitively working with the date of charms, in *Medioevo volgare da Montecassino all'Umbria* (Bari, 1971), vii–xiii, 93–110 (particularly 98, 99), 126–127.

72. Marco Vattasso, *Anedotti in dialetto Romanesco del sec. XIV* (Studi e Testi, 4) (Rome, 1901), 8, and *Per la storia del dramma sacro in Italia nel sec. XV e XVI* (Studi e Testi, 10) (Rome, 1903), particularly part II, "Le Rappresentazioni sacre al Colosseo," 71–89, 73–74; Brezzi, 413; Rome, Archivio di stato, Santo Spirito, B, 52. For the Blancis, see Rome, Vallicelliana, Archivio Orsini, II.A.I.46(ol.44), and cf. II.A.I.49(ol.47), a related 1270 disposition, and II.A.I.33(ol.31), the witness list for October 1.

73. Rome, Archivio di stato, Santo Spirito, B, 22; see also Friedrich Baethgen, "Quellen und Untersuchungen zur Geschichte der Päpstlichen Hof- und Finanzerwaltung unter Bonifaz VIII," *Quellen und Forschungen*, XX (1928–29), 209.

74. Hastings Rashdall, *The Universities of Europe in the Middle Ages*, ed. F. M. Powicke and A. B. Emden, II (Oxford, 1936), 28–31, 38–39. For a serious impression of learning and teaching at the curia, one should see Richard Mather's University of California Ph.D. dissertation on Matthew of Acquasparta now in process of publication by the German Institute in Rome; for libraries, see Mather, "The Codicil of Cardinal Comes of Casate and the Libraries of Thirteenth-century Cardinals," *Traditio*, XX (1964), 319–350, particularly 321; also 327 for the ranking of curial Latinists in the register of Urban IV. For

Notes

very recent work on cardinals' libraries, see Agostino Paravicini Bagliani, "Le biblioteche dei cardinali Pietro Peregrosso (†1295) e Pietro Colonna (†1326)," *Zeitschrift für schweizerische Kirchengeschichte,* LXIV (1970), 104–139. See Re, *Statuti,* 244–246, for the attempt to reestablish the *studium* and professors in Trastevere. For a stunning reevaluation of learning in thirteenth-century Italian cities, see Helene Wieruszowski, "Rhetoric and the Classics in Italian Education of the Thirteenth Century" (Collectanea Stephan Kuttner, I), *Studia Gratiana,* XI (1967), 171–207. For cardinals' households, in this connection, see Agostino Paravicini Bagliani, "Gregorio da Napoli, biografo di Urbano IV," *Römische Historische Mitteilungen,* XI (1969), 59–78, especially 69, and the same author's *Cardinali di curia,* throughout, but for variety within households particularly I, 158, and for chapel and chaplains, II, 478–495; for papal chaplains, see Reinhard Elze, "Die päpstliche Kapelle im 12. und 13. Jahrhundert," *Zeitschrift der Savigny-Stiftung für Rechtsgeschichte, Kanonistische Abteilung,* XXXVI (1950), 145–204. For a specific man of learning, see Paravicini's "Un matematico nella corte papale del secolo XIII: Campano da Novara († 1296)," *Rivista di storia della chiesa in Italia,* XXVII (1973), 1–32.

75. Gregorovius, *History,* V, 655–666. For a general statement about Byzantine influence, see Ernst Kitzinger, "The Byzantine Contribution to Western Art of the Twelfth and Thirteenth Centuries," *Dumbarton Oaks Papers,* XX (1966), 25–47, and particularly 45 for "the deliberate effort to revive the ancient native heritage . . . in Rome . . . [where] Torriti 'recreated' in the apse of S. Maria Maggiore an early Christian type of mosaic. . . ." On late thirteenth-century classicism and on Tre Fontane, see an important article by Carlo Bertelli, "L'Enciclopedia delle Tre Fontane," *Paragone,* N.S., XX, no. 235, Arte (1969), 24–49, and see also the same author's *"Opus Romanum,"* in the forthcoming collection of essays to be presented to Otto Pächt and published in *Kunsthistorische Forschungen.* For a rich miscellany of small plates illustrating Roman art at the end of the thirteenth century, see Antonio Muñoz, *Roma di Dante* (Milan and Rome, 1921).

76. It seems to me that all older general work on art in late thirteenth-century Rome is now superseded by Julian Gardner's unfortunately unpublished London Ph.D. thesis, "The Influence of Popes' and Cardinals' Patronage on the Introduction of the Gothic Style into Rome and the Surrounding Area, 1254–1305." The passages quoted here are from pages 103 and 145, but the work is valuable for every aspect of artistic activity and patronage considered here. A suggestion of the value of the total work can be gotten from one of its details published by the author as "The Capocci Tabernacle in S. Maria Maggiore," *Papers of the British School at Rome,* XXXVIII (1970), 220–230, and much more from his full very recent article, "Pope Nicholas IV and the decoration of Santa Maria Maggiore" in *Zeitschrift für Kunstgeschichte* (1973), which among much else includes interesting references to and plates from Nicholas III's Vatican palace. Roman mosaics have recently been cataloged and illustrated in Guglielmo Matthiae, *Mosaici medioevali delle chiese di Roma* (Rome, 1967): for Torriti at Santa Maria Maggiore, see I, 355–366 (dating, 355); for Cavallini at Santa Maria in Trastevere, see I, 367–368; for Rusuti at Santa Maria Maggiore, see I, 381–384; plates are in II. For Torriti, Cavallini, and Arnolfo, see White, *Art and Architecture,* 100, 97–99, 55, 61–63; see also

Notes

Hetherington, "The Mosaics of Pietro Cavallini"; Bertelli, "L'Enciclopedia," 35–36; the "Rusuti" dates are readjusted by Gardner, "The Influence," 244–250, and "Pope Nicholas IV," 28–33. Very recent work on Arnolfo's Roman period is to be found in an essay by Joachim Poeschke, "Betrachtung der römischen Werke des Arnolfo di Cambio," *Römische Quartalschrift*, LXVII (1972), 175–211, plates 8–24. See too Matthiae's *Pittura Romana del Medioevo*, II (Rome, 1966). For illustrations of the Tre Fontane "Coronation of the Virgin" and "Nativity," which are very closely related to the Torriti mosaics at Santa Maria Maggiore and which are still in *clausura*, see *Restauri della Soprintendenza alle gallerie e alle opere d'arte medioevali e moderne per il Lazio (1970–1971)* (Rome, 1972), plate 9. Figures closely related to the Tre Fontane fisherman are to be found at San Lorenzo fuori le mura, on the Pisano Fontana Maggiore (a February panel) at Perugia, and in copies after Giotto's Navicella, for example.

77. Gregorovius, *History*, V, 641; for a recent discussion of the tombs, see Mather, "The Codicil of Cardinal Comes," 324. The most fascinating aspect of the redating of the bronze Saint Peter, which has often been considered late antique, is its chemical analysis, for which see the two connected essays of Mario Salmi, "Il problema della statua bronzea di S. Pietro nella Basilica Vaticana," and of Bruno Bearzi, "Esame tecnologico e metallurgico della statua di S. Pietro," *Commentari*, XI (1960) 22–32; Salmi speaks of the *"tono antichizzante"* of the period, 26. It is perhaps misleading to imply that the effigy of Surdi presses itself upon the viewer, since the church in which it is preserved, Santa Balbina, has been closed pending another restoration for more than a decade; for its inscription see Forcella, *Iscrizioni*, 331, no. 1070.

78. For recent careful dating in a work not specifically devoted to this period, see Richard Krautheimer, *Corpus Basilicarum Christianarum Romae* (English edition, Vatican City, 1959), for example, II, 13, 42, 46, 183, 210, 276, 280; for bridges and gates, see Cecchelli, 199, 203; for an interestingly integrated guidebook to Christian Rome (to which Professor Stephan Kuttner drew my attention), see Noële Maurice-Denis and Robert Boulet, *Romée* (Paris: Desclée de Brouwer, 1963); for a guide to medieval remains in Rome, see Léon Homo, *Rome médiévale, 476–1420* (Paris, 1934), appendix, 291–323, "Les vestiges de la Rome medievale." Mr. Ronald Malmstrom's forthcoming work on thirteenth-century Aracoeli should change the look of this whole field. For the date of destruction of a Corso arch, together with an eleventh-century church, see C. Bertelli and C. Galassi Paluzzi, *S. Maria in Via Lata*, I (Rome, 1971), 13.

79. I have chosen the most common and likely, but not the only, translation of *spetiarius*, which could also mean a piecer in the cloth trade or a dealer in specie among the moneyers. For drawings of Matteo's and Giovanni's tombs see Muñoz, *Roma*, 381.

CHAPTER II
The Ideal City

1. See, for example, R. H. Hilton, *A Medieval Society* (Letchworth, New York, 1966), 7. For a description of the way in which thirteenth-century news was carried, see C. R. Cheney, "Gervase, Abbot of Prémontré," 39–40.

Notes

2. For the altar of the apostles, see *The Chronicle of Bury St. Edmunds, 1212–1301,* ed. Antonia Gransden (London, Edinburgh, 1964), 82. For Matthew's maps and plans, see Richard Vaughan, *Matthew Paris* (Cambridge, 1958), 235–250, particularly 247–248.

3. The reader will find parts of two books particularly helpful for the matter of this chapter, the first being the old standard work, the second a recent, brilliant summary and analysis: Arturo Graf, *Roma nella memoria e nelle immaginazioni del medio evo,* I (Turin, 1882), "I tesori di Roma," 152–181; Charles Till Davis, *Dante and the Idea of Rome* (Oxford, 1957), particularly 1–22, 33–34.

4. I am following the stories in Graf, particularly from the *Gesta Romanorum,* supplemented with William of Malmesbury in the *Gesta Regum:* Graf, 161–164. For the notarial date, see Rome, Archivio di stato, archivio del Collegio de' Notari Capitolini, vol. 1236, fos. 31ᵛ–32.

5. *Codice topografico,* III, 143–167: "Magistri Gregorii de mirabilibus urbis Romae," 145–147. This edition of the topographical texts of Rome, by Roberto Valentini and Giuseppe Zucchetti, includes a set of elaborate notes identifying objects and relating texts. The introductions attempt to survey earlier discussions of the texts and to date them. See too M. R. James, "Magister Gregorius de Mirabilibus Urbis Romae," *English Historical Review,* XXXII (1917), 531–554.

6. *Codice topografico,* III, 17–65: "La più antica redazione dei *Mirabilia,*" 32–33. The oldest redaction was put together in its present form and written in existing manuscripts before the end of the twelfth century. But it was probably composed, its parts assembled, by about 1140 at the latest.

7. *Codice topografico,* III, 40–42. See too Joseph van der Straeten, "Les Chaines de S. Pierre," *Analecta Bollandiana,* XC (1972), 413–424.

8. *Codice topografico,* III, 34–35; the vernacular is in *Codice topografico,* III, 116–136: "Le Miracole de Roma," 126–127. Readers may be more familiar with the slightly later version of this and other Roman stories in the *Legenda aurea.*

9. *Codice topografico,* III, 28–29; for Nero, see Graf, 332–361 (where his complexity in memory is exposed), but particularly, 332, 353–359. The quotation from Eclogue IV is verse 3. For the *exemplum* for sermons, see *Liber exemplorum ad usum praedicantium,* ed. A. G. Little (Aberdeen, 1908), 19–20.

10. *Codice topografico,* III, 24 (Saint Sebastian), 23 (*Quo vadis*). In general, for the genre of city descriptions, see J. K. Hyde, "Medieval Descriptions of Cities," *Bulletin of the John Rylands Library,* XLVIII (1965–1966), 308–340.

11. *Codice topografico,* III, 77–109: *Graphia Aureae Urbis,* 95–97 (and see *Mirabilia,* 31), 98–109. John White, "The Reconstruction of Nicola Pisano's Perugia Fountain," *Journal of the Warburg and Courtauld Institutes,* XXXIII (1970), 70–83, 80–81.

12. *Codice topografico,* 51–53.

13. *Codice topografico,* III, 77–79; Graf, *Roma,* 80–84. See Joseph R. Berrigan, "Benzo d'Alessandria and the Cities of Northern Italy," *Studies in Medieval and Renaissance History,* IV (Lincoln, Nebraska, 1967), 127–192,

Notes

particularly 132–136, for Benzo's attack upon the Janus legend. For the names Enea and Ottaviano, see, for example, "Regesto del Monastero di San Silvestro in Capite," *ASRSP,* XXII (1899), 529–531, nos. 70, 71, and Rome, Archivio di stato, San Silvestro, 69, 70.

14. Vogelstein, *Rome,* 1–3, 122–124; Graf, 91–92.

15. Quotation from *The Oracle of Baalbek,* ed. Paul J. Alexander (Dumbarton Oaks, 1967), 23.

16. A convenient source for imperial attitudes to Rome is Eugenio Dupré Theseider, *Idea imperiale di Roma nella tradizione del medioevo* (Milan, n.d.), particularly, in the introduction, pages 49–76 and, in the documents, for Frederick II, 173–196. On Frederick in general, one should read Ernst Kantorowicz, *Frederick the Second,* trans. E. O. Lorimer (New York, 1957), a magnificent book no matter how much one disagrees with it.

17. Kantorowicz, *Frederick,* 56, 107, 451 (the quotation)–452.

18. Kantorowicz, *Frederick,* 437–438, 448–450 (quotation 448–449). See also Paolo Brezzi, *Roma e l'Impero medioevale* (Bologna, 1947), 427–440, particularly plate XVII facing 432.

19. Kantorowicz, *Frederick,* 676; for the importance of an Italian city's *carroccio* (including Cremona's which was called "Bertha"), see Waley, *The Italian City-Republics,* (New York, Toronto, 1969) 138–140.

20. Kantorowicz, *Frederick,* 438 (for the shades), 451–452 (for the fiefs).

21. For Peter and Paul, see Nicholas III's Senatorial statute (Theiner, *Codex diplomaticus,* I, 216–218, no. 371, or *Les Registres de Nicolas III,* ed. Jules Gay (Paris, 1898), 106) or the urban statutes (Re, *Statuti,* 3). For the use of propaganda, see G. Ladner, "I Mosaici e gli affreschi ecclesiastico-politici nell'antico palazzo Lateranense," *Rivista di archeologia cristiana,* XII (1935), 265–292. For Innocent's sermon, see *Opera omnia,* IV, cols. 481–484, particularly cols. 481 and 484. For Nicholas, see Theiner and Gay, as above.

22. The forthcoming work of Charles Till Davis on Roman republicanism and late thirteenth-century thought and papal politics will, I think, drastically change our whole notion of the period; it will particularly affect our understanding of Nicholas III, who seems increasingly to have been a man, a Roman, and a ruler of the first importance. Professor Davis's first pertinent essay is called "Ptolemy of Lucca and Roman Patriotism"; it should be published shortly. For republicanism, see also Beryl Smalley, "Sallust in the Middle Ages," *Classical Influences on European Culture,* ed. R. R. Bolgar (Cambridge, 1971), 165–194. For the capitol as a wonder of the world, see H. Omont, "Les sept merveilles du monde au moyen âge," *Bibliothèque de l'école des chartes,* XLIII (1882), 40–59, 431–432, and particularly 47 for the sentence: "Primum miraculum est Capitolium Rome que totius mundi civitatum civitas est." On the movement of Jerusalem to Rome, see a work focusing on the seventeenth century but reaching back into the past: Irving Lavin, *Bernini and the Crossing of Saint Peter's,* College Art Association of America (New York, 1968), perhaps particularly 33–35 on "the process of what might be called 'topographical transfusion.'" Dante speaks of the Veronica in the *Paradiso,* XXXI, 204–206. There are good examples of earlier medieval (for example, ninth-century) pilgrims' cases for the relic earth of the Holy Land in the first room of the old

Vatican galleries of Christian antiquities between the Sistine Chapel and the Library. On another face of Christ in Rome in the later thirteenth century, the Mandylion, see Carlo Bertelli, "Storia e vicende dell'Immagine Edessena," *Paragone,* N.S. XXXVII, no. 217—Arte (1968), 3–37.

23. See G. Falco, "Il catalogo di Torino," *ASRSP,* XXXII (1909), 411–443, discussed in Robert Brentano, *Two Churches* (Princeton, 1968), 232.

24. Falco, 443.

25. For the feast of the snows, see van Dijk and Walker, *The Origins of the Roman Liturgy,* 376–377. For the indulgences, see G. Ferri, "Le carte dell' archivio Liberiano dal secolo X al XV," *ASRSP,* XXVII (1904), 145–202, 441–459, XXVIII (1905), 23–39, XXX (1907), 119–168, XXX, 120–121, of which the originals are in the Vatican library.

26. The poem is quoted by Graf, *Roma,* I, 42, and the pilgrims' hymn, Graf, 57. The whole first two chapters of Graf are pertinent (1–77). The theme of the foreigner's view of Rome is treated at some length in the first chapter of Brentano, *Two Churches;* see also John A. Yunck, "Economic Conservatism, Papal Finance and the Medieval Satires on Rome," *Mediaeval Studies,* XXIII (1961), 334–351, and George B. Parks, *The English Traveller to Italy,* I (Palo Alto, 1954); all give abundant references. For Rome from the point of view of Spain and the Spanish see Peter Linehan, *The Spanish Church and the Papacy in the Thirteenth Century* (Cambridge, 1971), 251–255.

27. *Vita Edwardi Secundi,* ed. N. Denholm-Young (London, Edinburgh, 1957), 46.

28. For Canterbury, see *Epistolae Cantuarienses,* ed. William Stubbs, Rolls Series (London, 1865), 191–192. For Walter of Chatillon, see *Moralisch-satirische Gedichte Walters von Chatillon,* ed. Karl Strecker (Heidelberg, 1929), particularly 19 (poem 2, line 13, the quoted line), 77 (poem 5, line 69, the quoted phrase), 5, 18, 25, 26, 70, 75, 110–112, 114 (particularly poems 1, 2, 5, 10, 11); see too Charles Witke, *Latin Satire* (Leiden, 1970), 233–266; I hope that my words will convince Mr. Richard Hunt that I am grateful for his having pushed me toward Walter. For the *de consideratione,* see Saint Bernard, *Opera omnia,* III, ed. J. Leclercq and H. M. Rochais (Rome, 1963), 381–493, 440 ("Quando hactenus aurum Roma refudit?"), 448–466, particularly 450, 456 (the scorpion's tail); see too, Saint Bernard, *Opera omnia,* I, ed. J. Mabillon, in Migne, *Patrologia Latina,* CLXXXII, cols. 422–431, 438–440, letters no. 238, 243, 244, particularly cols. 439, 442; Josef Benzinger, *Invectiva in Romam,* Historische Studien, 404 (Lübeck, Hamburg, 1968), generally valuable, but for Bernard, 83–91; Elizabeth Kennan, "The 'De Consideratione' of St. Bernard of Clairvaux and the Papacy in the Mid-Twelfth Century: A Review of Scholarship," *Traditio,* XXIII (1967), 73–115.

29. *Codice topografico,* III, 135.

30. I have provided examples of the problems of paying debts and have discussed at slightly greater length the rustics' fear in *Two Churches,* particularly 10–13, 43–44.

31. Brentano, *Two Churches,* 19–20.

32. Matthew Paris, *Chronica Majora,* ed. H. R. Luard, Rolls Series (London, 1872–1883), IV, 61. (The quotation is Proverbs 17:22.)

Notes

33. The phrases are quoted from Charles Till Davis, *Dante*, 33.

34. *Codice topografico*, 150–151 (Master Gregory), 65 (*Mirabilia*).

CHAPTER III
Who Ruled Rome?

1. Davis, *Dante*, 33.

2. Quotation from Sylvia L. Thrupp, "The City as the Idea of Social Order," in *The Historian and the City*, ed. Oscar Handlin and John Burchard (Cambridge, Massachusetts, 1963), 121–132, 125.

3. For the city seal, see, as an example, Bartoloni, *Codice diplomatico*, 206, and A. Theiner, *Codex diplomaticus dominii temporalis S. Sedis*, I (Rome, 1861), 248–251, no. 395; for the coinage, see Paolo Brezzi, *Roma e l'Impero medioevale* (Bologna, 1947), 488–490, and particularly plate XIV, facing 320 (misbound and mislabeled in the book as plate XII); and V. Capobianchi, "Appunti per servire all'ordinamento delle monete coniate dal senato romano dal 1184 a 1439," *ASRSP*, XVIII (1895), 417–445, XIX (1896), 75–123, particularly XVIII, 426–427 and plate I, and XIX, 75–84. See Eugenio Dupré Theseider, *Roma dal Comune di Popolo alla Signoria pontificia (1252–1377)* (Bologna, 1952), 40–42, 50–52, 252, and plate II, particularly for the moneys of Brancaleone and Charles of Anjou, and for the *grosso;* Mossa and Baldassari, *La vita economica*, plate 13; Louis Halphen, *Etudes sur l'administration de Rome au moyen-âge* (Paris, 1907), 82, n. 3. For a relatively recent bibliography particularly helpful for political history, see Waldemar Kampf's edition of Gregorovius (*Geschichte der Stadt Rom*) (Basel, 1954), particularly II, 908–920.

4. A. de Boüard, *Le régime politique et les institutions de Rome au moyen-âge, 1251–1347* (Paris, 1920), 159–165; Brezzi, 392; Waley, *The Italian City-Republics*, 76–77, 83–86, 132–135.

5. Gregorovius, *History*, V, 163.

6. Bartoloni, *Codice diplomatico*, 134–135, no. 82; see Waley, *The Italian City-Republics*, 60. In general, besides Bartoloni, for the senate, see Franco Bartoloni, *Per la storia del senato romano nei secoli XII e XIII* (Rome, 1946) [extract from *Bulletino dell'Istituto storico italiano per il medio evo*, LX]; A. Salimei, *Senatori e statuti di Roma nel medio evo* (Rome, 1935); Francesco Antonio Vitale, *Storia diplomatica dei senatori di Roma* (Rome, 1791); Luigi Pompili-Olivieri, *Il senato romano* (Rome, 1886); Halphen, particularly 60ff., and 167–179; Boüard, particularly 135ff., and 235–250. For relatively recent additions to the list of senators, see Pietro Gasparrini, *Senatori romani della prima metà del XIII secolo finora ignorati* (Rome, 1938), and my, "A New Roman Senator," *Quellen und Forschungen*, LIII (1973), 789–796, with the additional evidence of Petrus Petri Angeli Siniorilis active in 1261 from Ferri, *ASRSP*, XXX, 120–121.

7. In, for example, Gegorovius, Brezzi, Dupré Theseider (*Roma*), Halphen, Boüard; more specifically in W. Gross, *Die Revolutionen in der Stadt Rom 1219–1254* (Berlin, 1934). Daniel Waley, *The Papal State in the Thirteenth Century* (London, 1961) is constantly helpful and informative.

Notes

8. See particularly Halphen, *Etudes,* and Bartoloni, *Codice diplomatico* and *Per la storia.* The definition and significance of "money" in this sense will be clarified in the forthcoming work of Thomas Noel Bisson.

9. *Regesta pontificum romanorum,* II, ed. P. Jaffé (Leipzig, 1888), 81, no. 9606; Halphen, 69.

10. See, specifically, Friedrich Bock, "Le trattative per la senatoria di Roma e Carlo d'Angiò," *ASRSP,* LXXVIII (1955), 69–105. See too Dupré Theseider, *L'Idea imperiale,* 211–213; F. Liebermann, "Zur Geschichte Friedrichs II. und Richards von Cornwall," *Neues Archiv,* XIII (1887–1888), 217–222 (221–222 are interesting for the problem of *rioni*); and, most valuable, Karl Hampe, "Ungedrukte Briefe zur Geschichte König Richards von Cornwall aus der Sammlung Richards von Pofi," *Neues Archiv,* XXX (1905), 675–690, especially 686–688.

11. Dupré Theseider, *Roma,* 122; Vitale, *Storia diplomatica,* 137–141; *Registres de Clément IV,* ed. Edouard Jordan (Paris, 1893–1945), 351, no. 892; E. Martène and U. Durand, *Thesaurus novus anecdotorum,* II (Paris, 1717), cols. 96–98. Charles climbing is from Saba Malaspina, *Historia,* col. 864. M.-H. Laurent, *Le Bienheureux Innocent V,* Studi e Testi, 129 (Città del Vaticano, 1947), 418.

12. For an example of a "nervous" document, see Archivio segreto vaticano, Instrumenta Miscellanea, 108; see also Waley, 172–175; Boüard, *Le Régime,* 14; Dupré Theseider, *Roma,* 103–125.

13. Theiner, *Codex diplomaticus,* I, 216–218, no. 371; *Registres de Nicolas III,* ed. Jules Gay (Paris, 1898–1938), 106–108, no. 296; Friedrich Bock, "Il R(egistrum) super senatoria Urbis di papa Nicolò III," *BISI,* LXVI (1954), 78–113.

For Charles of Anjou's opposition to the election of Nicholas III and for his rather surprisingly intense vocabulary, see Friedrich Baethgen, *Ein Pamphlet Karls I. von Anjou zur Wahl Papst Nikolaus III.* [Sitzungsberichte der Bayerischen Akademie der Wissenschaften, Phil.-Hist. Klasse] (Munich, 1960); for use of Charles of Anjou's propaganda, see Partner, *The Lands of St. Peter,* 275, and particularly Charles Till Davis's forthcoming essay on Nicholas III and Ptolemy of Lucca.

14. Theiner, I, 248–251, no. 395; for the action of 1284 and the coalition of 1292, see Dupré Theseider, *Roma,* 231–232, 271.

15. Waley, *The Italian City-Republics,* 39; an article by Achille Luchaire, "Innocent III et le peuple romain," *Revue historique,* LXXXI (1903), 225–257, has been much used by historians dealing with this period. The major general narrative source for the dispute is *Gesta Innocentii III,* which is rich but confusing. For recent work on the *Gesta,* see Volkert Pfaff, "Die Gesta Innocenz' III und das Testament Heinrichs VI," *Zeitschrift der Savigny Stiftung für Rechtsgeschichte,* Kanonische Abteilung, I (1964), 78–126, 81–84.

16. *Gesta Innocentii III,* col. cxc. For an illustration of the Conti and Capocci towers, see Brezzi, plate XVIII, facing 464. Another bean-beech-cloth tower—this one called "faiolum"—seems to have existed in Orsini territory near the (now) Ponte Sant'Angelo in 1262: see Biblioteca Apostolica Vaticana, Archivio San Pietro in Vaticano, caps. 61, fasc. 225. For the distance con-

Notes

trolled by medieval weapons, see for example Lynn White, *Medieval Technology and Social Change* (Oxford, 1962), 103; Milton Lewine has in preparation a work on the church of the Madonna dei Monti in Rome which will explore the problem of weapons and distances between towers in connection with a Suburra tower.

17. *Gesta Innocentii III*, col. clxxxiii.

18. *Gesta Innocentii III*, col. clxxv–clxxxvi.

19. *Gesta Innocentii III*, col. clxxxvi–clxxxvii.

20. Gregorovius, *History*, V, 41–42, 164.

21. For the dying Boniface, see *Le Opere di Ferreto de' Ferreti Vincentino*, ed. Carlo Cipolla (Rome, 1908–1920) [Istituto storico italiano], I, 156. For Boniface's appointment of Fortebraccio and Riccardo, see *Les Registres de Boniface VIII*, ed. Georges Digard, Maurice Faucon, Antoine Thomas, Robert Fawtier (Paris, 1884–1939), IV, 57–58.

22. *Vita Gregorii IX* (*Liber censuum,* II), 25 cap. 19. Brezzi, 418 (S.P.Q.R.).

23. Karl Hampe, *Ein ungedrukter Bericht über das Konklave von 1241 im römischen Septizonium* (Heidelberg, 1913) [Sitzungsberichte der Heidelberger Akademie der Wissenschaften, IV, 3–134] 27–31; for the dictatorship of Matteo Rosso, see Brezzi, 441–451.

24. A study of Brancaleone (which I have never seen) is G. Rovere, *Brancaleone degli Andalò* (Udine, 1895). See the biographical notice by E. Cristiani, "Andalò, Brancaleone," in *Dizionario Biografico degli Italiani*, III (Rome, 1961), 45–48.

25. Matthew Paris, *Chronica Majora* ed. H. R. Luard, Rolls Series, (London, 1872–1883), V, 709.

26. Paris, V, 723.

27. Paris, V, 417–418, 662, 664, 699, 709, 723; for Belvoir, see Dupré Theseider, *L'Idea imperiale*, 197, 207–208.

28. Gregorovius, V, 310–311.

29. Paris, V, 723.

30. Boüard, *Le régime,* 14; Hampe, "Ungedrukte Briefe," *Neues Archiv,* XXX, 687. F. Liebermann, "Zur Geschichte Friedrichs II und Richards von Cornwall," *Neues Archiv,* XIII (1887–1888), 217–222, 221. Heinrich Finke, *Acta Aragonensia* (Berlin and Leipzig, 1908–1923), I, 15–17, no. 11 (16).

31. Theiner, *Codex diplomaticus,* I, 251. An extremely helpful model and guide for the study of citizenship in medieval Italian cities is now to be found in William M. Bowsky, "Medieval Citizenship: the Individual and the State in the Commune of Siena, 1287–1355," *Studies in Medieval and Renaissance History,* IV (Lincoln, Nebraska, 1967), 195–243. See too the suggestive essay by Peter Riesenberg, "Civism and Roman Law in Fourteenth-Century Italian Society," in *Economy, Society and Government in Medieval Italy,* 237–254.

32. Bartoloni, *Codice diplomatico,* 97–98, no. 61; Gay, *Nicolas III,* 107. See too Rome, Vallicelliana, Archivio storico capitolino, Archivio Orsini, II.A.II.29(ol.27), II.A.II.36(ol.34).

33. Rome, Archivio di stato, Santo Spirito in Sassia, Coll. B, 4; see too an Astalli in 1335 described as a citizen of Rome and now *habitator* in the *rione*

Ripa: Santo Spirito, B, 132. Technical meanings of *habitator* are discussed in Dina Bizzarri, *Studi di storia del diritto italiano* (Turin, 1937), 63–158, particularly 86ff.

34. Re, *Statuti*, 79, 274.

35. Re, 63.

36. Rome, Archivio di stato, San Silvestro, 54; see Re, 122–123.

37. See particularly Halphen, *Etudes*, 63, n. 2; for a discussion of the popular party in Rome, see A. de Boüard, "Il Partito popolare e il governo di Roma nel medio evo," *ASRSP*, XXXIV (1911), 493–512. The best guide for the early thirteenth century is the material within Bartoloni, *Codice diplomatico*.

38. Bartoloni, *Codice diplomatico*, 204–206, no. 128.

39. Halphen, 173; Boüard, *Le régime*, 70; Riccardo of San Germano, *Chronica*. See too Cesare de Cupis, *Regesto*, I, 106–119 (Archivio Orsini, II.A.III.16) for the *popolo* agreeing in 1312. See also Waley, *The Italian City-Republics*, 53, 61–62. See too Thierry de Vaucouleurs, *Vita Urbani IV*, in Muratori, *RIS*, III, 2 (Milan, 1734), col. 413.

40. The change from primitive representative assembly to one or two nobles is discussed with misleading overemphasis by Boüard, *Le régime*, 135–139, particularly 136, because of his romantic view of the earlier assembly.

41. See Gregorovius, *History*, V, 295–296; Halphen, *Etudes*, 64. In thirteenth-century Rieti, special councils seem, at least sometimes, to have been committees of general and special councils, as in a 1284 example recorded in Rieti, Archivio capitolare (communal collection), II.E.4.

42. Boüard, *Le régime*, 245.

43. Bartoloni, *Codice diplomatico*, 163–166, no. 99.

44. Bartoloni, *Codice diplomatico*, 161–162, no. 98.

45. Bartoloni, *Codice diplomatico*, 216–224, no. 136; the document dates the first action Tuesday March 5, but Tuesday was March 6 (neither scribe nor editor seems to have been bothered by the impossible sequence of days)—I assume that the day of the week would be better remembered than the date. For *anziani*, in general, see Waley, *The Italian City-Republics*, 186–187.

46. Rome, Archivio di stato, Santo Spirito, B, 43, 132; Biblioteca Apostolica Vaticana, Sant'Angelo in Pescheria, I, fo. 113. Rome, Vallicelliana, II.A.I.33(ol.31) for Magnifico Napoleone di Matteo Rosso, early; see too II.A.II.47(ol.44) and 48(ol.45); Waley, *The Italian City-Republics*, 165–167. The quick acceleration of self-conscious nobility is suggested by one of the documents which are connected with the Sant'Eustachio family and which are now preserved in the capitular archives at Rieti; it is Rieti, Archivio capitolare (communal collection), III.A.1 (from 1308). It is a variously interesting document: One of its Sant'Eustachios is also a canon of the church of Sant'Eustachio; two of its witnesses are the chamberlain of the "advocates and judges" and the rector of the "judges and advocates" of the City (Rome).

47. Antonio Rota, "Il Codice degli 'Statuta urbis' del 1305," *ASRSP*, LXX (1947), 147–162, 160–161. Re, *Statuti*, 172–173, 108–109; Saba Malaspina, *Historia*, col. 815; Boüard, *Le régime*, 298, 299, app. xiii; see also Re, 101–102, 104, 109–110. The importance of the distinction between horse

Notes

and foot is underlined by the division into categories of Ottobuono Fieschi's household legacies; see Federico Federici, *Della Famiglia Fiesca* (Genoa, n.d. [1641]), 131.

48. Saba Malaspina, cols. 834, 864; Finke, *Acta Aragonensia*, II, 614–618, no. 393, particularly 615, 616; Peter Partner, *The Lands of St. Peter: The Papal State in the Middle Ages and the Early Renaissance* (Berkeley and Los Angeles, 1972), 298. See too Waley, *The Italian City-Republics,* 200–218.

49. Halphen, *Etudes,* 61; for the Vico, see Halphen, 23; see also Carlo Calisse, *I Prefetti di Vico* (Rome, 1888), 18, also 3–5 (also in *ASRSP,* X (1887), 1–136, 353–594). I do not mean to imply that the Vico were not a powerful family in the thirteenth century.

50. Bartoloni, *Codice diplomatico,* 192–199, nos. 119–123. For the mosaic senators, see particularly Livario Oliger, "Due Musaici con S. Francesco della Chiesa di Aracoeli in Roma," *Archivum Franciscanum Historicum,* IV (1911), 213–251.

51. Bartoloni, *Codice diplomatico,* 190–192, no. 118; see also 188, no. 101.

52. Boüard, *Le régime,* 148; Bartoloni, *Codice diplomatico,* 201–203, no. 126.

53. Bartoloni, *Codice diplomatico,* 190–192, nos. 117–118.

54. Halphen, 21, 41, 48, 85; Boüard, *Le régime,* 149–151, 292; Bartoloni, *Codice diplomatico,* 137–141, no. 84 (139), 299–311, no. 130; Rome, Archivio di stato, Santo Spirito, B, 60.

55. Theiner, *Codex diplomaticus,* I, 365, no. 537.

56. See particularly Bartoloni, *Per la storia,* 3–10, but also his *Codice diplomatico,* throughout.

57. Bartoloni, *Per la storia,* 5, n. 2; also *Codice diplomatico,* 91, 93.

58. Bartoloni, *Codice diplomatico,* 216–224, no. 136 (224), 121–123, no. 75 (123), 204–206, no. 128 (205); Boüard, *Le régime,* 292. See too A. de Boüard, "Les notaires de Rome au moyen âge," *Mélanges d'archéologie et d'histoire,* XXXI (1911), 291–307, particularly 305, n. 1. For podestàs' bringing judges, see Waley, *The Italian City-Republics,* 86; and for a Sorrentine judge palatine in 1270, see Rome, Vallicelliana, Archivio Orsini, II.A.I.49(ol.47).

59. Halphen, *Etudes,* 75; Boüard, *Le régime,* 292.

60. Boüard, *Le régime,* 298–299, app. xiii.

61. Bartoloni, *Codice diplomatico,* 137–141, no. 84.

62. Bartoloni, *Codice diplomatico,* 137–141, no. 84.

63. Boüard, *Le régime,* 290–293, app. ix; see also Theiner, I, 260–261, no. 414; Halphen, 81; G. Tomassetti, "Del sale e focatico del comune di Roma nel medio evo," *ASRSP,* XX (1897), 313–368.

64. Schiaparelli, "I 'Magistri aedificiorum Urbis,'" 35–37, 52–53.

65. Schiaparelli, 26–27.

66. Schiaparelli, 33–35, 50–53.

67. Schiaparelli, 29–30, 37, 53, 46. For a private contract which selects the Masters as arbiters, see Rome, Vallicelliana, Archivio Orsini, II.A.I.50(ol.48).

68. For example, Rome, Archivio di stato, San Cosimato, 326.

69. Boüard, *Le régime*, 180–186; Tomassetti, "Del sale," 318; Theiner, *Codex diplomaticus*, I, 267–268, no. 426.

70. Boüard, *Le régime*, 297–298, app. xii; Gatti, *Statuti dei Mercanti,* 32; Bartoloni, *Codice diplomatico,* 211–213, no. 131.

71. Waley, 171.

72. Peter Herde, "Papal Formularies for Letters of Justice," *Proceedings of the Second International Congress of Medieval Canon Law* (Vatican City, 1965), 321–345, 340.

73. Rome, Archivio di stato, San Cosimato, 185.

74. Rome, Archivio di stato, San Cosimato, 199.

75. Rome, Archivio di stato, San Cosimato, 261.

76. Rome, Archivio di stato, Santo Spirito, B, 52; also 45.

77. Rome, Archivio di stato, San Cosimato, 259.

78. Rome, Archivio di stato, San Cosimato, 328.

79. Rome, Archivio di stato, San Cosimato, 322 (dated on dorse 1291, but by indiction 1290).

80. Rome, Archivio di stato, San Silvestro, B, 51.

81. Rome, Archivio di stato, San Cosimato, 233, 208 (connection in 1216), 307, 231, 232, 243, 301.

82. Re, *Statuti,* 90–93, 101, 182–183.

83. Re, 93.

84. Re, 99, 94, 125–126.

85. Re, 95–96, 123–124, 101–102, 109–110, 104–105.

86. Re, 172, 111, 113, 108, 107, 114, 114, 117.

87. See, for example, the edited statutes in *Statuti della Provincia Romana,* Fonti per la storia d'Italia, vols. 48 and 69.

88. Re, 125, 142, 140–141, 172, 170; for the terrible depredations of wolves later in 1443, see Pietro de Angelis, *L'Ospedale di Santo Spirito,* II, 61.

89. Re, 102–103.

90. In *The First Century of English Feudalism.*

91. Rieti, Archivio di stato, Fondo comunale, 4, 5.

92. For a hostile character sketch of Rainerio, see Kantorowicz, *Frederick the Second,* 584.

93. For renunciation, see, for example, Rome, Vallicelliana, Archivio Orsini, II.A.II.2. Rome, Archivio di stato, San Silvestro, 147; *ASRSP,* XXIII (1900), 120–121, no. 153. Also San Silvestro, 146; *ASRSP,* XXIII (1900), 118–120, no. 152. For an example of counselors (a Capoferro and a Malabranca to a Cenci), see Archivio Orsini, II.A.I.25(ol.24).

94. Rome, Archivio di stato, San Cosimato, 204.

95. Re, 54, 189.

97. Ira Marvin Lapidus, *Muslim Cities in the Later Middle Ages* (Cambridge, Massachusetts, 1967), 187. Of exposing documents, see, for example, Rome, Vallicelliana, Archivio Orsini, II.A.I.29(ol.27A), II.A.I.27(ol.26), II.A.I.46(ol.44), II.A.I.25(ol.24), II.A.I.49(ol.47); and see Halphen, *Etudes,* 84 and n. 1, for Martina. For Martino, see D. Waley, "Annibaldo Annibaldi," *Dizionario biografico degli Italiani,* III (Rome, 1961), 340–342, 341, where he speaks of Annibaldo as senator stopping the bad custom of the canons' of Saint

Notes

Peter's offering banquets to the popular magistrates, the *iudices Sancti Martini*.

98. Rome, Archivio di stato, Santo Spirito, A, 23; Santo Spirito, B, 55. For a reflection in written records of the actual involvement of Boniface VIII, see Heinrich Finke, *Aus den Tagen Bonifaz VIII* (Münster.i.W., 1902), xli–xlviii. For the hostile observer, see Finke, *Acta Aragonensia*, I, 119–121, no. 83 (121).

99. Strecker, *Moralisch-satirische Gedichte Walters von Chatillon*, 110–112 (poem 10); Ernest H. Kantorowicz, *Laudes Regiae* (Berkeley, 1958), 6 and n. 17 and 18.

CHAPTER IV
The Popes

1. Carroll Winslow Brentano has informed me that Urban IV also was considered to be a saint both in Perugia and Troyes. The best modern introductions to the papal chancery for this period are, in my opinion, Peter Herde, *Beiträge zum päpstlichen Kanzlei- und Urkundenwesen im 13. Jahrhundert,* 2nd ed. (Kallmünz, 1967), and the incisive introduction to C. R. and Mary G. Cheney, *The Letters of Pope Innocent III (1198–1216)* (Oxford, 1967), particularly, xi–xviii. They should be supplemented with one of the excellent editions of papal registers prepared by the French Schools of Athens and Rome, for example, *Les Registres d'Innocent IV,* ed. Elie Berger (Paris, 1884–1911), or *Les Registres de Boniface VIII,* ed. Georges Digard, Maurice Faucon, Antoine Thomas, and Robert Fawtier (Paris, 1907–1939), and particularly the essay by Fawtier in vol. IV. One should now also see O. Hageneder and A. Haidacher, *Die Register Innocenz' III* [Publikationen der Abteilung für historische Studien des österreichischen Kulturinstituts in Rom. Quellen] 1(1964). The best general introduction to the history of the papacy is probably Johannes Haller, *Das Papsttum* (Urach, 1950, Stuttgart, 1953). A much quicker introduction is available in Geoffrey Barraclough's recent and remarkable, *The Medieval Papacy* (London and New York, 1968). A serious discussion of a chronological segment of papal bibliography exists in Robert E. McNally's essay, "The History of the Medieval Papacy: a Survey of Research, 1954–1959," *Theological Studies,* XXI (1960), 92–132. Although Daniel Waley's *The Papal State in the Thirteenth Century* is not specifically a history of the popes, it is very revealing of them. English readers will find Horace K. Mann, *The Lives of the Popes in the Middle Ages* (London, 1902–1932), helpful and informative. The *Italia Pontificia* volumes edited by Paul Kehr and Walther Holtzmann, although they purportedly deal with a period prior to that of this book, are still, like the companion volumes for other countries, an invaluable introduction not only to the products of the papal curia but also to the archives in which they may be found.

2. For a convenient guide to the cardinalate, see Conrad Eubel, *Hierarchia catholica medii aevi,* I (Münster, 1913), 3–13, particularly, here, 9–10. For the meaning and history of the dignity, see Stephan Kuttner, "Cardinalis: the History of a Canonical Concept," *Traditio,* III (1945), 129–214. See too

Notes

Hans-Walter Klewitz, *Reformpapsttum und Kardinalkolleg* (Darmstadt, 1957) and Paravicini, *Cardinali di curia,* with its bibliography.

3. For Boniface, see T. S. R. Boase, *Boniface VIII* (London, 1933). This seems to me an insufficiently valued book, perhaps because it stresses neither the sort of specifically archival history nor the sort of legal learning which have been popular among scholars in recent years. It is a perceptive study in total personality, and as such, I now think, superior to all other biographies of thirteenth-century popes, although these include some major works like Achille Luchaire, *Innocent III* (Paris, 1904–1908). Thirteenth-century popes can now best be looked at, although seldom if ever in actual portrait likeness, in Gerhart Ladner, *Die Papstbildnisse des Altertums und des Mittelalters,* II: *von Innozenz II zu Benedikt XI* (Vatican City, 1970), which is vol. IV, ser. II of the Monumenti di antichità cristiana published by the Pontificio istituto di archeologia cristiana.

4. Boase, *Boniface VIII* 124, 258–259, 286–288; Heinrich Finke, *Aus den Tagen Bonifaz VIII* (Münster, 1902), il–l, and also particularly xlv; Pierre Dupuy, *Histoire du differend d'entre le Pape Boniface VIII et Philippes le Bel Roy de France* (Paris, 1655), 339; "Annales Genuenses Georgii Stellae," *RIS,* XVII (Milan, 1730), col. 1019.

5. Karl Hampe, *Ein ungedrukter Bericht über das Konklave von 1241,* 27–31; Horace K. Mann, *The Lives of the Popes in the Middle Ages,* XVI (London, 1932), 28; Saba Malaspina, *Historia,* cols. 783–874, 871–872· Boase, 29–41 (33 for the Stefaneschi translation). See too the conclave where *multi multa loquuntur,* the verse description of which is edited in Agostino Paravicini Bagliani, "Versi duecenteschi su un conclave del secolo XIII," *Miscellanea Gilles Gerard Meersseman,* Italia Sacra, 15 (Padua, 1970), 151–169. Although the Septizonium's remaining ruins were destroyed during the pontificate of Sixtus V (1585–1590), it can still be seen in Renaissance drawings; see for example in *Il Paesaggio nel disegno europeo del XVI secolo,* Mostra all'Accademia di Francia, Villa Medici, 1972–1973 (Rome, 1972), 158–161, nos. 108, 109 by Marten van Heemskerck.

6. Rose Graham, "Archbishop Winchelsey from His Election to His Enthronement," *Church Quarterly Review,* CXLVIII (1949), 161–175, 170.

7. Boase, 41; Arsenio Frugoni, "Laudi Aquilane a Celestino V," *Rivista di storia della chiesa in Italia,* V (1951), 91–99, 93.

8. Ptolemy of Lucca, *Historia Ecclesiastica, RIS,* XI (Milan, 1727), cols. 743–1242, 1200; R. Brentano, "'Consolatio defuncte caritatis': a Celestine V letter at Cava," *English Historical Review,* LXXVI (1961), 298–303; Boase, 45. For a Sulmona representation which, surely inadvertently, suggests this Celestine, see Ladner, *Die Papstbildnisse,* II, plate LVIIIa.

9. Saba Malaspina, *Historia,* col. 835; Ptolemy of Lucca, *Historia Ecclesiastica,* col. 1191. See the similar evaluation in "Alia continuatio" (of Martinus Oppaviensis), ed. Ludwig Weiland, *MGH,SS,* XXII (Hanover, 1872), 482.

10. *Chronicon de Lanercost,* ed. Joseph Stevenson (Edinburgh, 1839), 115; and see Salimbene de Adam, *Cronica,* ed. O. Holder-Egger, *MGH,SS,* XXXII (1905–1913), 618. See also the Lanercost verse in translation: *The Chronicle of Lanercost,* trans. Herbert Maxwell (Glasgow, 1913), 38.

11. *Liber censuum,* I, 117. For two particularly attractive visual representa-

Notes

tions of Honorius III, see Ladner, *Die Papstbildnisse,* III, plates XIV and XVb. See too Raoul Manselli, "S. Domenico, i papi e Roma," *Studi Romani,* XIX (1971), 133–143, 137.

12. Waley, *The Papal State,* 213; for the Crescenzi and Savalli, see Carlo Cecchelli, *I Crescenzi, i Savelli, i Cenci* (Rome, 1942).

13. *Registres d'Honorius IV,* ed. Maurice Prou (Paris, 1888), 267. For the swarming Boccamazzi, see I. Walter, "Giovanni Boccamazza," *Dizionario biografico degli Italiani,* XI, 20–24; M. T. Maggi, "Nicola Boccamazza," *Dizionario,* 24–25; *Necrologi e libri affini della provincia Romana,* ed. Pietro Egidi, I, 198/9, 200/1, 238/9, 240/1, 254/5, 266/7; Biblioteca Apostolica Vaticana, Santa Maria in Via Lata, cassetta 300–301A, no. 65 alias 909, no. 66 alias 239, no. 73 alias 380; Rome, Vallicelliana, Archivio storico capitolino, Archivio Orsini, II.A.I.51(ol.49).

14. The incident is used in F. M. Powicke, *Stephen Langton* (Oxford, 1928), 51.

15. Mann, *The Lives,* XVI, 28; Natalie Schöpp, *Papst Hadrian V* (Heidelberg, 1916), 309.

16. Canon Nicholas, *Clément IV* (Nîmes, 1910), 142. For Clement's splendid sepulchral head, see Ladner, *Die Papstbildnisse,* II, plates XXXIIa and XXXIIb.

17. Matthew Paris, *Chronica Majora,* V (London, 1880), 471; Jacopo da Varagine, *Cronaca di Genova,* II, ed. Giovanni Monleone (Rome, 1941), 399; Mann, XVI, 29. Federici, *Della Famiglia Fiesca,* 129, 132, 140; Santa Maria in Via Lata was planted by the fourteenth-century cardinal, Luca Fieschi.

18. "Continuationes breves chronici Martini Oppaviensis," ed. O. Holder Egger, *MGH,SS,* XXX, part 1, 711. For Ottobuono's foreign prebends, see Paravicini, *Cardinali di curia,* I, 360.

19. Etienne Georges, *Histoire du Pape Urbain IV et de son temps* (Arcissur-Aube, 1866), 2. Otto Schiff, *Studien zur Geschichte Papst Nikolaus' IV,* Historische Studien, V (Berlin, 1897), 11.

20. Decretals of Gregory IX: c.10, X, i, 9. For a quick English introduction to Innocent III and some of the problems of interpreting his thought and action, see James M. Powell, *Innocent III, Vicar of Christ or Lord of the World* (Boston, Massachusetts, 1963); for an even quicker introduction, and one that fits Innocent into a history of medieval theories of church and state, see Brian Tierney's brilliant *The Crisis of Church and State* (Englewood Cliffs, New Jersey, 1964), 127–138, or the introduction to C. R. Cheney and W. H. Semple, *Selected Letters of Innocent III* (London and Edinburgh, 1953), particularly ix–x.

21. *Gesta Innocentii III,* cols. xix–xxi.

22. *Gesta Innocentii III,* col. clxxxvii.

23. *Gesta Innocentii III,* col. lxxx.

24. Again, for a short, powerful discussion of Innocent's position, see Tierney, *The Crisis,* 127–131.

25. Migne, *Patrologia Latina,* CCXVII, col. 494; also cols. 465–466. For a clear appreciation of the antiquity of the topos *militia spiritualis,* see Hilarius Edmonds, "Geistlicher Kriegsdienst. Der Topos der militia spiritualis in der antiken Philosophie," *Heilige Uberlieferung* (supplement to *Beiträge zur Ge-*

schichte des alten Mönchtums und des Benediktinerordens) (Münster, 1938), 21–50.

26. Innocent III, *Opera omnia,* IV: Migne, *Patrologia Latina,* CCXVII, cols. 773–916, particularly 851–886; cols. 761–764; cols. 745–762. J. de Ghellinck, "Eucharistie au XIIe siècle en occident," *Dictionnaire de Théologie Catholique,* V, 1 (Paris, 1913), cols. 1233–1302, 1266–1267.

27. *Gesta Innocentii III,* cols. xviii, ccxiv.

28. *Gesta Innocentii III,* cols. cciv–ccx.

29. Lotharii cardinalis (Innocentii III) *De miseria humane conditionis,* ed. Michele Maccarone (Lucca, 1955), 16.

30. Migne, *Patrologia Latina,* CCXVII, col. 504.

31. Cheney, "The Letters of Pope Innocent III," 33–39 (particularly 33 for change, 35 especially referring to A. Fliche, and 36, 37, 39), 41. See too, Cheney and Semple, *Selected Letters,* xv. For what follows, see Helene Tillmann, *Papst Innocenz III* (Bonn, 1954), particularly 5. For the moustache, see Gerhart B. Ladner, "Eine Prager Bildnis-Zeichnung Innocenz III," *Collectanea Stephan Kuttner,* I, 25–35, 32–35 and fig. 1, particularly 34.

32. *Gesta Innocentii III,* col. lxxx.

33. *Chronicon Abbatiae de Evesham,* ed. W. D. Macray, Rolls Series (London, 1863), 152, 189.

34. Karl Hampe, "Eine Schilderung des Sommeraufenthaltes der römischen Kurie unter Innocenz III in Subiaco 1202," *Historische Vierteljahrschrift,* VIII (1905), 509–535, 528–535, particularly 534 and 535.

35. William of Malmesbury, *Historia Novella,* ed. K. R. Potter (Edinburgh, 1955), 11–12.

36. See, for example, Tillmann, 234–236.

37. Boase, *Boniface VIII,* 357, 360–362, 369.

38. For a directly opposite point of view, see Leonard E. Boyle, "An Ambry of 1299 at San Clemente, Rome," *Mediaeval Studies,* XXVI (1964), 345, where Father Boyle finds the accusations connected with the trials "ludicrous and unlikely." This in an essay rich in complex evidence about Boniface, particularly concerning his devotion to the Eucharist and his halo (332, 343–346). The public controversy about the trial has been a long one; a first step into it could be made through Antonio Corvi, *Il processo di Bonifazio VIII. Studio critico* (Rome, 1948). For Boniface's headwear (discussed in Boyle), see also Gerhart Ladner, "Die Statue Bonifaz' VIII in der Lateranbasilika und die Entstehung der dreifach gekrönten Tiara," *Römische Quartalschrift,* XLII (1934), 35–69. For Boniface iconography, see Ladner, *Die Papstbildnisse,* II, 285–340, and for particular and probably surprising pleasure, plates LXVIII and LXIXa and LXIXb. See too the specific collection of material in Clemens Sommer, *Die Anklage der Idolatrie gegen Papst Bonifaz VIII. und seine Porträtstatuen* (Freiburg i. Br., 1920).

39. Dupuy, *Histoire,* 541–542.

40. Dupuy, 527, 539–541.

41. Dupuy, 562–563, 529, 570–572, also 551.

42. Dupuy, 568, 338.

43. Dupuy, 526–527, 536, 550, 552.

44. Dupuy, 539, 560–561.

Notes

45. Dupuy, 339; Boase, *Boniface VIII*, 124; Finke, *Aus den Tagen*, xlv–l.

46. *Laude di fratre Jacopone da Todi*, ed. Giovanni Ferri (Rome, 1910) [Società filologica romana], 87, lauda 58, 11.19–20.

47. Dupuy, 570.

48. Tierney, *The Crisis*, 172–173. (My view of Boniface's personality is obviously very different from Tierney's.)

49. Finke, xxxix; Boase, *Boniface VIII*, 286; Gelasio Caetani, *Regesta Chartarum*, I (Perugia, 1925), 109–110; see Waley, *The Papal State*, 230–249, for Boniface's actual dealing with papal territories.

50. Boase, 161. See too the magnificent description of Boniface in 1301, "nothing but eyes and tongue in a wholly putrefying body," quoted from an Aragonese source in Partner, *The Lands of St. Peter*, 292.

51. See Villani, *Historie*, col. 397, for his intelligent Boniface, and Ptolemy of Lucca, *Historia Ecclesiastica*, col. 1203, for his impressive Boniface, and Jacopo da Varagine, *Cronaca di Genova*, II, 410, for his learned Boniface.

52. Gregorovius, *History*, V, 590; Boase, 346–348, 344, n. 5; Walther Holtzmann, "Zum Attentat von Anagni," *Festschrift Albert Brackmann* (Weimar, 1939), 492–507, particularly 495; "Relatio de Bonifacio VIII: de horribili insultatione et depredatione," ed. F. Liebermann, *MGH,SS*, XXVIII (Hanover, 1888), 621–626, 623. For the treasure see Emile Molinier, "Inventaire du Trésor du Saint Siège sous Boniface VIII," *Bibliothèque de l'école des chartes*, XLIII–XLIX (1882–1888); and see too Luisa Mortari, *Il Tesoro della Cattedrale di Anagni* (Rome, n.d.)

53. Boase, 182; "Martinus Oppaviensis Chronicon: Continuationes Anglicae Fratrum Minorum," ed. Ludwig Weiland, *MGH,SS.*, XXIV (Hanover, 1879), 253–259, 254–255 (earthquake). See Friedrich Bock, "Musciatto dei Francesi," *Deutsches Archiv für Geschichte des Mittelalters*, VI (1943), 521–544, particularly 533–536, for one of the interesting men connecting complexly and importantly Italian affairs and the court of France.

54. It seems to me that Father Boyle's remarks about Bolsena and the San Clemente ambry make an excellent starting point for considering Boniface and the Eucharist, "An Ambry of 1299," particularly 332.

55. See Ludovico Gatto, *Il Pontificato di Gregorio X (1271–1276)* (Rome, 1959) [Istituto storico italiano per il medio evo, Studi storici], 28–30 and the "Life" in Pietro Maria Campi, *Dell'historia ecclesiastica di Piacenza* (Piacenza, 1651–1662), II, 343–349; E. Martène and U. Durand, *Thesaurus novus anecdotorum*, II (Paris, 1717), cols. 96–97, and the *Registres de Clément IV*, ed. Edouard Jordan (Paris, 1893–1904), 253–268, nos. 691–703. The consideration of Celestine V should now begin with Edward Peters, *The Shadow King* (New Haven, 1970). For the Arezzo, visual Gregory X, Ladner, *Die Papstbildnisse*, II, plate XXXVI.

56. Boase, *Boniface VIII*, 4. For a more general examination of Alexander, see Salvatore Sibilia, *Alessandro IV (1254–61)* (Anagni, 1961); for a summing up of recent work on Alexander's family, see particularly 49–50.

57. Gregorovius, *History*, V, 514–515.

58. See Brentano, *Two Churches*, 3–61, and, for example, Rieti, archivio capitolare, VII.E.8.

Notes

59. Caetani, *Regesta Chartarum*, I, 76–86; for Caetani holdings, see Giuseppe Marchetti-Longhi, *I Caetani* (Rome, 1942), particularly 20–24. For Boniface's spending and its bibliography, see Peter Partner, "Camera Papae: problems of Papal Finance in the later Middle Ages," *The Journal of Ecclesiastical History*, IV (1953), 55–68, particularly 55; but see too, particularly for his method of collecting property, in this case around Cecilia Metella, Georges Digard, "Le domaine des Gaetani au tombeau de Cecilia Metella," *Mélanges G. B. de Rossi*, being Supplement XII to *Mélanges d'archéologie et d'historie* (Paris and Rome, 1892), 281–290, which includes references to the pertinent documents within Boniface's registers which Digard was then in process of editing. For the arrangements concerning the creation and erection of one of the statues of Boniface, the metal one in Bologna, see Maria Cremonini Beretta, "Il Significato politico della statua offerta dai Bolognesi a Bonifacio VIII," *Studi di storia e di critica dedicati a Pio Carlo Falletti* (Bologna, 1915), 421–431; and in connection with these statues, see again Boyle, "An Ambry of 1299."

60. Papal itineraries are not only clear from registers and from other extant papal letters, but also they have been outlined by their editors in the printed registers.

61. Theiner, *Codex diplomaticus*, I, 360–366; Friedrich Baethgen, "Quellen und Untersuchungen zur Geschichte der päpstlichen Hof-und Finanzerwaltung unter Bonifaz VIII," *Quellen und Forschungen*, XX (1928–1929), 114–237, 229.

62. "Barbarously involved" is from Gregorovius (V, 630) and springs from the irritation anyone must feel as he tries to read Stefaneschi's sort of verse in which, as Father Boyle has written very nicely, "sometimes, indeed, words are switched around like so many checkers, but without any regard to position or sense, for the sake of his not always happy hexameters" ("An Ambry of 1299," 339). For a very sympathetic treatment of Stefaneschi, see Arsenio Frugoni, "Il Cardinale Jacopo Stefaneschi, Biografo di Celestino V," *Celestiniana* (Rome, 1954), 69–124.

63. Baethgen, 207–210.

64. "Continuationes breves chronici Martini Oppaviensis," ed. O. Holder-Egger, 711; for Nicholas's translation to his sepulcher in his chapel on May 16, see Egidi, *Necrologi*, I, 212; the first numbering of days is correct, counting from Nicholas's election (November 25, 1277), not from his consecration (December 26, 1277): C. R. Cheney, *Handbook of Dates* (London, 1970), 38.

65. The quotation is from Tierney, *The Crisis*, 150; Gerard E. Caspary, "The King and the Two Laws: A Study of the Influence of Roman and Canon Law on the Development of Ideas on Kingship in Fourteenth-century England," an unpublished Harvard Ph.D. thesis; for Innocent seen in another light, see M. D. Lambert, *Franciscan Poverty* (London, 1961), 95–97. For Innocent's sepulchral head, see Ladner, *Die Papstbildnisse*, II, plate XXIIb.

66. See the discussion in Mann, *The Lives*, XVI, 23–24; but compare the very attractive image of the family that can emerge from Fieschi wills in Federici, *Della Famiglia Fiesca*, for example 131, Ottobuono's treasuring Innocent IV's own copy of his "Apparatus."

67. I have discussed this more thoroughly in "Innocent IV and the Chap-

Notes

ter of Rieti," *Studia Gratiana,* XIII [Collectanea Stephan Kuttner III] (1967), 383–410.

68. F. Pagnotti, "Niccolò da Calvi e la sua vita d'Innocenzo IV," *ASRSP,* XXI (1893), 7–120; the life is edited on 76–120.

69. By Herde in *Beiträge.*

70. In general, see Waley, *The Papal State,* and particularly here, 157.

CHAPTER V
The Natural Family

1. Gregorovius, V, 297, and through 300; also 45.

2. *Les Registres d'Honorius IV,* 577–583, no. 823, 578.

3. Livario Oliger, "B. Margherita Colonna," *Lateranum,* New Series, I (2) (1935), 83; its introduction and two lives are the source of almost all that follows on Margherita. For Sciarra, see Holtzmann, "Zum Attentat von Anagni," 497.

4. Oliger, 142–143 (and 211–212); compare *Confessions,* Book IX, chapter 10. See Peter Brown, *Augustine of Hippo* (Berkeley, 1969), 128–131, for the stay in Ostia.

5. Oliger, 143, 212–213.

6. Oliger, 152–153, 127; on the Veronica see, for example, Peter Brieger, *English Art, 1216–1307* (Oxford, 1957), 137, Brentano, *Two Churches,* 56–57, and Biblioteca Apostolica Vaticana, Archivio di San Pietro, V, 10, "85" (a 1272 letter of Gregory X ordering the canons to allow a private viewing by a countess and her entourage, a letter called a *licentia ostendendi vultum sanctum*); also on the Veronica, see Arsenio Frugoni, "Il Giubileo di Bonifacio VIII," *Bullettino dell'Istituto storico italiano,* LXXII (1950), 1–121, 23. The reference to Altruda's visits *ad limina,* of course, suggests the familiar devotion to the apostles Peter and Paul, but the description of Altruda's behavior, the *prolixitas,* and the varying time, surely seems to refer to a more sprawling and varying visitation of relics and churches, and not even just the addition of San Lorenzo: compare de Angelis, *L'Ospedale di Santo Spirito,* I, 101, 63. See André Vauchez, "Sainteté laïque au XIIIe siècle: La vie du bienheureux Facio de Crémone," *Mélanges de l'école française de Rome,* LXXXIV (1972), 13–53, particularly 39 for Facio's eighteen visits to the *limina apostolorum Petri et Pauli et aliorum sanctorum.*

7. Oliger, 212–213.

8. In general, see Oliger's introduction, and particularly 80–82, but for the name of Margherita's mother and for her genealogy, see Cesare de Cupis, *Regesto,* I, 45–51 (Rome, Vallicelliana, Archivio storico capitolino, Archivio Orsini, II.A.I.20), the will of the older Margherita's father, Giovanni Gaetani (or Giangaetano) di Orso di Bobone di Pietro, from 1232.

9. Oliger, 115.

10. Oliger, 132.

11. Oliger, 221.

12. Compare Dupuy, 29, and *Les Registres de Boniface VIII*, I, cols. 961–967, no. 2388, col. 962.

13. Oliger, 84, 213–214.

14. Waley, *The Papal State*, 214.

15. For a general introduction to the family, see Pio Paschini, *I Colonna* (Rome, 1955).

16. Paschini, 9–10.

17. I have particularly in mind the apse mosaic of Santa Maria Maggiore and Giovanni's senatorial mosaic now in the Palazzo Colonna. For Gregorovius's use of Petrarch's Colonna letters, see Gregorovius, *History*, VI, 306; and see Petrarch, *Epistolae de rebus familiaribus et variae* (Florence, 1859), I, 255 (Book V, Ep. 3). For a reward to a loyal family, see Giacomo's receiving Bartolomeo de Rocca, a canon of Rieti, as a member of his *consortium* of chaplains in 1308—moved, he says, by the continued loyalty of Bartolomeo's relatives: Rieti, archivio capitolare, VII.A.4. See *Regestum Clementis Papae V* (Rome, 1886), 323–324, no. 3535, for Clement's granting Giacomo the care of San Lorenzo in Lucina, from which his (Giacomo's) letter is dated. For Giovanni's connection with the Franciscans, see Herbert Grundmann, *Religiöse Bewegungen im Mittelalter*, Historische Studien: 267 (Berlin, 1935), 129–133.

18. Ludwig Mohler, *Die Kardinäle Jakob und Peter Colonna* (Paderborn, 1914), 216, within a general body of Caetani-Colonna material, 215–277. For Pietro and San Salvatore, see Pietro de Angelis, *L'Arcispedale del S. Salvatore* (Rome, 1958), 10. For Pietro and San Giacomo, see *L'Arcispedale di San Giacomo in Augusta* (Rome, 1955), 6. For Pietro's library, a significant part of which he admittedly bought already assembled, see Paravicini, "Le Bibliotheche dei Cardinali Pietro Peregrosso (†1295) e Pietro Colonna (†1326)," particularly 109, 125, 129, 131, 134–135.

19. See particularly Giuseppe Presutti, "I Colonna di Riofreddo (sec. xiii e xiv)," *ASRSP*, XXXIII (1910), 313–332, 328–331. On Colonna family divisions, see Waley, *The Papal State*, 213–214, Gregorovius, V, 541–542, and *Les Registres de Boniface VIII*, I, cols. 961–967, no. 2388, col. 963. On Colonna background and holdings, see, as Waley suggests, Richard Neumann, *Die Colonna und ihre Politik, 1288–1328* (Langensalza, 1916), 48–59. On family structure, see Oliger, 70–75.

20. By Oliger, 100.

21. *Cronica Fratris Salimbene de Adam ordinis minorum*, ed. O. Holder-Egger, *MGH,SS*, XXXII, part iii (Hanover and Leipzig, 1913),169–170.

22. *Les Registres d'Honorius IV*, 577–583, no. 823.

23. Cecchelli, *I Crescenzi*, 23–25; see too, A. Proia and P. Romano, *Il Rione S. Angelo* (Rome, 1935), 102–103; "Alia continuatio," *MGH,SS*, XXII, 482; Mann, *The Lives*, XVI (London, 1932), 361.

24. For Stefaneschi, see "Opus Metricum," in *Monumenta Coelistiniana*, ed. Franz Xaver Seppelt (Paderborn, 1921), 97; the Pandulfo bell is one of a pair of Saint Peter's Guidotto bells preserved together in the galleries. For the language of bells in medieval Rome and Lazio, see Alberto Serafini, *Torri campanarie di Roma e del Lazio nel Medioevo* (Rome, 1927), 76.

25. Gelasio Caetani, *Regesta Chartarum*, I, 57–58. For the loggia, see Biblioteca Apostolica Vaticana, Archivio San Pietro, caps. 61, fasc. 225. For a

Notes

318

plan and drawings of Monte Giordano, see Carlo Pietrangeli, *Guide rionali di Roma: Rione V—Ponte,* II (Rome, 1968), 33, 35, 37.

26. Rome, Vallicelliana, Archivio Orsini, II.A.III.1 (the act of 1300), II.A.II.38(ol.36) (sisters, Girardi, to brothers); for family action and involvement, see particularly II.A.III.27, but also II.A.II.36(ol.34), II.A.II.42(ol.40), II.A.II.47(ol.44), II.A.II.48(ol.45); for women, see II.A.I.29(ol.27A), II.A.II.42(ol.40), II.A.II.50(ol.47), II.A.III.27. For uncle and nephews, see Biblioteca Apostolica Vaticana, Archivio San Pietro in Vaticano, caps, 61, fasc. 225. Among cardinal priests of Santa Cecilia, Tommaso da Ocre lived next to his church but his predecessor Jean Cholet lived next to the church of San Crisogono, as their wills make clear; for a photographic copy and transcription of Cholet's will I should like to thank Agostino Paravicini Bagliani.

27. For Latino, see Angelus Maris Walz, *I Cardinali domenicani* (Florence and Rome, 1940), 17–18; Seppelt, *Monumenta,* 37; Salimbene, 436. For Matteo Rosso, see R. Morghen, "Il Cardinale Matteo Rosso," *ASRSP,* XLVI (1923), 271–372, 274–275. For Orsini and Santo Spirito, see Pietro de Angelis, *L'Ospedale,* II, 517, 545.

28. The classical monograph on Nicholas as cardinal is Richard Sternfeld, *Der Kardinal Johann Gaëtan Orsini (Papst Nikolaus III) 1244–1277* (Berlin, 1905).

29. Salimbene, 169–170; Dante, *Inferno,* canto xix, ll.70–71; Waley, *The Papal State,* 192; Boase, *Boniface VIII,* 246; see too the implications of Boniface VIII's letter condemning the Colonna in *Les Registres de Boniface VIII,* I, cols. 961–967, no. 2388, col. 962.

30. Waley, *The Papal State,* 192; Ferreto de' Ferreti, *Le opere,* I, 156. For the chapter of Saint Peter's, see Albert Huyskens, "Das Kapitel von St. Peter in Rom unter dem Einflusse der Orsini (1276–1342)," *Historisches Jahrbuch,* XXVII (1906), 266–290.

31. Seppelt-Stefaneschi, "Opus Metricum," 97; Frugoni, "Il Cardinale Jacopo Stefaneschi," 69–70, for Stefaneschi antiquity, and 104, for Stefaneschi manuscripts; for the Stefaneschi family, see also Frugoni's "La figura e l'opera del Cardinale Jacopo Stefaneschi (1270c.–1343)," *Atti della Accademia Nazionale dei Lincei,* series 8, V (1950), 397–424, and particularly 398; for the manuscripts, see also Gardner, "The Influence of Popes' and Cardinals' Patronage," 336–341. The term *mecenatismo* was applied to Stefaneschi by Muñoz, *Roma di Dante,* 404. For the clustering at Saint Peter's, see Huyskens, "Das Kapitel," 284, 285.

32. Francis Roth, *Cardinal Richard Annibaldi, First Protector of the Augustinian Order, 1243–1276* (Louvain, n.d.) [being extracts from *Augustiniana,* II–IV (1952–1954)], 3–4; Theiner, *Codex diplomaticus,* I, 118. For a convenient summary of Roman familial country-feudal activities, see Tomassetti, "Documenti feudali," particularly 296–304. Milton Lewine's forthcoming work on the Madonna dei Monti will deal with the tactical distance of these Annibaldi towers from the Colosseum.

33. Roth, 4; Waley, "Annibaldo Annibaldi," 341.

34. Roth, 5–6, 12–13; Waley, *The Papal State,* 307, 187; Walz, *I Cardinali,* 16 (for Annibaldo degli Annibaldi di Molara); Friedrich Bock, "Le Trattive per la senatoria di Roma e Carlo d'Angio," *ASRSP,* LXXVIII (1955),

Notes

69–105, 78; D. Waley, "Riccardo Annibaldi (Riccardo della Molara)," *Dizionario biografico degli Italiani*, III, 348–351. For the date of Riccardo's elevation, see Paravicini, *Cardinali di curia*, I, 143–144.

35. Roth, 14–15.

36. Roth, 9, 17–22, 25, 27, 31; Waley, *The Papal State*, 200–201; for the gossip about Ancher, see Salimbene, 170; for his murder, see B. M. Apolloni Ghetti, *Santa Prassede*, Le Chiese di Roma illustrate, 66 (Rome, 1961), 86–89; but for the obscurity of his life, see Paravicini, "Gregorio da Napoli," 75–76, n., and G. Mollat, "Anchier (Pantaléon)," *Dictionnaire d'histoire et de géographie ecclésiastique*, XII (Paris, 1914), col. 1514.

37. Roth, 24, 34.

38. Roth, 29.

39. Roth, 30, 29; Dupré Theseider, *Roma*, 314; Gelasio Caetani, *Domus Caetani*, I (San Casciano, 1927), 114, 146; *Regesta Chartarum*, I, 44–45, and 98–102, 114–132; Waley, *The Papal State*, 243; Gregorovius, *History*, V, 543; Boase, *Boniface VIII*, 164; "Annales Caesenates," *RIS*, XIV (Milan, 1729), 1115; "Annales Forolivienses," *RIS*, XXII (Milan, 1733), cols. 135–240. Giuseppe Marchetti-Longhi, "La carta feudale del Lazio," *Quellen und Forschungen*, XXXVI (1956), 324–327. V. de Donato, "Riccardo Annibaldi," *Dizionario biografico degli Italiani*, III, 351–352.

40. Roth, 42–49, 78–82, 97–101. The Aracoeli documents within the *Bullarium Franciscanum* do not suggest the involvement of Santa Maria del Popolo, but Annibaldi's position on the committee which got the Aracoeli for the Franciscans should be noted: *Bullarium Franciscanum*, I, ed. G. H. Sbaralea (Rome, 1759), 521–522, no. 288. On the other hand, other cardinals were also involved with the Augustinian Hermits, for example, see *Bullarium Franciscanum* I, 656–657, no. 475 (Cardinal Guglielmo Fieschi in 1253). For the Annibaldi scholar, see A. L. Redigonda, "Annibaldo Annibaldi," *Dizionario biografico degli Italiani*, III, 342–344. For Riccardo Annibaldi's chaplains, see Paravicini, *Cardinali di curia*, I, 153, 158–159.

41. Carlo Cecchelli, *S. Maria in Via*, English edition (Rome, n.d.), 19–21; Carlo Cecchelli, *I Margani, I Capocci, I Sanguigni, I Mellini* (Rome, 1946), 20–25; Vincenzo Forcella, *Iscrizioni delle chiese di Roma* (Rome, 1869–1884), II, 495, no. 1494, VIII, 357, no. 845, XI, 387, nos. 596, 597; Ragna Enking, *S. Andrea Cata Barbara e S. Antonio sull'Esquilino*, Le Chiese di Roma illustrate, 83 (Rome, 1964), 45–57, 109–114. Waley, *The Papal State*, 147–154; and, in general, Friedrich Reh, *Kardinal Peter Capocci, ein Staatsman und Feldherr des XIII Jahrhundert* (Berlin, 1933). See too Gardner, "Capocci Tabernacle." Berger, *Innocent IV*, III, 84, no. 5847. Ferri, *ASRSP*, XXVII, 166–167. Paravicini, *Cardinali di curia*, I, 300–306. For Valleranum, see too Archivio Orsini, II.A.II.2.

42. Pietro Egidi, *Necrologi e libri affini della provincia Romana*, I (Rome, 1908), 160. *Regesta Honorii Papae III*, ed. Pietro Pressutti (Rome, 1888–1895), II, 86, no. 4078, 472–473, no. 6203.

43. P. Fedele, "Tabularium S. Praxedis," *ASRSP*, XXVII (1904), 27–78, XXVIII (1905), 91–114, XXVII, 30. For the Savelli cloth, see Molinier, *Bibliothèque de l'école des chartes*, XLVII, 646–667, no. 1218.

44. Rome, Archivio di stato, Sant' Agostino, 6.

Notes

45. Rome, Archivio di stato, San Cosimato, 303. The Suburra will be more fully exposed in Milton Lewine's forthcoming work on the Madonna dei Monti.

46. Rome, Archivio di stato, San Silvestro in Capite, 156 ("Regesto," *ASRSP,* XXIII, 412–413, no. 165); Brezzi, *Roma,* 301; Rota, "Il Codice," *ASRSP,* LXX, 160–161.

47. Gasparrini, *Senatori Romani,* 13–18, 20–21; *Acta Sanctorum,* Maius, V (Antwerp, 1685), 85–100 (May 21).

48. Brezzi, 363; Caetani, *Regesta Chartarum,* I, 39–41, 44–45. But to the simple followers of Francis at least Giacomina did not seem romantically impoverished or decayed, but rather very noble and very rich: *de nobilioribus et ditioribus totius Urbis: The Writings of Leo, Rufino, and Angelo, Companions of St. Francis (Scripta Leonis, Rufini et Angeli sociorum S. Francisci),* ed. and trans. Rosalind B. Brooke (Oxford, 1970), 266–267.

49. Caetani, *Regesta Chartarum,* I, 226–227. The continuing importance of falling families like the Frangipane can be seen in their continued presence in the papal registers, and their indexes.

50. Caetani, *Regesta Chartarum,* I, 221–222.

51. Caetani, *Regesta Chartarum,* I, 235.

52. Rome, Archivio di stato, Sant'Alessio, 34.

53. Caetani, *Regesta Chartarum,* I, 61–64.

54. Rome, Archivio di stato, San Silvestro in Capite, 341. For a Roman proconsul, see D. Waley, "Annibaldo Annibaldi," *Dizionario,* III, 344–345.

55. Gelasio Caetani, "Margherita Aldobrandesca e I Caetani," *ASRSP,* XLIV (1921), 5–36; D. Waley, "Boniface VIII and the Commune of Orvieto," *Transactions of the Royal Historical Society* (London), Series 4, XXXII (1950), 121–139. The importance of marriage is readily apparent in the collection of documents within the Archivio Orsini: for a quick suggestion of its contents, see Cesare de Cupis, *Regesto,* I, 56, 64, 74–75.

56. Rome, Vallicelliana, Archivio Orsini, II.A.I.32(ol.30); F. Olivier Martin, et al., eds., *Les Registres de Martin IV* (Paris, 1935), 141–142, no. 337.

57. Re, *Statuti,* 175, 114, 93. For Giangaetano, see Biblioteca Apostolica Vaticana, Archivio San Pietro in Vaticano, caps. 61, fasc. 225.

58. Rome, Archivio di stato, San Cosimato, 188. Again see Herlihy, "Family Solidarity," and its terms and references.

59. See above, chapter 1, for Locrerengo; for Napoleone, see Biblioteca Apostolica Vaticana, Archivio San Pietro in Vaticano, caps. 64, no. 181, where this extract of a will now resides in a wrapper which is supposed to contain the will of Cardinal Stefaneschi.

60. *Les Registres de Boniface VIII,* I, col. 886, no. 2264.

61. Boüard, *Le Régime,* 52.

62. Proia and Romano, *Il Rione S. Angelo,* 102.

CHAPTER VI
The Spiritual Family

1. Giorgio Falco, "Il Catalogo di Torino delle chiese, degli ospedali, dei monasteri di Roma nel secolo XIV," *ASRSP,* XXXII (1909), 413–443 (and in *Codice topografico,* III, 291–318).

2. Biblioteca Apostolica Vaticana, Archivio San Pietro, V, 10 (Alexander IV, June 4, 1258); Rieti, archivio capitolare, III.B.4; Rome, Vallicelliana, Archivio Orsini, II.A.II.47(ol.44), II.A.II.48(ol.45), II.A.I.46(ol.44); *Liber censuum*, I, 565–567, II, 49. For a clear introduction to the problem of dealing with the fraternity, see Paul F. Kehr, *Italia Pontificia, I: Roma* (Berlin, 1906), 8–10. For the ten rectors, see Ferri, *ASRSP*, XXVIII, 24–25; for San Tommaso, see Carlo Pietrangeli, *Guide Rionali di Roma: Rione VII—Regola*, I (Rome, 1971), 58–60. The photograph is Gabinetto fotografico nazionale, E—16156.

3. See, for example, Augustus J. C. Hare, *Walks in Rome* (London, 1876), II, 404; Idelfonso Schuster, *Le Basilica e il Monastero di S. Paolo fuori le mura* (Turin, 1934), 274.

4. White, *Art and Architecture*, 62–63, 94–97; for the church in general, see Emilio Lavagnino, *San Paolo sulla Via Ostiense*, Le Chiese di Roma illustrate, 12 (Rome, n.d.); more specifically, John White, "Cavallini and the Lost Frescoes in S. Paolo," *Journal of the Warburg and Courtauld Institutes,* XIX (1956), 84–95. Schuster, 104, 106–107, 110, 139, plates 1, 5, 6, 12, 13, 15, 16, 18. Gardner, "Capocci Tabernacle," 225, and Otto Demus, whom Gardner cites (n. 34), for the Saint Paul's mosaicists and their probable Constantinople origins.

5. Dupuy, *Histoire*, 536; Schuster, 121–123, 127, 135, 145. But see Tommaso da Ocra's debt to Saint Paul's in Savini, "Il Cardinal Tommaso," 95.

6. Basilio Trifone, "Le Carte del monastero di San Paolo di Roma del secolo XI al XIV," *ASRSP*, XXXI (1908), 267–313, XXXII (1909), 29–106, XXXI, 294–300, no. 16. In general, on Italian monasticism, see Gregorio Penco, *Storia del monachesimo in Italia* (Rome, 1961). One may look also at Giulio Salvestrelli, "Lo Stato feudale dell'abazia di San Paolo," *Roma,* I (1923), 221–231, 419–431.

7. Trifone, "Le Carte," *ASRSP*, XXXI, 305–311, no. 24. One can compare the monastery partially described in these documents with the roughly contemporary one in *The Customary of the Benedictine Abbey of Eynsham,* ed. Antonia Gransden.

8. David Knowles, *The Religious Orders in England,* I (Cambridge, 1950), 281–283. For reform of Saint Paul's, see Schuster, 102.

9. Evelyn Underhill, "A Franciscan Mystic of the Thirteenth Century. The Blessed Angela of Foligno," *Franciscan Essays by Paul Sabatier and Others* (Aberdeen, 1912) [British Society of Franciscan Studies, Extra Series, I], 88–107, particularly 95, 101, 102; J. R. H. Moorman, *A History of the Franciscan Order* (Oxford, 1968), 263–264; see also Brentano, *Two Churches,* chapter 3. See note 14, this chapter.

10. The problems are exposed in M. D. Lambert, *Franciscan Poverty* (London, 1961), 1–30: "The Problem of Saint Francis"; but one should at least be aware of J. R. H. Moorman, *Sources for the Life of St. Francis* (Manchester, 1940), and M. Bihl, "Contra duas novas hypotheses prolates a Ioh. R. H. Moorman, *"Archivum Franciscanum Historicum,* XXXIX (1946), 3–37; a reassembled Francis, after criticism, is presented in the first chapter of Moorman's history, and a most recent and more optimistic study of some of the most im-

Notes

portant sources exists in the introduction to Rosalind Brooke, *The Writings of Leo, Rufino and Angelo.*

11. Brooke, 234–235 (the cicada).

12. Brooke, 21, and, for example, 94–95, 124–125, 264–269, 274–275, 296–299. (Not all of the pertinent material is of course included in the Brooke texts.)

13. Brooke, 212–215. The theme of Francis as romantic is gracefully developed in A. G. Little, "The Seventh Centenary of St. Francis of Assisi (1226–1926)," *Franciscan Papers, Lists and Documents* (Manchester, 1943), 1–150; for the round table and Roland quotations, see Brooke, 212–213, 214–215; the Lancelot who is suggested here is a character, of course, from the world of R. W. Southern, *The Making of the Middle Ages* (London, 1953).

14. Lester K. Little has been much concerned with the theme of Franciscan poverty in a newly moneyed society; he is in process of presenting his conclusions in a series of publications, the first of which is Lester K. Little, "Pride Goes before Avarice: Social Change and the Vices in Latin Christendom," *American Historical Review,* LXXVI (1971), 16–49.

15. The skin and the fire are Brooke, 178–179. On poverty, see Father Cuthbert, "Saint Francis and Poverty," *Franciscan Essays* (Aberdeen, 1912), 18–30, 24; Lambert, *Franciscan Poverty,* 38–67.

16. Lester K. Little will write at length on the jingling, see note 14; for Lazarus and Dives, see Brieger, *English Art,* 132 n. 2.

17. Gino Sigismondi, "La 'Legenda Beati Raynaldi' le sue fonti e il suo valore storico," *Bollettino della deputazione de storia per l'Umbria,* LVI (1960), 5–111, 42.

18. For these scholars, see John W. Baldwin, *Masters, Princes, and Merchants, the Social Views of Peter the Chanter and his Circle* (Princeton, 1969).

19. For a recent learned and intelligent discussion and defense of Bonaventure's approach, see Rosalind B. Brooke, *Early Franciscan Government* (Cambridge, 1959), particularly 272–275; see also Lambert, 137, 140, and Moorman, *History,* 260–261.

20. See Moorman, *History,* 5, 29, and particularly 18–19 for doubts about Francis's Roman activity; see too Brooke, *The Writings,* 248–253, 258–259. "Mirror of Perfection," chapter cxii, and Brooke, *The Writings,* 266–269, 274–277, for the quotation see 266–267; I have very slightly altered the wording of Mrs. Brooke's translation to make this clause intelligible out of context; see also Brezzi, *Roma,* 302. See too W. R. Thomson, "The Earliest Cardinal-Protectors of the Franciscan Order," *Studies in Medieval and Renaissance History,* IX (Lincoln, Nebraska, 1972), 27–39. For Dominic in Rome, see Manselli, *Studi Romani,* XIX, 133–143.

21. This is discussed more fully with bibliographical references in Brentano, *Two Churches,* chapter 4. See Moorman, *History,* 209 n. 1 and *Bullarium Franciscanum,* III, ed. D. A. Rosso (Rome, 1765), 244–246, for the rather exotic case of nuns exiled from "Romania" living in a Roman house who were allowed to transfer themselves from a Damianite rule (from Franciscan) to a Dominican rule (rule of St. Augustine, customs of S. Sisto). For Gregory IX and the nunneries—Santa Bibiana, Sant'Agnese, Sant'Andrea in Biberatica, Santa

Maria in Campo Marzio, San Cyriaco, Santa Maria de Massima, see *Les Registres de Grégoire IX,* I, cols. 551–558, no. 932.

22. A. F. C. Bourdillon, *The Order of the Minoresses in England* (Manchester, 1926), 5 (from the *Legenda*). Clare and the orders are treated at some length in Moorman, *History,* particularly 32–39, 205–215.

23. Paschal Robinson, "St. Clare," *Franciscan Essays* (Aberdeen, 1912), 31–49, 41: I have changed the wording of the translation slightly.

24. Bourdillon, 6; see too the similar remark, "by the fourteenth century," in Moorman, *History,* 406; see, on the matter of rule, poverty, and property, Bourdillon, 2–3; Livario Oliger, "De origine regularum Ordinis S. Clarae," *Archivum Franciscanum Historicum,* V (1912), 181–209, 413–447, particularly 193 (familial emphasis), 209 (example of difficulties with rule), 414–429 (struggle with property, complexity); Robinson, 34, 41, 44, 47; *Bullarium Franciscanum,* I, ed. G. H. Sbaralea (Rome, 1759), 263–267, 476–483, 671–678; II, ed. G. B. Colombino (Rome, 1761), 477–486, 509–521, and particularly 477, 478, 479 (clothes), 480, 509, 510, 511 (clothes).

25. For relics and feasts, see Giovanni Giacchetti, *Historia della venerabile chiesa et monastero di S. Silvestro de Capite di Roma* (Rome, 1629), 17–49, 67–76. See too a recent work, J. S. Gaynor and I. Toesca, *S. Silvestro in Capite,* Le Chiese di Roma illustrate, 73 (Rome, 1963).

26. For examples of these names, see San Silvestro in Capite, 85(82), 77(75), 92(88), 91(87). The following pages are so filled with references to San Silvestro documents that the repetition of even a normally abbreviated proper reference to them would become tedious to the eye, particularly because there is a double reference to most documents. Most of the documents are edited in V. Federici, "Regesto del monastero di San Silvestro in Capite," *ASRSP,* XXII (1899), 213–300, 489–538, XXIII (1900), 67–128, 411–447. Most of their originals exist in Rome, Archivio di stato, San Silvestro in Capite. In the references employed here (henceforth cited as San Silvestro), numbers not in parentheses refer to the numbers of the documents in Federici's edition (without reference to volume or page); numbers within parentheses refer to the numbers of documents within the fond San Silvestro in Capite in the Archivio di stato.

27. Giacchetti, 43; A Serafini, *Le torri campanarie di Roma e del Lazio nel medioevo,* 216–217; see particularly Richard Krautheimer, "A Christian Triumph in 1597," in *Essays in the History of Art presented to Rudolf Wittkower,* ed. Douglas Fraser, Howard Hibbard, and Milton J. Lewine (London, 1967), 174–178, 175–176, nn. 13, 18.

28. San Silvestro, 52(49), 105(101), 106(102), 93(89), 103(100), 70(69), 79(76), 95(91), 58(55), 137(132), 138(133), 128(124), 129(125), 134(129), 136(120), 149(142), 151(144), 167(158), 124(119), 148, 154, 168.

29. San Silvestro, 73(70), 74(72), 82(79), 99(95), 106(102).

30. San Silvestro, 84(81), 86(83), 91(87), 92(88), 113(108).

31. San Silvestro, 142(136), 158(151), 128(124), 150, 168.

32. San Silvestro, 99(95), 102(98), 105(100), 132(128), 74(72), 92(88), 58(55), (52), (53), (35), 65; see Federici, *ASRSP,* XXIII, 443–444.

Notes

33. San Silvestro, 85(82).
34. San Silvestro, 87, 88(84).
35. San Silvestro, 126(122).
36. San Silvestro, 122(118).
37. San Silvestro, 164(155); see too Gregorovius, *History*, V, 491.
38. San Silvestro, 38(35).
39. San Silvestro, 49(46), 70(69).
40. San Silvestro, 80(77), 84(81), 92(88), 96(93), 103(100), 110(107), 112(106), 117(111), 121(116), 124(119), 125(121), 128(124), 136(120), 137(133), 143(137), 146(139), 149(142), 151(144), 158(151), 162(154), 96(93).
41. San Silvestro, 128(124), 131(127), 149(142), 162(154).
42. San Silvestro, 90(86), 67(65), 154.
43. San Silvestro, 67(65), 138(132), 145(138).
44. San Silvestro, 53(50), 104(99).
45. San Silvestro, 107(104), 147(141); the feast of Saint John now celebrated by San Silvestro is the appropriate one of the *decollatio* (August 29), but I think that the documents' repeated talk of Saint John in the Summertime means not this feast but rather the great general feast of the Nativity of Saint John (June 24).
46. San Silvestro, 171.
47. San Silvestro, 164(155), 167, 168(158)(159), (161), 170.
48. *Les Registres de Clément IV*, ed. Jordan, 684, no. 139. *Le Registre de Jean XXI*, ed. E. Cadier (Paris, 1898–1960), 23, no. 62, 16, no. 43; *Les Registres de Nicolas III* (Paris, 1898), col. 194, no. 510; *Registres de Martin IV* (Paris, 1901–1913), 134, no. 313, 171, no. 415; San Silvestro, 161, 163, 169, 170.
49. San Silvestro, 106(102).
50. Giacchetti, *Historia*, 47–49; see also Federici, *ASRSP*, XXII, 231 (with the year 1277).
51. San Silvestro, 173(160); *Les Registres d'Honorius IV*, col. 104, no. 121.
52. Oliger, "B. Margherita Colonna," 84.
53. San Silvestro, (221), 180(165a), 184(173).
54. San Silvestro, 192(178), (227), (230).
55. But see André Callebaut, "Saint François et les privilèges, surtout celui de la pauvreté concédé à Sainte Claire par Innocent III," *Archivum Franciscanum Historicum*, XX (1927), 182–193.
56. San Silvestro, 183(171).
57. San Silvestro, 174(161), 178(164), 175(162), 176, 182, 189.
58. San Silvestro, 174(161).
59. Oliger, "B. Margherita Colonna," 203; San Silvestro, 161, 163.
60. San Silvestro, 187, 189(176), (187), (207), (227).
61. Livario Oliger, "Documenta originis Clarissarum Civitatis Castelli, Eugubii (a 1223–1263), necnon statuta Monasteriorum Perusiae Civitatisque Castelli (saec. XV) et S. Silvestri Romae (saec. XIII)," *Archivum Franciscanum Historicum*, XV (1922), 71–102, 99–102.
62. Oliger, "B. Margherita Colonna," 30–32; *Bullarium Franciscanum*, IV, ed. D. A. Rosso (Rome, 1768), 456, 468.

Notes

63. Oliger, "B. Margherita Colonna," 31–32; *Bullarium Franciscanum*, V, ed. Conrad Eubel (Rome, 1898), 8–9; Gregorovius, *History*, VI, 306.

64. San Silvestro, (227), (193); Rome, Archivio di stato, 30 Not. Cap., Uff. 33: vol. 37, Angelus Cesius (20 Ott. 1558), fo. 1257r; Falco, "Il Catalogo," 427; Oliger, "Documenta," 102.

65. San Silvestro, (187). For the eggs, see Gaynor and Toesca, *San Silvestro*, 67.

66. Rome, Archivio di stato, Santi Cosma e Damiano in Trastevere (San Cosimato) [henceforth cited as San Cosimato], 178, 170–213, 230. The problem of *conversi*, although never specifically *conversi* of this sort, is repeatedly approached in *I Laici nella "Societas Christiana" dei secoli XI e XII*: Pubblicazioni dell Università Cattolica del Sacro Cuore: Miscellanea del Centro di Studi Medioevali, V (Milan, 1968). Two essays, particularly, might be looked at in this connection: Cosimo Damiano Fonseca, "I conversi nella communità canonicali," 262–305, and Jacques Dubois, "L'institution des convers au XIIe siècle. Forme de vie monastique propre aux laïcs," 152–176.

67. San Cosimato, 170, 172, 204, 212, 223; see too P. Fedele, "Carte del monastero dei SS. Cosma e Damiano in Mica Aurea," *ASRSP*, XXI (1898), 459–534, XXII (1899), 25–107, 383–447, particularly XXI, 459–494 —this edition did not reach the thirteenth century.

68. San Cosimato, 188.

69. San Cosimato, 192.

70. San Cosimato, 218, 175.

71. For example, San Cosimato, 158, 200, 205.

72. San Cosimato, 329 (compare with 180), 173, 178, 315.

73. San Cosimato, 174, 193, 201, 202.

74. San Cosimato, 186.

75. San Cosimato, 198, 215, 216.

76. San Cosimato, 214. *Bullarium Franciscanum*, I, 50–51, no. 37.

77. San Cosimato, 238; see too *Bullarium Franciscanum*, I, 137–138, no. 143 (transcription of the supposed original papal letter in Milan, where the different date is surely due to a misreading of Kalends as day—15 Kl' September = 18 August). *Liber censuum*, 18–19. For the affair with the Camaldolese, see *Bullarium Franciscanum*, I, 112–113, no. 113 and *Les Registres de Grégoire IX*, ed. Lucien Auvray (Paris, 1896–1955), I, col. 809, no. 1450.

78. San Cosimato, 230, 252.

79. San Cosimato, 258.

80. San Cosimato, 306. *Bullarium Franciscanum*, I, 249–252, no. 274, 257, no. 284; *Les Registres de Grégoire IX*, II, cols. 1103–1108, nos. 4478–4481, cols. 1187–1189, nos. 4647–4650.

81. San Cosimato, 252, 384, 310, 303, 199, 334; Falco, "Il Catalogo," 441.

82. San Cosimato, 243, 231, 232, 233.

83. San Cosimato, 295, 307, 308, 291, 303.

84. San Cosimato, 240, 250.

85. San Cosimato, 278, 280, 281, 277, 262, 292, 287.

86. San Cosimato, 296, 275, 286, 284, 196.

87. San Cosimato, 185, 246, 252, 244, 261, 260, 279.

88. San Cosimato, 257.

Notes

89. F. Camobreco, "Il Monastero di S. Erasmo sul Celio," *ASRSP, XXVIII* (1905), 265–300, 288–290. *Les Registres de Grégoire IX,* I, col. 176, no. 54; Monaci, "Regesto . . . di Sant'Alessio," 162.

90. Roth, *Cardinal Richard Annibaldi,* 46; P. S. L. (Lopez), "De origine conventus Romani Stae. Mariae de Populo, ordinis S. Augustini," *Analecta Augustiniana,* IX (1921), 71–75, 72; Rome, Archivio di stato, Sant'Agostino, 3, 4. For the elaborate business of getting the Franciscans into the Aracoeli, see *Bullarium Franciscanum,* I, 521–522, no. 288, 530–531, no. 304, 545, no. 330, 556–558, no. 346, and particularly 616–618, no. 418; *Les Registres d'Innocent IV,* II, 147, no. 4848, III, 96, no. 5897; and see earlier, chapter 5, note 40.

91. Margaret R. Toynbee, *S. Louis of Toulouse and the Process of Canonisation in the Fourteenth Century* (Manchester, 1929), particularly 111–117.

CHAPTER VII
Last Wills and Testaments

1. The connection between Theodoric and the Theater of Pompey has been recalled recently in Caecilia Davis-Weyer, *Early Medieval Art, 300–1150* (Englewood Cliffs, New Jersey, 1971), 52–53; see also Cassiodorus, *Variae,* IV, no. 51, in *MGH, Auctores Antiquissimi,* XII (Berlin, 1894), 138–139.

2. For Maitland, who thought specifically of Blake's angels, see Ermengard Maitland, *F. W. Maitland, A Child's-Eye View,* Selden Society (London, 1957), 6. For the making of wills, see Michael M. Sheehan, *The Will in Medieval England* (Toronto, 1963), particularly 192–195.

3. Acton in the "Inaugural Lecture on the Study of History," in *Essays on Freedom and Power,* ed. Gertrude Himmelfarb (Cleveland, Ohio, 1955), 32.

4. The lines from "La Vie de Saint Alexis," are 606–607, page 33, in the edition of Gaston Paris (Paris, 1903). For the evangelical revival see particularly M.-D. Chenu, *Nature, Man and Society in the Twelfth Century,* trans. Jerome Taylor and Lester K. Little (Chicago, 1968), and especially "The Evangelical Awakening," 239–269; Lester K. Little, "Pride Goes before Avarice: Social Change and the Vices in Latin Christendom," *American Historical Review,* LXXVI (1971), 16–49; Tadeusz Manteuffel, *Naissance d'une hérésie* (Paris, 1970), particularly for Saint Alexis, 13 ("a veritable hymn to the glory of voluntary poverty"), 33–37 for Norbert and Robert; the essays within *I Laici nella "Societas Christiana" dei secoli XI e XII:* Relazioni del X Congresso internazionale di scienze storiche, III: Storia del medio evo (Milan, 1968) — for example, Herbert Grundmann, "Eresie e nuovi ordini nel secolo XII," 356–402, 377–389, particularly 380, and Gerd Tellenbach, "Il Monachesimo riformato ed i laici nei secoli XI e XII," 118–142; Ernst Werner, *Pauperes Christi* (Leipzig, 1956), 19–27; "Pauvres Catholiques," in *Cahiers de Fanjeaux,* II (1967), 207–272; for Norbert, recently recalled in connection with a pertinent miracle, see John F. Benton, *Self and Society in Medieval France: The Memoirs of Abbot Guibert of Nogent,* 238–239; see too the "Vita Norberti," ed. Roger Wilmans, in *MGH,SS.,* XII, 663–703; for the

Notes

modern quotation, see Gordon Leff, *Heresy in the Later Middle Ages* (Manchester, 1967), 2. For thirteenth-century combinations of saintly virtue there were, of course, much older precedents as Fortunatus's Radegunda and Christ Himself make clear.

5. For Francis with parsley and the cicada, see Brooke, *Scripta,* 296–299, 234–235.

6. *Mission to Asia,* ed. Christopher Dawson (New York, 1966), 140–141.

7. Frugoni, *Celestiniana,* 56–67, 57.

8. See particularly the 1957 catalog of the *Mostra di sculture lignee medioevali* for the Museo Poldi Pezzoli in Milan, but also the general work, Paul Thoby, *Le Crucifix* (Nantes, 1959), and especially Adriano Prandi, "L'Espressione del dolore e della morte attraverso una serie di crocefissi del Museo di Cividale," in *Il Dolore e la morte nella spiritualità dei secoli XII e XIII,* Convegni del Centro di studi sulla spiritualità medievale, V (Todi: Accademia Tudertina, 1967), 367–380, and particularly plates 14 and 15; also in this richly pertinent collection, see Raoul Manselli, "Dolore e morte nell'esperienza religiosa catara."

9. Rome, Archivio di stato, Santo Spirito, A, 22.

10. Beryl Smalley, *The Study of the Bible in the Middle Ages* (Notre Dame, Indiana, 1964), 285; E. Delaruelle, "Influence de Saint François d'Assise sur la piété populaire," *I Laici,* 449–466, 452.

11. Moorman, *History of the Franciscan Order,* 259.

12. Gino Sigismondi, "La 'Legenda Beati Raynaldi,' le sue fonti e il suo valore storico," *Bolletino della deputazione di storia patria per l'Umbria,* LVI (1960), 5–111, 31, 41–44.

13. J. de Ghellinck, "Eucharistie au XIIᵉ siècle en occident," *Dictionnaire de Théologie Catholique,* V. 1 (Paris, 1913), cols. 1233–1302; Migne, *Patrologia Latina,* CCXVII, cols. 773–916; S. J. P. van Dijk and J. Hazelden Walker, *The Origins of the Modern Roman Liturgy* (Westminster, Maryland, 1960), 360–365; V. L. Kennedy, "The Moment of Consecration and the Elevation of the Host," *Mediaeval Studies,* VI (1944), 121–150, 123, 133, 148–150, and "The Date of the Parisian Decree on the Elevation of the Host," *Mediaeval Studies,* VIII (1946), 87–96, 94; the quotation is from Joseph A. Jungmann, *The Mass of the Roman Rite* (London, 1959), 91, see also 422–427, 425–426.

14. Ernest W. McDonnell, *The Beguines and Beghards in Medieval Culture* (New Brunswick, 1954), especially 299–319; F. Bonnard, "Bolsena," *Dictionnaire d'histoire et de géographie ecclésiastique,* IX, cols. 679–680; Andrea Lazzarini, *Il Miracolo di Bolsena* (Rome, 1952), particularly 65–71, 83–86; Dawson, *Mission* 180; Brooke, *Scripta,* 226–229; Kajetan Esser, *Das Testament des heiligen Franziskus von Assisi* (Münster, Westphalia, 1949), 101; Boyle, "An Ambry," 329.

15. *The Book of Margery Kempe,* ed. Hope Emily Allen, Sanford Brown Meech, et al., Early English Text Society (Oxford, 1940), see particularly 47.

16. Rome, Archivio di stato, Santo Spirito, B, 131.

17. For the grail, see Lizette Andrews Fisher, *The Mystic Vision in the Grail Legend and in the Divine Comedy* (New York, 1917).

Notes

18. *Acta Sanctorum,* May, V, 85–100 (May 21), 88–90; Ferdinand Schevill, *Siena,* ed. William M. Bowsky (New York, 1964), 266.

19. The thirteenth-century building history of Aracoeli is the subject of a forthcoming work by Professor Ronald Malmstrom of Williams College.

20. Rome, Archivio di stato, Sant'Alessio, 39.

21. For an example of names, see Rome, Archivio di stato, Santo Spirito, B, 99 (the nuns of Santa Maria in Campo Marzio in 1335).

22. For example, Rome, Archivio di stato, Sant'Alessio, 31.

23. Compare Herbert Grundmann, *Religiöse Bewegungen im Mittelalter,* Historische Studien, 267 (Berlin, 1935), 5.

24. Biblioteca Apostolica Vaticana, Archivio San Pietro, Caps. 64, no. 181.

25. Egidi, *Necrologi,* I, 262–263.

26. Egidi, *Necrologi,* I, 260–261, 262–263.

27. Egidi, *Necrologi,* I, 242–243, 228–229; see, for example, Pio Pecchiai, "I segni delle case di Roma nel medio evo," *Archivi,* XVIII (1951), 231, 233.

28. Egidi, *Necrologi,* I, 222–223, 278–279, 220–221, 97, 101.

29. Egidi, *Necrologi,* I, 194–195, 198–199, 200–201, 238–239, 240–241, 254–255, 266–267, 336–337, 222–223, 268–269, 278–279, 234, 236, 260; 240–241; 246–247.

30. Egidi, *Necrologi,* I, 242–243, 212–213, 97.

31. Egidi, *Necrologi,* I, 200–201, 278, 226–227, 198–199, 262–263, 264–265, 192–193, 184–185; see also Pecchiai, "I segni."

32. Egidi, *Necrologi,* I, 198–199, 214–215, 272–273, 174–175, 196–197, 210–211; and see Grimaldi, "Opusculum de Sacrosancto Veronicae Sudario . . ." (Archivio San Pietro, H, 3), particularly fos. 36r–52r. Bertelli, "Immagine Edessena," 3–5.

33. Worcester, Saint Helen's, Register of Bishop G. Giffard, fos. 345^v–347.

34. Rome, Archivio di stato, San Cosimato, 198.

35. Rome, Archivio di stato, San Cosimato, 311; see, for example, Santo Spirito, B, 10, and San Cosimato, 237. For a nicely illustrated and learned introduction to the pyx, see S. J. P. van Dijk and J. Hazelden Walker, *The Myth of the Aumbry* (London, 1957), particularly 33–35, frontispiece, plates 1, 5, 6.

36. Huelsen, *Le Chiese di Roma,* 212, 455–456; *Les Registres de Grégoire IX,* I, 400, no. 636.

37. See, for example, Gilles Gerard Meersseman, "I Penitenti nei secoli XI e XII," *I Laici,* 306–339, 325.

38. For example, Archivio Segreto Vaticano, Instrumenta Miscellanea, 145.

39. Rome, Archivio di stato, Sant'Agostino, 5. For the Hungarian cardinal's selection of Santa Balbina as his place of burial, see Paravicini, "Un frammento," 181 and n. 9. The will of Pietro's son Adwardus also survives (Archivio Segreto Vaticano, Celestini, 18), written in July 1296, after Pietro's death; it repeats his benefactions selectively. It includes Grottaferrata, and asks burial in a chapel to be made at Aracoeli.

40. The order of Saint Silvestro Guzzolini of Osimo; see Paulus Weissenberger, "Die ältesten Statuta Monastica der Silvestriner," *Römische Quartalschrift,* XLVII (1939), 31–109; see also Gregorio Penco, *Storia del*

Notes

Monachesimo in Italia, Tempi e Figure, 31, (n.d.), 301–305 (and for the Celestinians 305–312); for the Crucifers at San Matteo, see Ferri, *ASRSP,* XXX,
141–142, no. 97, in which a 1296 gift refers to the order specifically. For San
Giacomo's undistinguished position at the beginning of the pontificate of Innocent III see Hageneder and Haidacher, *Die Register Innocenz' III,* I, 417–419,
no. 296; for its privileges, including burial privileges from Honorius IV, see
Prou, *Honorius IV,* cols. 225–226, 230, nos. 293, 300, both from February 5,
1286; the latter allows San Giacomo to receive profits from usury, etc., up to a
limited value, from penitents. The power of San Giacomo to attract burials is
witnessed by the recorded burial there of a fourteenth-century canon of Saint
Bartholomew's on the Island; both it and the fourteenth-century burial of a
miller call attention to San Giacomo's site by the river; for these burials, see
Forcella, *Inscrizioni,* VI, 324, nos. 1063, 1065. For redecoration at Grottaferrata,
see Carlo Bertelli, "La Mostra degli affreschi di Grottaferrata," *Paragone,* no.
249, Arte (1970), 91–101, particularly 100–101, and plates 40–49.

41. Worcester, Saint Helen's, Register of Bishop G. Giffard, fos.
345ᵛ–347; see too Sheehan, *The Will,* 258–265. For Ottobuono Fieschi's
related preference for the friars, see Federici, *Della Famiglia Fiesca,* 129.

42. Rome, Vallicelliana, Archivio storico capitolino, Archivio Orsini,
II.A.I.48(ol.46).

43. Vallicelliana, Archivio Orsini, II.A.I.21(ol.20).

44. Rome, Archivio di stato, Santo Spirito, B, 107; Vallicelliana, Archivio
Orsini, II.A.II.16(ol.15); San Silvestro in Capite, *ASRSP,* XXIII, 426–428,
no. 171.

45. Felice Nerini, *De templo et coenobio Sanctorum Bonifacii et Alexii*
(Rome, 1752), 446; Gelasio Caetani, *Regesta Chartarum,* 39–41; Vallicelliana, Archivio Orsini, II.A.I.54(ol.53).

46. Oliger, "B. Margherita Colonna," 174; Rome, Archivio di stato,
Sant'Agostino, 6; Egidi, *Necrologi,* 142; Hirschfeld, "Genuesische Dokumente,"
130–131; David Herlihy, *Medieval and Renaissance Pistoia* (New Haven,
1967), 249.

47. Biblioteca Apostolica Vaticana, Sant'Angelo in Pescheria, I, 2, fos.
68–70; for different conclusions about age suggested by fifteenth-century de-
mographic-statistical material from Pistoia, see Herlihy, *Pistoia,* 78–93, particularly 89.

48. Sant'Angelo in Pescheria, I, 2, fos. 58–61ᵛ.

49. See Sant'Angelo in Pescheria, I, 2, 64ᵛ–66; Rome, Archivio di stato,
Sant'Agostino, 26.

50. Biblioteca Apostolica Vaticana, Sant'Angelo in Pescheria, I, 1, fos.
81–81ᵛ, 52ᵛ, 140–142, 117–118; Pietro de Angelis, *Ospedale,* II, 617–620;
for the madonna of Sant'Angelo, see Ilaria Toesca, "L'Antica 'Madonna'
di Sant'Angelo in Pescheria a Roma," *Paragone* 227 (1969), 3–18 and plates
1–5b.

51. Biblioteca Apostolica Vaticana, Sant'Angelo in Pescheria, I, 1,
78ᵛ–79, 164ᵛ–165; 118ᵛ–119; Rome, Archivio di stato, San Silvestro in
Capite, 193.

52. Biblioteca Apostolica Vaticana, Sant'Angelo in Pescheria, I, 3, 88.

53. The quotation "only for recounting" refers to Francis's talk of those

Notes

who gain credit "only for recounting" the deeds of Charles, Roland, and Oliver, and his comparison of them with "us" who tell stories of the saints' wonders: Brooke, *Scripta,* 214–215; the second Franciscan reference is to the friar's talk of Francis's selling of his sackcloth to the lord and the *multi baldaquini et panni de serico* ("many rich brocades and silken cloths") which would later cover his body: Brooke, *Scripta,* 260–263.

54. Ilaria Toesca, *Una croce dipinta romana* (Rome, 1966), who should not, however, be held responsible for the suggestions about the identity of the anonymous saints. In connection with this cross and its piety one might also look at the same author's "Opere d'Arte del territorio di Tivoli restaurate dalla Soprintendenza alle gallerie," *Atti e Memorie,* XLII (1969), 197–202. For the Bible which Nicholas III gave to Aracoeli, see *Il libro della Bibbia* (Vatican City, 1972), 41, no. 77; it is Vat. lat. 7797.

55. For the evocative description of Constantius II's entry into Rome, taken from Ammianus, see Ernst H. Kantorowicz, *Laudes Regiae* (Berkeley, 1958), 66.

56. Gregorovius, *History,* V, 466. Ludovico Frati, *Libro di cucina del secolo XIV* (Livorno, 1898), 15, for the *fritata* recipe; I do not have evidence for its use in thirteenth-century Rome itself, but mean here to suggest the specific act of eating a specific dish.

57. One should think of the little Veronica in connection with surviving souvenirs from other centers, like those from Canterbury preserved at the Guildhall in London, or those still available, like the little pens from Loreto, through whose viscous liquid the holy house sails.

Index

Index

Index

Index

Index

Index

Index

340